OFFSHORE PIONEERS

BROWN & ROOT AND THE HISTORY OF OFFSHORE OIL AND GAS

OFFSHORE PIONEERS

BROWN & ROOT AND THE HISTORY OF OFFSHORE OIL AND GAS

JOSEPH A. PRATT

TYLER PRIEST

CHRISTOPHER J. CASTANEDA

Gulf Publishing Company
Houston, Texas

OFFSHORE PIONEERS

BROWN & ROOT AND THE HISTORY

OF OFFSHORE OIL AND GAS

Gulf Publishing Company
Book Division
P.O. Box 2608 □ Houston, Texas 77252-2608

10 9 8 7 6 5 4 3 2 1

Library of Congress Cataloging-in-Publication Data

Pratt, Joseph A.
Offshore pioneers : Brown & Root Marine and the history of offshore oil and gas / Joseph A. Pratt, Tyler Priest, Christopher J. Castaneda.
p. cm.
Includes bibliographical references (p.) and index.
ISBN 0-12390-838-8 (alk. paper)
1. Brown & Root (Firm)—History. 2. Offshore oil field equipment industry—United States—History. 3. Offshore gas field equipment industry—United States—History. 4. Ocean mining—History. 5. Oce: **0-12390-838-** g—History. I. Priest, Tyler. II. Castaneda, Christopher James, 1959–
HD9569.B76P7 1997
338.7'6817665'0973—dc21 97-34516
CIP

All photographs courtesy of Brown & Root, Inc., unless otherwise noted.

Cover photograph of BP's Highlands One under tow along Scottish coast from Highlands Fabricators to the Forties field in 1974.

Dedication

To the thousands of Brownbuilders
who made this book possible.

Contents

Foreword

During the first half century of the offshore oil and gas industry, Brown & Root Marine has been a pioneer in the design and construction of offshore platforms and pipelines. Through applied engineering, the industry has moved into deeper waters and harsher environments, providing an important case study of the technological forces driving the energy economy. The history of a single company, in this case Brown & Root, reveals the human dimensions of this broad process of change. Real people made practical choices based on their best understanding of the technical possibilities and the environmental constraints. When these choices proved wrong, they returned to the drawing board and made adjustments. Their efforts produced astonishing long-term advances in offshore technology.

The process of change is of more than historical interest. As we write in 1997, the offshore industry as a whole is in the midst of a wave of profound technological change that is redefining what is possible, while at the same time dramatically lowering the cost of the recovery of oil and gas from great depths. Our history seeks to provide a perspective on the last fifty years that will be useful as context for understanding the possibilities of the next fifty.

This vision of the history project originated with Dr. Jay Weidler, senior vice president of Brown & Root Energy Services. With almost thirty years of experience in the design and analysis of offshore structures, he approaches the challenges of the current era from a technical perspective informed by history. As the sponsor of our history project, he supported our work in every way, from helping to design the project, to serving as an invaluable "interpreter" on technical matters, to reading and editing the chapters. Ms. Billie McMahon, his executive assistant, managed the project. She facilitated our work by locating research materials and photographs and by coordinating interviews with current and former Brown & Root employees. She also read the entire manuscript several times and helped keep the project on schedule.

Fifty-three interviews of Brown & Root personnel were central to the completion of the history. Joe Rainey, a retired company executive, carefully and tirelessly conducted most of these interviews, and his hard work provided one of the key sources for reconstructing the past. Suzanne Mascola transcribed these interviews. Those interviewed were: Mark Banjavich, Dirk Blanken, Charles L. "Charlie" Buck, J.S. "Sage" Burrows, Marshall Cloyd, A.B. Crossman, Peter S. "Pete" Cunningham, Benny L. Davis, Adi Desai, L.E. "Bo" Driggers, Jamie Dunlap, H.S. "Frank" Frankhouser, Leon Garrett, Anthony V. Gaudiano, W.R. "Bill" Golson, Hugh Gordon, Joe E. "Buddy" Hoke, S.J. "Stan" Hruska, J.D. Irons, A.R. Jackson, Allen C. Johnson, Reeves Kolb, Joe C. Lochridge, John G. Mackin, Jr., Basil Maxwell, P.H. "Pat" Moore, Frank Motley, Jim Noake, Clyde Nolan, C.D. "Chuck" Osborn, Tim Pease, W. B. "Ber" Pieper, Gene Raborn, W.R. "Rick" Rochelle, Heinz Rohde, N.S. "Nat" Self, D.K. Smith, Harris P. Smith, T. "Bo" Smith, John E. Sole, Bill Stallworth, Lawrence A. Starr, Ed Tallichet, Dr. W.H. Tonking, R.A. Turrentine, Ken Wallace, Delbert Ward, Bob Weatherly, Jay Weidler, and R.O. Wilson. Many of those interviewed also provided useful historical materials from their personal records.

Discussions with people outside of Brown & Root gave us a broad perspective on the evolution of aspects of the offshore industry. Especially useful were interviews with Pat Dunn (Shell Oil, retired), Griff Lee (J. Ray McDermott, retired), Malcolm Sharples (Noble Denton and the American Shipping Bureau), and Jim Winfrey (Exxon, retired). Each graciously shared valuable historical documents with us.

We should also note here the value of the technical articles and reports written by specialists from throughout the industry and published in such places as the proceedings of the annual Offshore Technology Conference and various industry journals such as the *Engineering-News Record.* Trade journals such as *Offshore,* the *Oil & Gas Journal,* and *World Petroleum* also regularly reported on technical issues.

Special thanks go to George R. Bolin for the use of memorabilia from his father, long-time Brown & Root executive L.T. Bolin. Valuable research assistance came from Sethuraman Srinivasan, Jr., Jeffrey M. Crawford, and Bruce Andre Beauboeuf. Marsha Attaway and Kathy Groccia kept the office going while Billie worked on this book. Dell Avery and Barbara Fletcher from the Information Resource Center at Brown & Root provided guidance in using reference materials. Joyce Alff, our editor at Gulf Publishing, displayed admirable patience as we wrote and rewrote the manuscript. Leonard LeBlanc graciously allowed us to use historical materials from *Offshore.*

In addition to Jay Weidler and Billie McMahon, Rick Rochelle, R.O. Wilson and Zelma Branch read all or most of the manuscript. Dirk Blanken, Clyde Nolan, Bill Stallworth, and Stan Hruska carefully read and reread multiple drafts of chapters. Alfonso Barnetche, Mark Banjavich, Jamie Dunlap, Tony Gaudiano, Joe Lochridge, John Mackin, Basil Maxwell, Tim Pease, Ber Pieper, Joe Rainey, David Smith, Harris Smith, Larry Starr, Ken Wallace, John McClellan, Andre

Mangiavacchi, and Jerry Slaughter reviewed sections of the manuscript. A footnote at the beginning of each chapter lists those who made noteworthy contributions to particular chapters.

Funding for research and writing was provided by Brown & Root. This project also drew upon research for a general history of offshore oil and gas in the Gulf of Mexico, funded in part by the University of Houston Energy Laboratory and the Cullen Chair in History and Business at the University of Houston.

While we thank all of those who helped us complete our history, the final responsibility for its contents rests, of course, with the co-authors.

Joseph A. Pratt
Richard Tyler Priest
Christopher J. Castaneda

Introduction

A Half Century of Innovation

Brown & Root, a Houston-based engineering and construction firm, built the platform from which Kerr-McGee Oil Industries drilled the first producing well beyond the sight of land in the Gulf of Mexico. The well came in on November 14, 1947, a date that marks the birth of the modern offshore oil and gas industry. Before 1947, companies had extracted oil from underwater fields, but these deposits were located primarily in protected inland waters. Oil had been produced offshore in the Gulf of Mexico before 1947, but always in sight of land. Kerr-McGee's well, in eighteen feet of water, ten and a half miles from the Louisiana shore, went a step beyond those previous developments. The Kermac 16 stood in the open waters of the Gulf, exposed to the fury of waves and wind. The success of this platform in producing oil from beyond the horizon heralded a new era of technological innovation that subsequently spread to offshore provinces throughout the world.

Brown & Root Marine has been a pioneer in the innovation of offshore technology since the 1930s. Through the engineering, construction, and installation of platforms and marine pipelines, the company helped make offshore oil and gas a critical component of the world's energy supply. From negligible production in 1947, the total production of offshore oil grew steadily to account for about 14 percent of the world's supply in 1974 to about 33 percent in 1996. That year, the output of offshore oil reached about 21 million barrels per day out of a total world production of 63 million barrels per day. Over the same fifty year period, worldwide natural gas production rose to around 228 billion cubic feet per day, the energy equivalent of almost 40 million barrels of oil. Offshore gas accounted for 20–25 percent of this total.

Brown & Root was one of the original members of what became a full-fledged offshore industry. From the early days, those working offshore formed a fraternity of sorts. The various parts of the offshore system—exploration, production, transportation, and supply—moved forward together. Oil and gas companies, engineering and construction concerns, and oil service and supply firms did whatever was necessary to operate in deeper waters. As the offshore industry matured in the Gulf of Mexico, firms like Brown & Root branched out to apply and modify technologies in other marine environments. Brown & Root personnel and vessels traveled around the world in search of opportunities and undertook projects in almost every locale that attracted attention offshore, from the Gulf of Mexico to the Persian Gulf, from the North Sea to the South China Sea. The fiftieth anniversary of the first oil production from an out-of-sight-of-land well marks an appropriate point to reflect on the history of the offshore industry and the companies that drove it forward.

Numerous companies contributed to the development of an offshore oil and gas system in the Gulf of Mexico. With its gradually sloping deltaic plain and vast expanse of shallow water, the northern Gulf of Mexico was an ideal proving ground for those who saw potential offshore. Building large, freestanding structures in open waters, however, had never been attempted before. Marine construction thus required trial-and-error methods and close interchange between engineers, fabricators, and construction crews. Engineers adapted prevailing technologies to new demands, making incremental improvements that nudged offshore operations into deeper waters. But not all innovation was gradual. Like bolts from the blue, new processes such as mobile drilling and computer-assisted design radically changed existing practices and depth capabilities.

Innovation allowed the offshore industry to redefine what was technically and economically possible. Responses to basic environmental and economic imperatives spurred on technological change. Environmental challenges were often paramount. Engineers had to design structures that were appropriate for particular field characteristics and that could withstand waves, winds, soil movements, earthquakes, ice, and deepwater. The price of oil and gas was another constraint on the industry; material and construction costs set parameters around innovation. The capacity of available equipment and fabrication sites also shaped fundamental design choices. As field sizes, water depth, and pace of development increased, designs often outran the ability of existing equipment to lift, transport, and install massive structures. The creation of larger fabrication yards and vessels to handle them were essential for retrieving oil and gas from more demanding environments. Less obvious, but perhaps more important in the long-term, was the development of another type of sophisticated equipment: computers and programs that greatly enhanced structural design and analysis. The history of Brown & Root Marine is, to a great extent, the history of technological innovation arising from the interplay between the physical environment, commercial conditions, and equipment.

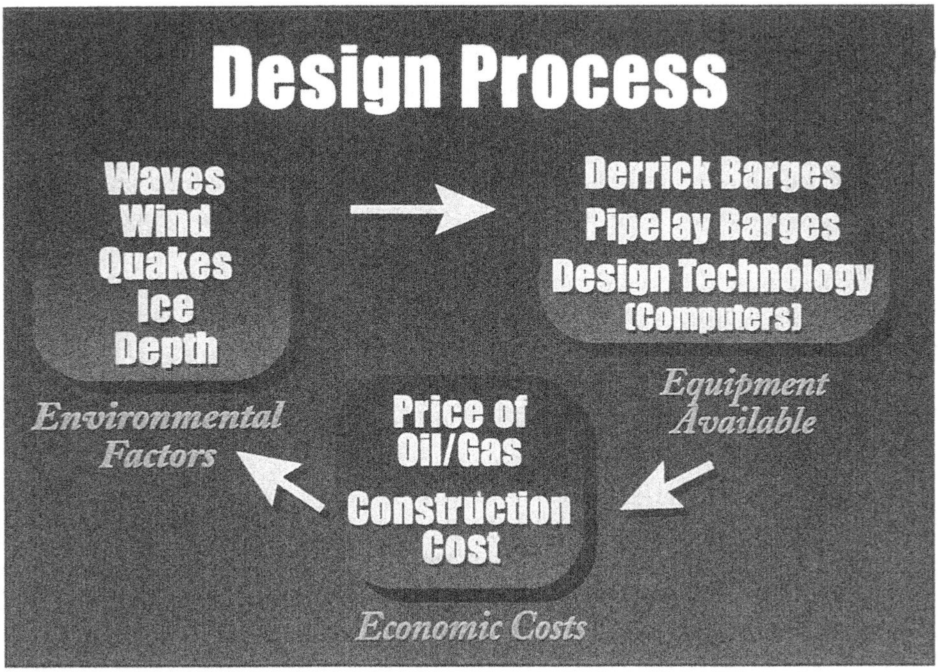

Design Process. Courtesy of University of Houston Media Services.

Given Brown & Root Marine's wide-ranging activities, a comprehensive history of all its offshore projects could not easily be contained in a single book. This volume focuses on major technological developments and the company's approach to innovation. It is divided into three parts: the development of a basic technology for offshore production in the Gulf of Mexico, the response of the company to the extreme challenges posed by regions outside the Gulf, and Brown & Root's participation in the development of the North Sea. This method of organization stresses thematic and geographic coherence over chronological order. Technologies were often developed for particular locations and concurrent with other innovations. Some movement back and forth through time in the narrative, therefore, is unavoidable.

Part I examines the creation of a modern offshore industry in the Gulf of Mexico. From the 1930s through the early 1950s, the industry adapted onshore technology and equipment for use in shallow waters. A growing understanding of the impact of wind and wave forces on offshore structures led to the development of standard design criteria. New equipment and procedures for offshore development resulted in larger and sturdier structures in deeper waters. Permanent onshore facilities were built to fabricate these structures, and a cluster of vital service industries also emerged. In these formative years of offshore development,

Brown & Root established a leading position in building platforms and laying pipelines. The company joined many others in developing a mature technical system which, by the early 1960s, was able to deliver oil and gas from water depths of 200 feet.

In the 1970s and 1980s, an offshore construction boom, set off by higher oil prices and by declining discoveries onshore in the United States, pushed platforms into progressively deeper waters in the Gulf of Mexico. Large investments in equipment and improved design capabilities during the 1960s prepared Brown & Root for the new challenge. The size and sophistication of the company's vessels and equipment had increased dramatically. Moreover, Brown & Root had become proficient in one of the most significant aspects of offshore innovation, the advanced application of computers in designing structures for deeper waters and harsher environments. As fields beyond the edge of the continental shelf were prospected, Brown & Root produced important innovations to bring them into play.

While developing and improving the technology needed to place platforms farther out in the Gulf of Mexico, the offshore industry also moved around the world in search of opportunities. Brown & Root took on demanding jobs in many countries, helping to bring much needed offshore oil and gas onstream. Some of this work involved transplanting basic technology to new locations, although geopolitical considerations often made such work far from routine. Other times, Brown & Root confronted particularly difficult engineering and construction problems that required it to innovate on a grand scale. Part II discusses the company's responses to selected "extreme challenges" during the 1960s and 1970s, including Venezuela and the Middle East, Project Mohole, commercial diving, earthquakes and ice offshore California and Alaska, and the crash development of Mexico's Bay of Campeche.

In the relatively tranquil waters of Venezuela and the Middle East, Brown & Root employed well-tested technology and equipment from the Gulf of Mexico. But these new environments also presented unique challenges that required new methods and modifications to basic technology. Beginning in the 1950s, Brown & Root assumed a prominent position in Venezuela's Lake Maracaibo, designing various kinds of structures to meet the lake's particular marine conditions. As the company entered other countries, it took on a new speciality in supertanker terminals. Especially important was the Persian Gulf, whose giant oil fields justified path-breaking departures in the scale and cost of terminals and pipelines.

In its enthusiasm to test the limits of offshore construction, Brown & Root Marine reflected the mentality of the entire company. Founders Herman and George Brown had built an ambitious construction company eager to take on extraordinary challenges. The company's growth and achievements during and after World War II earned it a national reputation in engineering as well. Marine engineering and construction became an increasingly important part of the diversified, multinational Brown & Root organization. In 1962, Brown & Root was acquired by the Halliburton Company, another diversified firm whose traditional strengths

Offices on Clinton Drive, Houston

in oil services complemented the offshore work of Brown & Root. That same year, Brown & Root was selected to manage the National Science Foundation's "Project Mohole," an experiment to drill through the earth's crust to the mantle. Although Project Mohole's celebrated quest to explore "inner space" was ultimately aborted by the U.S. government, it nonetheless produced valuable scientific and engineering advances that were applied to marine drilling. In the 1960s and 1970s, Brown & Root pushed the exploration of inner space in a different way, through a subsidiary company called Taylor Diving & Salvage. Taylor Diving pioneered innovations in diving that facilitated marine construction and pipelining in deeper water. The accomplishments of Project Mohole and Taylor Diving helped map out entirely new technological frontiers for the offshore industry.

Beginning in the 1960s, Brown & Root tackled other difficult challenges as the offshore industry ventured into harsher environments. In California, the company had to come up with offshore design criteria that factored in the effects of earthquakes. This experience led to a series of demanding projects in the Cook Inlet in southern Alaska, where the combination of earthquakes and destructive ice floes, coupled with frequent tidal changes, demanded fundamental changes in platform jacket designs. Brown & Root later designed and built facilities for the Prudhoe Bay field on Alaska's North Slope, where arctic conditions required adjustments in all aspects of the company's operations.

In the late 1970s, Brown & Root became involved in another exceptional development, this time in the warmer waters off Mexico. Mexico's national oil company wanted to bring large offshore fields in the Bay of Campeche into production quickly, and Brown & Root played a major role in helping Mexico to increase offshore production from virtually zero to nearly two million barrels per day by the early 1980s. In the midst of this effort, the company also responded to a spectacular well blow-out at Ixtoc with creative measures to contain the oil spill.

Part III reviews the extremely demanding technological challenges posed by oil and gas development in the unforgiving waters of the North Sea. Brown & Root was one of the first construction companies to enter these waters. During the 1960s and early 1970s, it laid all the gas pipelines and built most of the platforms in the southern North Sea. Brown & Root was also a chief participant in developing the first major North Sea oil deposits at the Ekofisk and Forties fields in the early 1970s. These projects marked turning points both for North Sea oil and for Brown & Root Marine. As project manager for the installation of four giant platforms at the Forties, the company stretched its technical and managerial resources to new heights. Few other Brown & Root Marine projects offered the same measure of risk and reward.

Through its growing British subsidiary, Brown & Root expanded its involvement in the North Sea, laying pipelines and installing platforms in the roughest seas yet encountered by the offshore industry. Rising oil prices into the mid-1980s paved the way for the exploitation of oil and gas in areas previously deemed uneconomical. Along with production from Alaska and Mexico, North Sea crude altered the world's oil supply system in the wake of the OPEC embargo. As an important contributor to offshore development in all three areas, Brown & Root helped international oil companies meet the challenge of the world oil crisis.

The severe downturn in oil prices in the mid-1980s presented a much different challenge for Brown & Root Marine and the entire offshore industry. Brown & Root joined other companies in a gut-wrenching reexamination of traditional practices. Severe cutbacks, consolidations, and an elevated concern for the bottom line transformed the industry, allowing a return to profitability for those companies that survived. As the demands of the offshore industry became more technologically advanced and more expensive to meet, Brown & Root Marine faced hard choices, including a shift in focus from its traditional strength in construction toward its growing expertise in design and engineering. In the 1990s, the consolidation of the energy services of Brown & Root and its parent company, Halliburton, strengthened the capabilities of each. The thrust by some companies into extremely deep waters of the Gulf of Mexico in the late 1990s portends a new era of offshore expansion made possible through the aggressive application of cost-cutting technologies. This exciting new era seems to be only the latest example of the offshore industry responding to environmental and economic constraints with technological innovations.

Brown & Root always has prided itself on its "can do" attitude in overcoming such constraints. Initial setbacks or failures rarely detered the search for a practical engineering solution. The offshore industry has given acclaim to Brown & Root for its record of innovation and reputation for completing demanding work on time and on budget. Representatives from the offshore industry have met annually since 1969 at the Offshore Technology Conference (OTC) in Houston, where they exchange technical information, make deals, and renew old acquaintances. In 1994, the OTC awarded Brown & Root its Distinguished Achievement Award for Companies, which recognized the firm's "sustained and continuing contributions to the design and construction of offshore facilities in many physical environments and under a wide range of technical and commercial conditions." Since Brown & Root's individual achievements were possible only in the context of the evolution of the industry as a whole, its history is also a partial history of the offshore fraternity to which it belongs.

PART 1

The First Challenge: Creation of the Modern Offshore Industry in the Gulf of Mexico

CHAPTER 1

Before the Dawn

The search for oil and gas has known few boundaries. Until the 1930s, however, one such boundary throughout the world was the point at which land met ocean. Before that time, the petroleum industry had found innovative ways to develop oil fields beneath the waters of shallow lakes and protected bays. It had created elaborate systems of piers to extend wooden fingers from the land out into the Pacific Ocean in California. But the winds and waves of the open seas had blocked the development of offshore oil or gas. In the 1930s several oil companies decided it was time to explore the possibilities for offshore exploration and production. For a time most of their activity came offshore California, but in the late 1930s, several Texas-based oil companies made the first forays into the the Gulf of Mexico, and Brown & Root went with them.[1]

The Great Depression and an unprecedented glut of oil had dampened the prospects of the oil industry as a whole, much less the prospects for a risky new venture into what were literally untested waters. Nonetheless, a handful of companies looked offshore, drawn by economic incentive to find large oil and gas fields and curiosity to find out what was out there. Before World War II halted their efforts, the successes and failures of these pioneers introduced the industry to the challenges to be expected by those who hoped to conduct operations offshore.

A Construction Company in Search of Opportunities

Brown & Root began as a road building company in central Texas. Herman Brown had entered this business in 1914 after his graduation from Temple High School. When his employer faced bankruptcy several years later, Herman received the company's meager assets in lieu of back wages. At the tender age of 22, Herman Brown became the proud owner of several mule teams and the rudimentary equipment used to grade roads in this era. He set out with a vengeance to build a successful road construction company. Hard work and the timely financial assistance of his brother-in-law, Dan Root, helped Brown & Root survive and prosper in the 1920s. In 1922 Herman's younger brother George, who had a degree in mining engineering from the Colorado School of Mines and several years experience with Anaconda Copper, joined the firm. The brothers gradually became established road builders in central Texas.

The Great Depression curtailed most road building, forcing the Brown brothers to scramble for work. George had moved his office to Houston in 1926, and in an effort to keep their crews busy, Brown & Root at one point in the 1930s even hauled that city's garbage. The company turned the corner in 1936, when it won the contract to construct the Marshall Ford Dam (later renamed the Mansfield Dam) on the Colorado River above Austin. This project propelled Brown & Root to a new status as a significant regional construction firm while also expanding its engineering capacities.

Being in Houston in the 1930s meant being near the center of the nation's oil and gas business. The Gulf Coast of Texas had long been a focal point of exploration and refining, in part because of the prolific production from early wells near the salt domes so prevalent along the coast. Inevitably, exploration moved out from land into bays and the shallow water offshore. Galveston Bay presented an obvious target of opportunity for Houston-based oil companies, and the first geological survey of the bay was conducted in 1933 using the torsion balance system.[2] Earlier, in 1928, a submarine had been used to survey parts of the Gulf of Mexico using the gravity method, which was supplemented by early efforts at coring in 1934.[3] By the late 1930s, several prospects had been identified either entirely offshore or out in the water connected to onshore finds.[4] Petroleum geologist Michael Halbouty captured the enthusiasm for prospects offshore in an article entitled "Probable Undiscovered Stratigraphic Traps on Gulf Coast" in *World Petroleum* in June 1938. This article included discussion and drawings of possible traps in the shallow waters of the Gulf of Mexico.

Encouraged by such studies, several major oil companies began drilling in Galveston Bay and off the Texas and Louisiana coasts in the late 1930s. Looking past the temporary problems of the 1930s, these companies decided to look offshore in the hopes of finding large fields of the sort that had become increasingly difficult to find onshore in the United States. Their competitive instincts launched them offshore in search of new fields. Most major companies already had refineries on the Gulf Coast that would provide a ready market for oil from offshore in the Gulf of Mexico. The risks were high—offshore work would entail higher costs and extreme challenges in adapting existing technology to new conditions. But several companies with strong ties to the Houston area, notably the Humble Oil & Refining Company and the Pure Oil Company, became offshore pioneers in the late 1930s, and Brown & Root joined them as a contractor.

Testing the Waters

Brown & Root had no previous experience with offshore construction, but the Browns had never hesitated to try new things. When questioned about Brown & Root's lack of experience in building dams before it took on the Marshall Ford project, George Brown responded: "To be road-builders you have to know about concrete and asphalt. You have to learn something about bridges. Once you learn these things, it's a small step, if you are not afraid, to pour concrete for a dam."[5] The Browns fostered an entrepreneurial, "can do" attitude in their company. When opportunities near Houston opened up in the early years of offshore development, Brown & Root jumped at the chance to apply its general construction skills to the task of helping oil companies create the facilities needed to produce oil from under water.

The company took its first step out into the water on a project for Humble Oil & Refining Company in Galveston Bay in the late 1930s. Until then, Brown & Root's experience in oil-related construction consisted of small "set up" jobs to prepare drilling sites, and constructing roads in the newly discovered Conroe, Texas, oil field about thirty miles north of Houston. Oilfield construction had evolved as a specialized industry separate from the general construction practiced by Brown & Root. As the company sought to apply its experience in general construction to the needs of offshore oil development, Brown & Root got in on the ground floor of what would become a thriving industry.

Humble Oil's move into Galveston Bay was a logical extension of its traditional leadership in exploring for oil on the Gulf Coast. The largest oil company in Houston, Humble had been incorporated in 1917 to develop properties in the Houston area. Standard Oil of New Jersey acquired a controlling interest in Humble in 1919, cementing its status as the leader in the southwestern oil industry. In the early 1930s Humble had developed a major new field at Friendswood, a small town near the western bank of Galveston Bay. When Humble became interested in exploring the bay after another company discovered oil there,[6] it called on Brown & Root for help. Brown & Root had developed a relationship with Humble by doing some hard construction jobs all across southern Texas and Louisiana. Whatever work Humble offered, the struggling construction company accepted, including moving rigs and equipment in the Friendswood field.

Humble and Brown & Root sought the least expensive means of recreating onshore conditions for drilling. If a land-like drilling site could be created out in the water, Humble could get on with the task at hand. No special construction was required to protect against rough waters or the threat posed by hurricanes, since the bay was relatively calm and at least partially protected from high waves and wind.

Industry experience drilling in inland lakes and protected bays proved instructive. Oil had been produced in Caddo Lake in northern Louisiana since 1911 by drilling from platforms built out of cypress trees from along the lake's shore. In the 1920s and 1930s Lake Maracaibo in Venezuela sprouted a forest of platforms and derricks. In these protected waters, much of the equipment used in drilling could be housed on tender vessels or even onshore, reducing the need for the construction of large platforms. In the bayous and marshes of "inshore" Louisiana, however, drillers tried any means to create a solid surface from which to work. They tried drilling barges, submersible barges resting on "mats" constructed from oyster shells packed on the bottom of the bayou, and platforms supported by wooden pilings at the end of plank roads or canals dredged through the bayous. Humble and its affiliates within Standard Oil of New Jersey (now Exxon), if not Brown & Root, had experience with all of these approaches to drilling in bays, and the company tried several of them in its search for oil in Galveston Bay.

In retrospect, Brown & Root's earliest ventures offshore seem singularly unremarkable, even in the context of contemporary activities in Venezuela and Louisiana. But its work in Galveston Bay for Humble Oil laid the foundation for the evolution of Brown & Root's marine operations. There, the company drove its first piles for an offshore platform. To aid in this work, Brown & Root recruited Frank Motley, who had experience in pile driving for a bridge company. Motley later recalled, "When I first started working for Brown & Root, we had no men with offshore experience or technical know-how. All we had was ambition."[7]

In addition to building pile-supported drilling platforms, Brown & Root prepared sites for drilling barges in the bay. Basil Maxwell, one of the the company's original offshore workers, recalls: "We would drive a 40–50 foot two-pile trestle . . . to tie the barge to . . . And if the bottom wasn't right, we would either add oyster shell or take out a little if we had to." Building mats was not, however, to remain an important part of the company's marine operations. Brown & Root's onshore experience in piledriving and heavy construction made it a natural for the construction of drilling platforms requiring piles.

In Galveston Bay the company also laid its first submarine pipeline, entering another area of the offshore business that became an enduring strength of the company. In 1936 Brown & Root laid a short pipeline for Humble Oil from an offshore tank to onshore facilities. The pipeline was welded onshore, floated to the site using barrels for support, and sunk to the bottom of the bay. Frank Motley was interested in the practical problems raised by underwater pipelines, and he developed several major innovations for laying pipe offshore. He led the way in developing a practical machine for burying pipelines by using high pressure jets to dig the sea bottom out from under the submerged pipe, creating a trench into which the pipeline could sag under the guidance of a sled pulled along the pipe by a barge.[8]

This innovation, as well as Brown & Root's first efforts driving piles offshore and building platforms, was a lasting legacies of the company's initial foray into Galveston Bay. This early venture allowed the company to develop and test new construction techniques in relatively calm, shallow, and protected

Early pipelaying

waters before facing the more demanding environment of the Gulf of Mexico. Brown & Root came away with a good relationship with an important new client, a nucleus of "marine" specialists with at least a modicum of practical experience, and an inside look at an emerging new field of design and engineering.

As Humble and other companies poked holes in Galveston Bay in a disappointing search for substantial deposits of oil and gas, they took a hard look across the Bolivar Peninsula to the Gulf of Mexico. This narrow peninsula had long seemed a promising prospect for exploration. Humble joined other oil companies in staking out leases off this segment of the upper Texas Gulf Coast. Before World War II the first offshore wells in Texas waters were drilled in this region, and Brown & Root worked with Humble Oil to complete these wells at McFadden Beach up the coast from High Island, a small coastal community located on a salt dome some thirty miles from the Texas-Louisiana state line.

The exploratory wells in this area took a half step into the Gulf of Mexico, retaining a tenuous connection to land in the form of a trestle that reached several thousand feet out into the Gulf. In the spring of 1938 Brown & Root constructed this trestle, which was widened at several points to create derrick foundations. Off the end of the trestle approximately a mile from shore, a separate drilling platform was constructed at a cost of about $38,000. A train track built on the trestle carried equipment and supplies for the rigs, and a boat serviced the separate platform from the end of the pier. A plank road

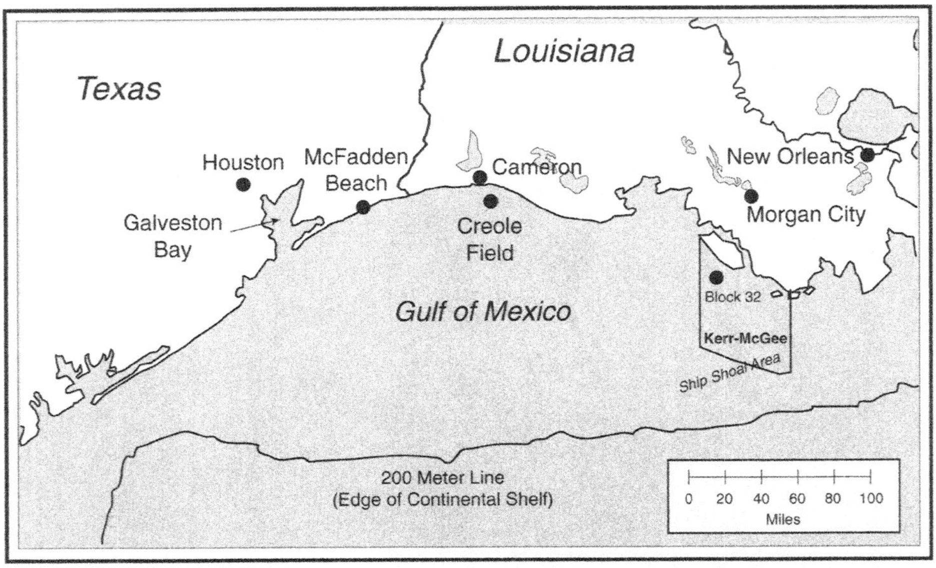

Early Gulf of Mexico

over the sandy beach facilitated movement onto the trestle. This trestle arrangement was similar to the piers that had extended drilling out into the Pacific Ocean along the southern California coast at the turn of the twentieth century. Beginning in the late 1920s, extensive trestle systems had also been constructed offshore from Baku into the Caspian Sea in the Soviet Union. Unlike the Pacific Ocean and the Caspian Sea, however, the Gulf of Mexico never witnessed the use of trestles over a long period of time.

The experience of Humble and Brown & Root at McFadden Beach suggests why this was so. Hundreds of wooden piles driven into the soft soil of the Gulf supported the trestle and the rigs that sat on it. A separate 50-foot by 90-foot platform built off the end of the pier in 18 feet of water also was supported by wooden piles driven into the ocean floor. When this well proved dry, Brown & Root constructed a second, larger platform in 15 feet of water.[9]

Within nine months of completing the facility, problems of this approach had surfaced. The wooden piles provided a shaky foundation for drilling. They were weak structurally, and they had not been driven deeply enough into the ocean floor to provide a solid base capable of supporting the pipe and equipment needed for deep drilling. Even creosoted wood proved vulnerable to ocean borers, whose attacks on the pilings further weakened the structure. In August 1938 a hurricane damaged the platform, the trestle, and the onshore support facilities. One eye-witness remembered it as "pretty ferocious. In fact, we lost a rig from the trestle one time. A hurricane came along and knocked the piles out from under it." Such problems convinced Humble to give up on the McFadden site, where it had not yet found sufficient oil to justify the expense and the uncertainties of its first venture offshore.

McFadden Beach pier

The episode served as a warning of the difficulties of extending onshore or even inshore technology into the Gulf of Mexico. Wooden structures would be hard pressed to withstand oceanic conditions or support the heavy weights required to drill to great depths. Missing was the basic knowledge about waves, weather, and soil needed to make informed engineering decisions about design and construction. The cost of building trestles and platforms for exploratory drilling was sobering; a dry hole meant the loss of most of the investment in these facilities, which were not easily salvaged. The railroad track on the trestle symbolized one of the great challenges: the installation and servicing of structures built at sea and not connected to land.

Out Into the Gulf

Brown & Root had one subsequent opportunity to address such questions before World War II. In a path-breaking project for Pure Oil and Superior Oil beginning in 1937, Brown & Root constructed a mammoth wooden platform complex for the Creole field about one mile out in the Gulf south of Cameron, Louisiana. In March 1938 this site produced the first oil in the open waters of the Gulf of Mexico. This was both the climax of the developments in offshore drilling in the 1930s and the forerunner to the dramatic surge in such activity after World War II. While putting the oil industry out into the Gulf, it also revealed the need for fundamental technological changes if the industry was to stay there.

Pure Oil became interested in exploring the extreme western coastal region of Louisiana in the mid-1930s as part of the general development of the marshlands of south Louisiana. However, the company had great difficuties exploring the area around Cameron. Lake Charles, some forty miles to the north, was the closest point served by a good highway. The traditional transporation link between Lake Charles and Cameron had been steamboat service via Calcasieu Lake, and the first road between the two towns opened only shortly before Pure Oil began its explorations. Reflection seismograph surveys conducted in 1936 from swamp buggies indicated salt domes in the area east of Cameron. The prospect seemed to extend beyond land between Cameron and the mouth of the Mermentau River. Pure and Superior Oil went in together to lease 7,000 acres onshore and 33,000 offshore. After onshore test wells came up dry, the companies turned their attention to the offshore lease. The companies decided to move farther offshore to drill what became known as Number 1 State Gulf of Mexico well.[10]

Brown & Root was a logical choice to build the platform for Pure's project. Its work with Humble Oil had made it one of only a handful of companies in the region with experience in offshore construction. Its headquarters in Houston was within about a hundred miles of the drilling site, which was only about seventy-five miles up the Gulf coast from the Humble site off McFadden Beach. In addition, Brown & Root had personal connections into Pure Oil through Judge James Elkins, whose Houston-based Vinson, Elkins law firm represented Pure Oil. George and Herman Brown were friends and frequent business associates of Elkins, and this tie no doubt helped smooth the way for Brown & Root to become the contractor in the Creole field.

The drilling site was only about a mile from the nearest point onshore, but it was ten to twelve miles away by water from Cameron, the logical supply point. In practice, this site presented all of the problems of a location out of sight of land. Hines Baker, president of Humble Oil & Refining in the late 1940s, felt that "to oil men with twenty years of development and experience in drilling in protected waters, moving out to the open Gulf was a natural and logical step." He noted that the Creole field "utilized techniques developed in swamplands and barge operations to open a new phase of drilling methods."[11] Just how much was borrowed from previous practices and how much was new could be seen in the detailed descriptions of the project published in professional journals by I.W. Alcorn, one of the Pure Oil engineers who designed it. His description and the remembrances of Brown & Root employees who worked on the project also provide benchmarks for measuring subsequent progress in offshore design.

The companies involved wanted a sturdy foundation for drilling and production with a life of at least twenty years. They considered building an island of sheet steel piling, driving wooden pilings and then using concrete footings around the top of the piles to extend steel columns out of the water, and building a long trestle. A very large wooden platform seemed to provide the sturdiest, most dependable, and least expensive alternative, and construction began in June 1937.

Brown & Root took a general design and adapted it to conditions in the open Gulf. The original plan called for a system of platforms supported by 300 creosoted wooden piles made from southern pine approximately eleven inches in diameter. The platform system was to be 180 feet by 320 feet, with

room for a derrick, drilling mud, pipe racks, water tanks, and power supply but no quarters for the working crew. This ambitious project called for a large artificial island on which everything needed to drill a well could be placed.

The water depth of about fifteen feet allowed for the use of flat-top barges for driving the piles, which were pounded through about six feet of silt and 28 feet of hard pan sand. This was Brown & Root's first experience installing a platform from floating equipment. One of the company's employees later recalled this "stickbuilding" process used to construct a platform while battling the elements: "We drove all the piles and staged them with boards (and) tied them together. We had a double row of the largest piles in an array for the derrick foundation." The rough waters of the open sea introduced new challenges, some big and some small: "If we dropped anything, we went down and got it, without scuba gear."

Design adjustments were made as construction went forward. The original plan included the use of a steam powered rig and called for the building of a separate platform attached by a pier to the main platform to house four large fuel tanks. Switching to a diesel-electric rig eliminated the need for this separate platform, reducing the number of piles needed by 45. A second set of changes had nothing to do with economic efficiency: "We had a hurricane that knocked out everything, all the pilings except the large ones we had for the drilling platform. So, we go back and offset the design a little bit to miss the piles we had put in and drove some more." Such "construction on the run" was the sort of thing that Brown & Root had always relished; the company took great pride in its ability to find practical solutions to construction problems.

One difference in onshore and offshore construction was readily apparent. Government regulation of all structures in navigable waterways required greater concern for safety and control of pollution. The War Department through the Army Corps of Engineers regulated the seaworthiness of coastwise tugs and barges used in the ocean. Their requirements for offshore operations ranged from minor regulations, such as special lighting on the derrick and the installation of buoys around the platforms, to major ones, such as devices to eliminate the loss of oil into the water and new provisions for shutting in a well in an emergency.[12]

The Creole project demonstrated how economic considerations encouraged technological innovations. Engineers were learning as they went along about the costs of offshore work, and they were quick to search out savings that would make offshore work more competitive with onshore work. For example, the switch from a steam to a diesel-electric rig was driven by economies to be gained through "slim-hole" drilling, which drilled a smaller hole for exploratory wells. Greater savings came from an experiment with directional drilling. The central platform for the Creole field had cost about $70,000, and a similar expenditure for separate platforms for the operations of each developmental well in the field would make the total cost of the project prohibitive. Instead, Pure Oil and Superior Oil embarked on a directional drilling program from the central platform, which was enlarged to accommodate the operations of eleven producing wells. These wells ultimately reached out from the central platform in a circle that encompassed almost 300 acres. Although described as "unusual" in 1940, the use of directional drilling subsequently became the norm for offshore operations.[13]

The engineers who designed the platform for the Creole field and the workers who constructed it identified and partially answered key questions about offshore structures. Before driving the piles, they made and recorded what were, for the time, careful core samples of the soil that would support the platform. Their use of hundreds of piles resulted from careful analysis of the weight-bearing capacities of the wooden piles, not from some sense that if enough piles were driven, the structure was bound to stand up. They calculated the benefits of round piles compared to flat ones, arguing that "the resistance (i.e. drag) of a flat surface having a width equal to that of the

Creole platform

diameter of the pile, is probably five times that of a round surface."[14] They could not, however, apply this insight by using tubular joints, since prevailing construction technology used metal connections to hold together wooden structures. On such key issues as soil composition, structural strength, and the need for rounded surfaces in structures exposed to waves, Brown & Root and its partners made obvious progress in the Creole field.

The most significant design limitation of this era was the total lack of knowledge about the effect of hurricanes on offshore structures in the Gulf of Mexico. Little data on wind speeds or wave heights existed. Engineers designed the Creole platform to withstand 150 mile-per-hour winds, but they knew that waves, not wind, were the central concern. They protected against waves by adding piles, cross-bracing the piles on the north and south sides of the platform, and building the deck high enough to prevent waves breaking on the platform. But with decks only 12–15 feet above mean sea level, one of the designers acknowledged the inevitability of serious damage in a hurricane. In 1938 he advised other designers of offshore platforms to prepare for the worst a storm could offer by placing the decking "so that if in the event of sufficiently severe storm waves should strike the bottom, the decking would be washed off without further damage to the structure."[15] The limits of this approach became evident in future decades, when the original platform had to be substantially rebuilt to withstand the force of hurricanes.

The other obvious limitation of this early platform design was the lack of living quarters for the crews. Despite its giant size, the platform was not truly self-contained or self-sufficient. This reflected legitimate fears of the dangers hurricanes posed for offshore workers in an era with slow and unreliable transportation to and from the rig and no effective early warning system for storms. The lack of quarters created hardships for those who operated the platform. Each day workers commmuted from the onshore base of operations at Cameron to the platform, a ten- to thirteen-mile trip each way. Rough seas on the one- to two-hour journey tested the mettle of even the most experienced boat riders. Frequently, they had to turn back after reaching the platform, since rough seas often made it impossible for workers to climb from the boat up the rope ladder to the platform. This proved costly for the men and the companies, since the wells could not produce without a crew.[16]

The difficulties of commuting to and from the platform revealed a practical problem that surpassed design issues. If a company such as Brown & Root hoped to develop a substantial business in the construction of offshore structures, it needed the services of companies in related businesses that had not yet seen sufficient opportunities for profits to risk entering the Gulf. In these earliest years of offshore development, almost no dependable supply or service companies operated in the Gulf. Workers were carried to and fro by chartered shrimp boats, which also towed out flat-top barges and at times even carried equipment to the platform when barges were not available. Heavy fog often blanketed the region. Given the lack of communications between the platform and land crews, supply boats had little option except to turn off their engines and listen for the sounds of the platform. Obviously, offshore operations could not progress much beyond the point reached in 1938 without the development of better means of communication and new ways to deliver services and supplies to the platforms.

Such limitations led one engineer who worked on the Creole well to describe it as "an onshore operation offshore."[17] However, an initial investment of about $150,000 created a complex of eleven wells that quickly produced 1,500 to 2,000 barrels of oil per day. Over the thirty years after the first well came in, this field produced approximately four million barrels of oil.[18] The Creole field stands as a historic symbol of the possibility of profitable operations out in the Gulf.

Yet the tentative steps taken at the Creole field and with other pioneering initiatives before World War II have been largely forgotten. This reflects, in part, the limitations inherent in the application of onshore technology to offshore operations, but bad timing also explains why this venture attracted so little attention from either contemporaries or historians. The first day of production from the Creole platform, March 18, 1938, was a crowded one in oil history. On that date the expropriation of the properties of U.S. and British companies in Mexico captured newspaper headlines around the world, riveting the attention of oil executives and government officials on the issue of sovereignty over oil deposits. On that same day came news of the successful completion of the discovery well for the first major field in Saudi Arabia. Although this event escaped wide public notice at the time, it has since been noted by historians as the beginning of an epoch in oil history, the movement of the center of oil production to the Middle East. The coming of World War II overshadowed all other events in the late 1930s. When the war finally ended, even those who had helped develop the first fields in the Gulf became too busy developing new fields to take much time to remember the prewar roots of the offshore industry.

Brown & Root nonetheless reaped lasting benefits from its early start in the Gulf in the 1930s. One careful study concluded that "approximately twenty-five wells were drilled from conventional pile foundations in shallow water off the Gulf Coast from 1937 to 1942."[19] In its work for Humble Oil and Pure Oil, Brown & Root constructed the platforms used to drill about half of these wells. The company had staked out leadership in a new type of construction. It had taken a seat-of-the-pants approach, developing solutions to practical problems as they arose. Working with no proven models for the construction of offshore structures and little assistance in the provision of supplies and services, Brown & Root and its clients made a start at defining key design and construction issues and attempting to devise practical solutions by adapting existing technology. World War II suspended this promising evolutionary process, diverting the energies of Brown & Root into war production and taking the company's key offshore personnel into the armed services for the duration. As workers completed the expansion of the platform and the drilling of directional wells in the Creole field, they kept a wary eye out for the periscopes of German submarines. Years would pass before workers could put the war behind them and return to the job of creating a new industry out in the Gulf of Mexico.

Notes

1. This chapter draws from the interview with Basil Maxwell. For the best general history of the offshore industry, see Hans Veldman and George Lagers, *50 Years Offshore* (Delft, Holland: Foundation for Offshore Studies, 1997). Also, see the historical anniversary issues of *Offshore,* including the twenty-fifth anniversary issue in September 1979, the fortieth anniversary issue in April 1994, and the fiftieth anniversary of the industry in May 1997.
2. "There Is Where the Oil Is," *Brownbilt* (Fall 1972): 8.
3. "Marine Exploration is Coming of Age," *World Petroleum* (March 1954): 76.
4. "Drilling Wells Off Shore in Texas Bays and Inlets," *Brownbuilder,* 75th Anniversary issue, 114.
5. Chris Castaneda and Joseph Pratt, *Builders: The Brown Brothers of Houston* (College Station: Texas A & M University Press, forthcoming). Also, Jeffrey L. Rodengen, *The Legend of Halliburton* (Fort Lauderdale, Florida: Write Stuff Syndicate, 1996).
6. Henrietta Larson and Kenneth Porter, *History of Humble Oil & Refining Company: A Study of Industrial Growth* (New York: Harper & Brothers, 1959).
7. "A Pioneer Among Pioneers," *Brownbilt* (Fall 1972) 20–22.
8. "There Is Where the Oil Is," 8.
9. Jack S. Toler, "Offshore Petroleum Installations," *Proceedings of the American Society of Civil Engineers* (September 1953): 289–3; Griff C. Lee, "Offshore Structures: Past, Present, Future and Design Considerations," Offshore Exploration Conference, New Orleans, Feb. 14–16, 1968, 2.
10. I.W. Alcorn, "Marine Drilling on the Gulf Coast," *Drilling and Production Practice* (American Petroleum Institute: 1938), 40–45; I.W. Alcorn, "Derrick Structures for Water Locations," *Petroleum Engineer* (March 1938): 33–37; *Pure Oil News* (March 1938), as reported in "First Well Drilled in Gulf of Mexico Just 25 Years Ago," *Offshore* (October 1963): 17–19.
11. "Marine Production Gains Place in Oil Industry Picture," *World Petroleum* (March 1948): 44.
12. Alcorn, "Marine Drilling," 46; Alcorn, Derrick Structures," 36–7.
13. C.D. Lockwood's Reference Report, Southern Louisiana, 1940, 97.
14. Alcorn, "Derrick Structures," 36.
15. Alcorn, "Derrick Structure," 36; Alcorn, "Marine Drilling," 46.
16. "First Well," 17–19.
17. "First Well," 18.
18. These figures are taken from Lockwood, 97; "First Well," 19; and Alcorn, "Marine Drilling," 46.
19. Toler, "Offshore Petroleum Installations," 289–3.

CHAPTER 2

Beyond the Horizon

World War II stopped the move offshore in its tracks, as mobilization for a war on two fronts became the top priority of the nation, the oil industry, and Brown & Root. Even after the end of the war—indeed, to the end of the Korean War in 1953—military demands and corresponding steel shortages constrained the growth of the offshore industry. During these same years, state and federal governments bickered over ownership of offshore lands. Before the Tidelands Act in 1953 established a legal framework for leasing, disputes over rights to the tidelands shut down most activity offshore. Yet despite military and legal constraints in the years from 1941 to 1953, the offshore oil industry made notable advances. As companies first entered the Gulf, they identified the technical challenges they would have to overcome to create an efficient technical system for offshore oil and gas production. The competitive race to build such a system then began in earnest, as companies explored alternatives in designing and building offshore facilities.[1]

Offshore—In Ships

The Brown brothers' focus from 1941–1945 was on war production. Brown & Root participated in a three-company consortium that built a $90 million naval air station at Corpus Christi, Texas, and in 1941 the Browns took charge of a small contract to build four PC boats (Patrol Crafts or subchasers) in Houston. Upon entering into this project, George Brown admitted "We had never seen a ship built, but we decided we could build one, even though we did not know a bow from a stern."[2] This first venture in shipbuilding led to contracts to build subchasers, destroyer escorts, LSIs (landing ship infantry), and LSMs (landing ship medium, a vessel used to transport tanks and other heavy equipment). By the end of the war, Brown Shipbuilding Company, which was organized separately from Brown & Root, had built more than 350

Brown Shipbuilding—Greens Bayou during World War II

combat ships, more than 300 landing craft, and smaller numbers of pursuit craft, rocket-firing boats, and salvage boats. While constructing ships valued at more than $500 million, Brown Shipbuilding employed as many as 25,000 workers. The Browns established a reputation for completing their defense contracts on budget and on time, and they received favorable national publicity, five Army-Navy "E's," and a Presidential Citation for their efforts.

This work enhanced the Browns' regional and national reputations and pushed their construction activity to a new scale. To construct these ships as rapidly and efficiently as possible, the Browns acquired a 180-acre track of land at Greens Bayou, about ten miles east of Houston with access to the Gulf of Mexico through the Houston Ship Channel. After the war Brown & Root used a segment of the site for its Greens Bayou yard, which was used to support both offshore and onshore oil and gas-related construction. Even more important to the company's future were the more than one hundred engineers originally hired by Brown Shipbuilding. After the war, the Browns retained many of these engineers, who became the core of an engineering design group that helped transform what had been essentially a construction company into a construction and engineering company.[3]

With its newly enhanced engineering capacity, Brown & Root expanded and diversified aggressively after World War II. The Browns built roads, dams, military installations, petroleum and chemical plants and other industrial facilities, onshore pipelines, and offshore oil and gas platforms and pipelines. The company's total assets grew from $6 million in 1946 to $27 million in 1954 to $61 million in 1962. By the late 1960s, Brown & Root regularly appeared at or near the top of the *Engineering News-Record's* annual listing of the nation's largest contractors.[4]

Although Brown & Root took on projects all over the world after the 1940s, it retained a particular interest in projects near its home base in Houston. Speaking to an audience of McGraw-Hill editors in 1952, Herman Brown called the Southwest his "first love," remarking that "you can't do business all over an area for 38 years without feeling you have a certain proprietory interest in the locality. . . . Every rooster is king of his own dung-hill, and in the Southwest Brown & Root can simply do a job cheaper than most of its competitors."[5] Oil, gas, and petrochemicals fueled an economic boom in the Southwest after World War II, and Houston-based Brown & Root became one of the biggest players in the specialized construction required by these closely related industries.

Brown & Root benefitted from its experience in oil and gas construction, its close ties to major companies, and its Houston location—home of the national headquarters for the major natural gas transmission companies, many independent oil producers, and regional offices and refineries for most major oil companies. Brown & Root's leadership in oil and gas-related construction became one of its defining characteristics, which encouraged the growth of a specialized marine division with an emphasis on offshore oil and gas work.

The company developed an extremely close tie with the natural gas pipeline business that expanded after the war. The Brown brothers led a group of investors that acquired the Big Inch and Little Big inch pipelines from the federal government in 1947 and then used these lines to create the Texas Eastern Transmission Company. The inch lines had been built during the war to carry crude oil and petroleum products from the oil fields and refineries of the Southwest to the northeastern United States. With their purchase and conversion to carry natural gas, the Browns entered one of the nation's fastest-growing industries in the 1950s and 1960s, the transportation of natural gas from the fields of the Southwest to the markets of the Northeast.[6] They also helped create a major customer for Brown & Root, which completed an estimated $1.3 billion in construction work for Texas Eastern from 1947 to 1977,[7] with numerous projects offshore in the gas fields of the Gulf of Mexico.

The post-war era witnessed a boom for both the oil and gas industries. Demand skyrocketed. The United States became a net importer of oil in the late 1940s, and fear of growing dependence on imports made offshore oil production attractive as a domestic alternative to foreign oil. Gas had long been the neglected stepchild of oil production, but the emergence of a pipeline system capable of moving southwestern gas more than a thousand miles to markets created an opportunity to build gathering lines for gas out into the Gulf to connect offshore gas discoveries into the emerging national gas pipeline grid.

Aggressive expansion of offshore oil and gas could not occur in the post-war years, however, until the tidelands issue was resolved. Before the war, offshore drilling sites had been leased from state governments, although federal agencies such as the War Department and the Coast Guard had regulated the design and construction of these and other structures built in the nation's navigable waters. The states and the federal government asserted conflicting claims of authority to lease these submerged lands, and legal warfare raged. With millions of dollars in leasing bonuses at issue, compromise was unlikely. Oil executives lobbied hard for the "states' rights" position favored by most in the industry. Before legislation resolved this issue in 1953, court decisions in 1950 essentially shut down expansion in the Gulf, but not before new offshore projects had explored the effectiveness of several approaches to drilling wells in the oceans.

The Challenges of the Open Sea

Several companies led the way into the Gulf in the late 1940s. Humble Oil was the early leader, but other large integrated companies with operations on the Gulf coast joined in the search for offshore oil and gas. Such companies as Shell Oil, Magnolia (the Texas affiliate of Standard of New York or Mobil Oil), and the Texas Company all recognized the potential of the Gulf of Mexico. These worldwide leaders in the oil industry entered promising new producing areas, but the Gulf was particularly attractive to them because they had major refineries on the Houston Ship Channel and in the Beaumont-Port Arthur area.

Other companies led by Kerr-McGee, Pure Oil, Superior Oil, and the California Company also entered the Gulf of Mexico after World War II. Still others followed, including Standolind, Continental Oil, Phillips Petroleum, Sun Oil, Standard Oil of Texas, and the Texas Eastern Production Company.[8] Each had its own incentive for investing in the risky waters of the Gulf, but as Dean McGee of Kerr-McGee said, "We had strong competition from the large companies to find productive (onshore) leases. We realized finally that we had to go somewhere else." Betting its future on the potential of offshore production, Kerr-McGee "decided to get out of the land-drilling business . . . and went to the Gulf."[9] Few other small companies placed such faith on the future of offshore production, but those who entered the Gulf in the late 1940s had calculated their prospects for making major finds offshore and found them to be greater than their prospects for making significant new finds in the more thoroughly explored lands onshore.

But rewards would come only if efficient systems for retrieving offshore oil could be developed. The oil companies that looked offshore and the construction and service companies that supported their work faced severe challenges in the open sea, with little experience to point their way. The construction of seawalls and piers had created an understanding of the basic forces affecting the design and construction of such onshore structures exposed to the seas. But other than a few lighthouses built in shallow waters off the Florida Keys, engineers found no useful models for designing structures capable of providing sturdy, durable working spaces in the open sea. No one had ever needed to work for extended periods on artificial islands in the ocean. Basic engineering principles could suggest starting points, but practical experience offshore would be required to test these principles and create technologies for the new offshore environment.

In most respects, the Gulf was an excellent place to perfect new technology for offshore production. Its gentle slope meant shallow water far out from shore and a relatively flat bottom. Its waters were often calm, and it was near centers of oil-related activities, where specialized construction and services could emerge from existing companies.

The big drawback to exploring in the Gulf was hurricanes. These storms rose quickly and unpredictably from June through November, sweeping through the Gulf with harsh winds and high waves. Designers had to consider not only the force exerted by the normal action of winds and waves over the projected life of an offshore structure, but also the extreme forces that might be exerted if a major hurricane developed.

The early designers had little systematic data on hurricanes. Engineers could fortify offshore structures against severe hurricane winds, but the unpredictable forces exerted by hurricane-related waves presented more difficult design problems. Before companies invested as much as $1 million in building an offshore platform, they needed an answer to one question in particular: how high should the deck of a platform be in different depths of water to avoid having waves break on top of it, washing it away or damaging it?

The experiences with hurricanes before World War II suggested a short answer to this question: decks should be higher than those that had been built off McFadden beach and in the Creole field. This answer supplied only a low range—decks fifteen feet or less above the mean water level of the Gulf were too low—without helping to establish the upper range for a safe deck. Engineers seeking designs that would be both cost-effective and sturdy enough to survive a hurricane could only guess how high was high enough.

When designers sought expert opinions supported by data and research on wave action, they discovered little. A handful of specialists in universities and weather forecasting supplied estimates of probable wave heights in the Gulf, but, given the lack of concrete measurements, these were little more than guesses. In October 1947, for example, an authoritative paper excerpted in *The Oil and Gas Journal* stated that "in 100 feet of water or less, waves will seldom exceed twenty feet in height and even this size of wave will be exceptional." This article suggested that "the deck of an open structure erected offshore in the Gulf should therefore be raised at least twenty feet above the still water line."[10] Another oceanographer ventured a higher number in 1947, estimating that a 32-foot wave was the maximum that could develop in the Gulf.[11] At about this time, Magnolia was building a platform in the Gulf with a deck height of twenty feet. Others who followed the best available advice at the time paid for it later when waves much higher than predicted destroyed many of this first generation of "low rider" platforms.

One big problem was the lack of basic data about the behavior of hurricanes in the Gulf. Government agencies had made little progress in tracking hurricanes, which were as yet unnamed and referred to simply by date and location. The Gulf Coast was, of course, no stranger to the devastating impact of hurricanes. The worst hurricane-related disaster in American history had killed more than 6,000 people in Galveston in 1900, and numerous storms subsequently had struck the coast near the areas where offshore exploration was spreading along the Louisiana and Texas coasts. But hurricane trackers had not kept accurate or detailed records of the waves created by such storms nor of the frequency of the most severe storms.

The early experts based their projections on the likelihood of a "twenty-five year hurricane," that is, the largest storm projected to hit the region in a twenty-five year period. Oil executives accepted the projections with a sort of wishful thinking, assuming that these forecasts would probably not occur during the twenty-five year life of platforms and if they did, the damage would no doubt be inflicted on other companies' platforms. Such thinking underpinned the decisions made by companies to build platforms in the Gulf with deck heights in the 20–30 foot range in the late 1940s.

Soil conditions in the Gulf were also unknown. Structures heavy enough to support deep drilling would need to rest on a firm foundation, but great soil variations were readily apparent, expecially in the comparison of conditions from the mouth of the Mississippi River with those offshore in other areas. Soil samples were vital, since offshore construction would need to withstand the horizontal forces exerted by waves as well as the vertical load forces traditionally considered in onshore construction. Short of a systematic soil survey in the Gulf and greater experience in driving piles and operating in soils off different parts of the Gulf Coast, careful analysis at each new site and overbuilding were the best protections against soil problems. Of course, in a new construction environment, all potential problems could not be anticipated by engineers. No one predicted, for example, that giant mudslides along the ocean floor near the mouth of the Mississippi River could simply devour platforms and pipelines during storms.[12]

Installing a structure in the open sea also posed materials, equipment, and design problems. The "stick building" approach of the pre-war period had involved driving hundreds of wooden pilings and then constructing massive platforms on top of them. Bracing consisted of more wood above the water level and perhaps metal strapping down under the water line. If offshore construction was to keep pace with the demands of the oil industry in the late 1940s, construction companies would have to develop a new generation of heavy equipment far superior to what existed at the time.

Giant barges capable of lifting extemely heavy loads at sea would be needed, as would larger hammers to drive piles. Shortages of materials in the immediate post-war years turned the attention of oil companies and construction companies to war surplus vessels that could be refitted for heavy lifting and carrying at sea. Brown & Root made do by converting a surplus Navy yard freight vessel into the *Herman B* in honor of Brown & Root's founder, Herman Brown. It was initially equipped as a construction barge with a lifting capacity of 35 tons. In the years before "purpose-built" vessels were designed for specialized construction and pipeline work, this multipurpose offshore workhorse and a second such converted craft, the *L.T.Bolin,* carried the load for Brown & Root's offshore construction efforts.

Design problems were often related to equipment capabilities. It seemed clear that steel should replace wood in the fabrication of platforms. The limited offshore lifting capacity and small size of pile driving equipment forced the use of relatively small diameter metal tubing of only about 12" in diameter for pilings. Such small tubes required designs using many pilings, creating the steel equivalent of the wooden forests of piles which had supported early wooden rigs. Not until equipment capacities increased could a new generation of platforms supported by a few large steel piles evolve. As steel pilings grew larger, and taller platforms were needed for deeper water, engineers asked the logical question: how much fabrication could be done onshore to minimize weather-induced delays and other difficulties inherent in stick-building out at sea? Such an approach promised to lower the cost and raise the quality of offshore structures. But breakthroughs in onshore fabrication awaited the development of specialized fabrication yards and specialized equipment to install prefabricated structures at sea.

The combined design challenges posed by waves, soil conditions, and installation at sea presented barriers to the rapid expansion of the offshore industry. Such challenges were heightened by the lack of established firms capable of delivering the goods and services required in the construction and operation of offshore structures. As companies addressed these challenges, they had to make informed investment decisions amid considerable uncertainty. In the years from 1945 to 1953, adventurous companies pushed ahead with the construction of a generation of largely experimental offshore structures while aggressively studying the engineering issues raised by this new environment for oil exploration and production.

The Race Is On

During the race into the Gulf after World War II, three approaches to offshore development emerged: (1) small platforms with tender vessels; (2) larger self-contained platforms; and (3) mobile drilling vessels for exploration combined with permanent platforms if large deposits of oil and gas were discovered. Each of these systems had its strengths and weaknesses, its advocates and critics. None was a clear winner in the years before 1954, but the competition among companies using these approaches accelerated the development of the technology needed for offshore development.

Brown & Root reinforced its pre-war reputation as an offshore pioneer by building the first successful small platform with tender. This project won acclaim for the company as the builder of the platform for the first out-of-sight-of-land producing well. It also gave Brown & Root a role in the development of an innovative approach to offshore operations that became, for a time, the most common technological system for finding and producing oil in the Gulf.

Dean McGee of Kerr-McGee Oil Industries and George Brown of Brown & Root are generally credited with leadership in the development of Kermac 16, as the original small platform with tender came to be called. In 1946 Kerr-McGee paid $10,000 to the state of Louisiana for offshore leases. The most promising seemed to be Ship Shoal Block 32 located some ten and a half miles from the shore off Terrebonne Parish and twenty-one miles southeast of Eugene Island. Phillips Petroleum Company purchased a half interest in this lease and Stanolind Oil Company purchased a three-eighths interest, with Kerr-McGee retaining a one-eighth interest as operator of the project. Kerr-McGee selected Brown & Root to undertake construction. The choice of designs was a calculated gamble aimed at reducing the costs of exploration by using a tender vessel to carry much of the equipment and personnel needed to drill the well, thereby reducing the size and cost of the platform constructed for the rig itself. The economic calculations were clear and convincing: the early small platforms cost only $200,000 to $300,000, about one-fourth the cost of some of the larger self-contained platforms constructed in the late 1940s.[13] If tenders proved affordable and effective, this approach promised to alter the basic cost structure of offshore exploration.

The 38-foot by 71-foot platform was dramatically smaller than most previous offshore drilling platforms. By way of contrast, it had less than one-twentieth the area of the Creole platform built by Brown & Root in 1938. This allowed the deck to be supported by only sixteen steel pilings, which were 24 inches in diameter. These 140-foot long piles were driven 104 feet into the ocean floor, and they were braced with 9-3/4" pipe to a point two feet above mean low tide. The main drilling deck was only 38 feet by 58 feet, with the remainder of the platform's area accounted for by a small landing platform built off one side of the steel piles and supported by six timber pilings. This design provided room on the drilling deck for only the barest of essentials: the substructure and derrick, rotary table, shale shaker, drilling engines, bay tank for water and fuel, small mud tank, and an auxiliary mud pump. In a revolutionary change from past practices, the tender barge held everything else needed for drilling.[14]

This was no ordinary barge. Kerr-McGee had acquired for this job several war surplus YF Navy barges, each 260 feet long with a beam 48 feet wide. With a rated cargo capacity of 2,300 long tons, these big, rugged barges could be purchased easily and cheaply immediately after the war. The conversion

Kermac 16 under construction

process was simple. Workers stripped out almost everything originally installed on the barge, including the engine, and installed the specialized equipment needed to build and operate a rig. At a cost ranging from $700,000 to $1 million to convert each YF Navy barge, a functional offshore working space could be created. Once towed out and moored at the work site, these tender vessels housed the mud pits and pumps, dry mud, cement and chemicals, pipe racks, logging equipment, diesel fuel, drilling water, and the quarters and galley for thirty to forty workers.

Mooring a mammoth barge near a small platform required a strong mooring system capable of protecting the platform from the barge in rough seas. The Kermac 16 used 19 wooden pilings tied together by steel H-beams to create three "dolphins" to buffer the platform from the barge. An anchoring system sought to corral the barge to minimize the possibility that it would swing around and smash into the platform in high seas. Not easily solved was the problem of transferring equipment, including pipe, from the tender of the platform in rough waters. On an average of 30 to 60 days per year, bad conditions forced such tenders to stay away from their platforms, producing an uncomfortably high number of lost days due to weather.[15] In severe storms, the tender would be pulled away from the rig with a tug to ride out the storm.

Brown & Root faced a demanding job of installing the Kermac 16 in the open sea more than 52 miles from the onshore support site at Berwick, Louisiana. The lack of specialized equipment large and heavy enough to perform the work promised to make a difficult job almost impossible. Kerr-McGee offered to lease to Brown & Root one of the YF Navy barges it had purchased. This was the vessel that Brown & Root converted into the *Herman B.* Its size and lifting capacity facilitated the fabrication of the platform in the open sea by providing a large, sturdy surface for driving piles and installing the deck and drilling equipment, which included a 129-foot derrick.

Tender barges solved one problem that had plagued previous offshore efforts by providing living quarters for crews at the construction site. Instead of making a daily commute back to shore, workers began to work seven days on and seven days off, with standard shifts of twelve hours per day. Kerr-McGee had also purchased several war surplus air-sea rescue boats to service its offshore operations. Though far from perfectly adapted as supply and personnel boats in the Gulf, these craft were clearly superior to the leased shrimp boats they replaced.[16]

The initial well on the Kermac 16 site was spudded in on September 9, 1947; oil began to flow on October 14; and the well was completed on November 14. The lack of barges capable of transporting oil to shore delayed regular production until May 1948. Although the well is correctly celebrated as the first producer out-of-sight-of-land, several other wells being drilled from large, self-contained platforms began to flow in the Gulf of Mexico at roughly the same time.

Considerations of which projects deserve recognition as "firsts" in offshore operations miss one of the more significant aspects of the Kermac design. The fact that it was out-of-sight-of-land was of only symbolic importance; given conditions off the Louisiana coast, numerous previous offshore projects had been built and operated without ready access to land. The platform's design, not its location, was its key contribution, for the small platform with tender greatly lowered the cost of offshore exploration for a brief, but critical, time in the years before 1954. This design can be understood as a transitional step toward the advent of the mobile drilling vessel. The tender could, after all, be towed on to another site if the well drilled from the platform was a dry hole. In addition, the original platform could be salvaged, moved, and re-erected at another location for roughly sixty percent of its original cost.[17] In comparison to the large, self-contained platforms which had dominated offshore exploration in the Gulf, the small platform with tender was much more flexible and less expensive in drilling exploratory wells, since much of the cost of construction at an offshore site was not lost in the event of a dry hole.

Because of its cost advantages over large platforms, this "semi-mobile" approach found growing popularity from 1947 to 1953. According to Dean McGee, this technology "proved so successful at lowering costs and speeding up construction that this method is now [1949] being used by most companies drilling in the open waters off the Louisiana coast."[18] Companies experimented

Small platform with tender-Kermac 16

with this approach by expanding the deck to allow for more directional drilling, using even smaller platforms without pipe racks for the initial exploratory well with room to expand the platform if warranted by results, and including an engine in the tender vessels to make them self-propelled.[19]

Limitations to this approach became evident when technological breakthroughs made possible better alternatives. The costs of the small platform were not wholly recoverable if oil was not found. Despite much effort to design good mooring, a giant barge in the open sea could not be used effectively in high winds and waters. Particularly troublesome were problems in moving pipe and drilling mud from the tender to the platform, which led to lost work days during drilling. Dean McGee voiced other concerns: "It is not practical for use in very shallow waters and its practicability is yet to be determined for water depths of more than sixty feet."[20] Despite such limitations, the small platform with tender vessel was much used in offshore exploration through the early 1950s, and it remained an option for drillers in the Gulf of Mexico well into the 1960s. A National Petroleum Council study in 1965 found 30 tenders still in use.[21]

Brown & Root's role in the development of this innovative design brought favorable publicity to the fledgling offshore company. In 1982 the American Petroleum Institute honored George Brown for his contributions to the Kermac 16 project: "It was your vision that led to the design and construction of the first producing offshore drilling platform thirty-five years ago. This effort pioneered what has since become extensive exploration, drilling, and production of petroleum in waters all over the world."[22] Brown & Root obviously shared in the spotlight, for its engineers and construction crews had turned their boss's "vision" into a reality. The project showed the practical value of a new approach to offshore exploration.

As small platforms captured much of the attention of the offshore industry, Magnolia and Humble Oil made heavy investments in larger, self-contained platforms for their initial post-war explorations in the Gulf. These ambitious projects demonstrated that self-contained platforms could not compete with other approaches to exploratory drilling. Yet steady improvement in the design and construction of self-contained platforms continued, driven by the knowledge that such structures would be needed for efficient offshore development and production of large offshore fields originally discovered by other, less expensive methods of drilling.

The Magnolia Petroleum Company entered the Gulf dramatically in 1946, leasing five large blocks offshore from the state of Louisiana for several hundred thousand dollars and building a large self-contained platform to explore the most promising prospect. The company began construction work in May 1946 at a site ten miles southeast of Eugene Island off Terrebonne Parish. Although not considered out-of-sight-of-land because it was visible from a marsh island off the coast, this project marked a substantial step by a major company determined to become a force in offshore development. The completion of the platform and the commencement of drilling operations in August 1946 vaulted Magnolia into the lead in the quest for new oil fields in the Gulf.

The project incorporated one signficant advance, the use of steel pilings. Magnolia used hundreds of wooden piles to support the structure, but it also added more than fifty 15-1/2-inch steel pilings. This project marked both the birth of the age of steel and the death of the extensive use of wood in pilings offshore. An unusually large crew of 64 worked both onshore and off to build the platform, with quarters provided on an old Mississippi River steamboat moored at a wharf on Eugene Island. The crews traveled back and forth to a 77-foot by 173-foot platform mounted 20 feet above the water. Obviously, Magnolia's civil engineering department, which designed and constructed the platform, had decided to do it right, investing the funds necessary to launch its company offshore in style. The handiwork would have merited note among the "firsts" for the offshore industry but for one crucial detail—it failed to find oil. Magnolia salvaged what it could and searched for a new lease to explore and a new, less costly approach to exploratory drilling.[23]

As the Magnolia project wound down in April 1947, other even more ambitious self-contained platforms moved toward completion. The most expensive of these was Humble's Grand Isle Block 18 platform six miles south of Grand Isle, Louisiana, in about 45 feet of water. In addition to the water depth attained, the size and design of this project were new. With a total area of 39,000 square feet, this dual platform had a lower deck at 34 feet and an upper deck at 48 feet above the Gulf. The derrick floor was sixty feet above the ocean. The structure supporting these decks incorporated two advances that became the norm in offshore platform construction: it was all steel and it was built using templates fabricated onshore.

The logic behind the use of these templates or "jackets" was overwhelming, and their widespread adoption altered the history of offshore platform construction. J. Ray McDermott, a New Orleans-based competitor of Brown & Root, helped develop this critical innovation, which strengthened offshore platforms and solved problems inherent in the stickbuilding approach. McDermott first used templates in 1947 in building a large, self-contained platform for Superior Oil Company at about the same time that the W. Horace Wallace Company, another competitor to Brown & Root, used this approach on the Humble platform.[24] The advantages included easier installation, stronger underwater bracing, and lower costs from onshore fabrication. The use of templates quickly caught on in platform construction throughout the Gulf.

At the time, Humble was the most active oil company in the Gulf, and it recognized the advantages in a standardized approach to the manufacture and use of offshore templates. Humble developed a system of "template towers," each with four jacket legs sixteen inches in diameter arranged in a square ten feet wide. Humble used twenty-five such squares in the construction of Grand Isle 18.[25] For smaller platforms, Humble simply used fewer of the towers. Companies originally viewed prefabricated templates as a way to cut costs while providing a more efficient guide for the driving of piles than the "batter piles" traditionally used to guide piles into place. In practice, however, the jackets also provided structural strength, and they soon became a standard part of most platforms not constructed in shallow waters.

The Humble design included living quarters on the platform, an arrangement that became the norm after improved transportation and communication enabled platforms to be evacuated more rapidly, reducing the danger of storms. Companies experimented with the placement of quarters, with some even choosing to build separate platforms connected to the working deck by bridges to allow workers to live with a bit of distance between them and the noises and dangers of the working platform.

Humble's platform included room for seven wells. The design of the decks also incorporated a large separator, metering equipment, and a water-distilling unit. It expanded the practical meaning of self-contained by incorporating space for more of the equipment needed for efficient production of oil and gas. Upon its completion in March 1948, Humble's impressive dual platform at Grand Isle 18 illustrated the state-of-the-art in the construction of self-contained exploratory platforms that could be easily transformed into effective production facilities once oil had been found.

The scale and sophistication of this design, however, carried an unprecedented price tag of $1.23 million. A large company such as Humble Oil could absorb the risks entailed in the building of Grand Isle 18, and the investment would prove profitable if the platform produced substantial quantities of oil and gas.

But with the memory of Magnolia's recent failure still fresh, executives at Humble no doubt swallowed hard when the first test well produced a dry hole. Despite better results from subsequent drilling, Humble decided to take a less risky approach to its next wave of exploratory drilling. In 1948 it purchased 19 war surplus LSTs (Landing Ships Tank) for use as tenders. In its next nine efforts to locate oil deposits in the Gulf, it used smaller platforms tended by converted LSTs, and most other companies took the same path.

The great disadvantage of the self-contained platform in this era was its cost. Exploration was a gamble, and every company looked for ways to avoid an ante of $1 million. Kermac 16 had shown them an option for lowering costs. Using the large self-contained platform in the race to find new offshore oil and gas was like driving a new Cadillac in a demolition derby. Even first prize might not cover the high cost of entry.

The missing link in the system of technology needed for more efficient offshore development was a truly mobile drilling rig for exploration. These advances began to emerge in the late 1940s, most notably the development of a submersible drilling barge that could be used in shallow water. One of the first, John Hayward's *Breton Rig 20* borrowed heavily from the older barges used for drilling in the muck at inshore Louisiana sites. In the early 1950s, Hayward began to convince skeptics that his barge could slash the costs of exploratory drilling by moving from location to location with little set-up cost.[26] His work paved the way for a revolution in mobile drilling that sparked an offshore boom in the 1950s and 1960s.

The Offshore Industry Before the Tidelands Act of 1953

By 1953 the offshore industry had forged a technological system to produce competitively priced oil from offshore fields. In place were systems for finding petroleum deposits, an early version of metal jackets for platform construction, methods for installing platforms at sea, a prototype mobile drilling rig, and many of the specialized services and supplies needed for offshore operations.

The creation of this technological framework after World War II helped define the industry's business structure. At the center were the major oil companies whose decisions to enter the Gulf pulled along a closely related cluster of other industries. By the early 1950s, Brown & Root had established its niche within this cluster. Capitalizing on its early entry into the Gulf and its work on Kermac 16, it had emerged as a leader in platform construction. It had also taken its first steps toward leadership in laying underwater pipelines, a specialized business that would become increasingly important both to the offshore industry and to Brown & Root.

Even in these early years of platform construction in relatively shallow water, it was clear that only a handful of companies could survive in this small market which required heavy investments in specialized equipment. Brown & Root had made do in offshore construction with the *Herman B* and the *L.T. Bolin,* converted war surplus barges it had acquired from Kerr-McGee.

Initially the 35-ton lifting capacity of these barges was adequate. Brown & Root's competitors used roughly the same boats. W. Horace Williams used a converted war surplus LST with a 50-ton crane mounted on it to build Humble's Grand Isle 18 platform; J. Ray McDermott completed its first offshore construction with a World War I barge using a floating harbor crane on deck. But the wave of the future was clear in 1949 when McDermott commissioned the first derrick barge built specifically for offshore use. Its lifting capacity of 150 tons was a breakthrough for offshore construction. By 1953, McDermott already had a second unit with a 250-ton capacity.[27] The race to offer bigger and better offshore construction equipment had begun. Such expensive advances in construction equipment were vital to the expansion of the offshore industry, since a prerequisite for the evolution of larger jackets fabricated onshore for use in deeper waters was the capacity to install these structures at sea.

Brown & Root's early investment in such equipment and its pioneering work on several of the first platforms in the Gulf assured that it would have a leading role in designing, fabricating, and installing new platforms. As the industry moved from stick-building out in the Gulf to the onshore fabrication of ever larger jackets, Brown & Root and McDermott emerged as leaders in this area. They gradually increased their capacity to build these specialized

Early prefabricated jacket and deck

structures, making use of the largest diameter steel tubing available to them. By 1953 both companies neared the point at which the scale and complexity required to fabricate offshore structures justified the heavy investment needed to build specialized onshore fabrication yards.

They would need such yards to keep pace with the growth of the offshore industry. According to one authoritative observer, from 1947 through 1951 seventeen oil companies had invested more than $260 million for leasing, exploration, and development offshore in the Gulf of Mexico. These years witnessed the drilling of 242 wells, 130 of which produced oil, gas, or gas-condensate from twenty producing areas. By January 1, 1953, approximately seventy drilling platforms had been erected in the Gulf out to water depths of seventy feet.[28] Offshore Louisiana led the way, producing about 20 million barrels of oil from 1947–1952.[29] Texas trailed far behind, with only 19 wells drilled offshore and about 600,000 barrels of oil produced through November 1954.[30] The production figures were not impressive compared to total U.S. production or to onshore Louisiana or Texas, but they represented a noteworthy achievement for an industry still searching for the right technology to succeed in a new and challenging environment.

While operating offshore in these years, the oil industry studied the broad issues posed by wave forces and soils. In October 1949, a small hurricane struck several platforms offshore of Texas, causing more than $200,000 in damages to one platform which had been built 26 feet above the normal water level. A nearby platform with a deck 33 feet off the water suffered only minimal damage. Study of the damage suggested that waves had crested 8 to 14 feet higher than had been predicted by experts. After this storm, a consortium of oil companies, universities, and consulting firms launched a research program on hurricanes and their potential effect on offshore structures.[31] Humble Oil had initiated its research on waves and weather forecasting techniques, and the leading expert on these issues, Walter Munk of the Institute of Geophysics and Scripps Institute of Oceanography, University of California, had published data on wave action in the Gulf. In 1949 offshore engineers representing twelve oil companies met in Houston at the Offshore Operators' Committee to discuss wave action at "great length."[32] They hoped to trade research notes on wave action, experience with hurricanes, and practical measures for minimizing the danger to personnel and equipment. The industry was coming to a consensus on several issues, including the need for shared research and a general principle for platform construction: when in doubt, raise the deck.

The industry also cooperated in exchanging information about soil composition. Company experts published papers on soil analyses of borings in the Gulf and on the results of pile tests.[33] Others analyzed soil conditions and catalogued information about soils, steadily reducing one source of uncertainty facing the offshore industry.

By the early 1950s, the industry had overcome most of the challenges first encountered in its move offshore after World War II. A National Petroleum Council (NPC) study of offshore technology in 1953 noted that most of the individuals who had contributed to offshore developments "had little or no previous experience with marine operations in open waters, so their assignment involved considerable research . . . to determine the most economical methods of finding and producing oil from structures located in offshore areas." The NPC report argued that offshore costs would remain "relatively high as compared to operations on land," but that some of the costs offshore "will be reduced by the economies resulting from large-scale operations and by the development of more efficient equipment for offshore drilling."[34] The report described an industry whose accomplishments had prepared it for a boom in which rapid expansion would lower costs, creating opportunities for further expansion and technological change.

The boom awaited the resolution of legal and political disputes over leasing rights in the tidelands, which had sharpened since World War II. The discovery of offshore oil before the war had first raised the issue of whether the federal or state government held the authority to lease lands underneath the sea. The federal government had moved to forestall any international questions about sovereignty over tidal lands with a proclamation by President Truman in September 1945 asserting the authority of the United States over the subsoil of the Continental Shelf contiguous to the land area of the United States. The coastal states nonetheless continued to lease offshore tracks of land. In 1947 Texas asserted its claim to the entire Continental Shelf off its shore; Louisiana claimed control over lands up to 27 miles off its coast, and it defined its coast by drawing its boundary as far out as possible from the marshy expanse of south Louisiana. The two states auctioned offshore leases and launched aggressive political and legal campaigns to defend their control.

The federal government struck back in a series of suits against California, Texas, and Louisiana. The Supreme Court ultimately ruled in all three cases that paramount rights over submerged lands belonged to the federal government. The Texas and Louisiana decisions came down on the same day, June 5, 1950, blocking the further leasing of offshore lands until a new system of controls could be established. In the next three years, offshore development slowed dramatically, as all involved parties awaited the creation of a leasing system acceptable to both the states and the federal government.

The final act in this drama was played out in national politics, as the tidelands became one of the hottest issues in the 1952 presidential campaign. After much posturing on all sides, Congress enacted the Tidelands Act of 1953, which validated all state leases issued before the Supreme Court decisions in 1947 and 1950 and established a system of shared control in which the federal government had the authority to lease all lands outside of a given distance from the shore. Louisiana and most other states were given control of submerged lands out to three miles from their shorelines. Because of historical claims, Texas and Florida were given control out to three leagues or about ten miles. This compromise pleased almost no one, but after the resolution of a wave of new suits and countersuits, the Tidelands Act ultimately established a new framework for offshore leasing.

A round of new auctions for offshore leases displayed the industry's optimism about prospects offshore. This optimism was grounded in the years before 1953, when the industry had assembled the technological framework within which the economic puzzle of offshore development could be put together. In the decades after 1953, an offshore boom in the Gulf of Mexico would allow the industry to complete the puzzle.

Notes

1. This chapter draws from an interview with G.C. Lee.
2. "Can-do attitude leads Brown brothers to shipbuilding," *Brownbuilder-75th Anniversary:* 13.
3. Joseph Pratt and Chris Castaneda, *Builders: The Brown Brothers of Houston* (College Station: Texas A & M University Press, forthcoming).
4. Pratt and Castaneda, *Builders.*
5. Pratt and Castaneda, *Builders.*
6. Chris Castaneda and Joseph Pratt, *From Texas to the East* (College Station: Texas A & M University Press, 1993).
7. Pratt and Castaneda, *Builders.*
8. These lists are taken from a variety of sources; including James W. Calver, "Louisiana Offshore Operations," *World Petroleum* (November 1954): 68; Thomas Mireur, "Texas Offshore Exploration," *World Petroleum* (November 1954): 70.
9. Dean McGee as quoted in "McGee Recalls First Well 40 Years Ago," unidentified newspaper clipping in Brown & Root Archives.
10. F.R. Harris and Harry Gard Knox, "Important Considerations in Marine Construction," American Institute of Mining and Metallurgical Engineers, Technical Publication No. 2324, October 1947; F.H. Harris and H.G. Knox, "Marine Construction," *The Oil & Gas Journal* (October 18, 1947) 131.
11. Griff C. Lee, "Offshore Structures: Past, Present, Future and Design Considerations," paper presented at Offshore Exploration Conference, New Orleans, Feb. 14–16, 1968.
12. Lee Interview; Jack S. Toler, "Offshore Petroleum Installations," *Proceedings of American Society of Civil Engineers,* September 1953, 289–5–6; M.B. Willey, "Engineering Characteristics of the Gulf Coast Continental Shelf," American Institute of Mining and Metallurgical Engineers, Technical Publication No. 2323, October 1947, 1–11.
13. Most of the information on costs in this section is from National Petroleum Council, *Technology of Offshore Operations on the Gulf Coast Continental Shelf* (Washington: 1953).
14. The primary source for most of this information is Dean McGee, "A Report on Exploration Progress in Gulf of Mexico," *Drilling* (May 1949) 50, 53, 117, and 120. See, also, John Samuel Ezell, *Innovations in Energy: The Story of Kerr-McGee* (Norman: University of Oklahoma Press, 1979): 147–170.

15. NPC, *Technology of Offshore,* 21.
16. "Man, Oil and the Sea," *Offshore* (October 1972): 60.
17. NPC, 1953 Study, 21.
18. McGee, "Exploration Progress," 117.
19. Jack Toler, "Offshore Petroleum Installations," 189–6–8.
20. McGee, "Exploration Progress," 117.
21. NPC, "Impact of New Technology on the U.S. Petroleum Industry, 1946–1965," 210.
22. API Award Ceremony, 1982 Annual Meeting, as reported in "Ex Brown & Root Chairman to Receive Special Award," *Houston Post,* (November 8, 1982).
23. "Magnolia Testing Offshore Formations in the Gulf," *World Petroleum* (March 1947): 60–61; McGee, "Exploration Progress," 42–44, API Papers, 1948.
24. M.B. Willey, "Structures in the Sea," *The Petroleum Engineer* (November 1953): B-43.
25. Pierce Shannon, "Lower Cost Drilling Platforms Reduce Off-Shore Expenses," *Drilling* (May 1950): 66–67; Dean McGee, "Exploration in the Gulf of Mexico," 49; and Mercer H. Parks and James C. Posgate, "Drilling and Producing Operations on the Continental Shelf," API Papers, 1948, 111.
26. McGee, "Exploration Progress" *Drilling* (May 1949): 120.
27. Lee, "Offshore Structures," 2.
28. Toler, "Offshore Petroleum Installations," 289–2–3.
29. James W. Calvert, "Louisiana Offshore Operations," *World Petroleum* (November 1954): 68.
30. Thomas Mireur, "Texas Offshore Exploration," *World Petroleum* (November 1954): 71, 73.
31. Willey, "Structures in the Sea," B-41; R.C. Farley and J.S. Leonard, "Hurricane Damage to Drilling Platform," *World Oil* (March 1950): 85–92; and "Mr. Gus Goes to Sea," *World Petroleum* (December 1954): 37.
32. Farley and Leonard, "Hurricane Damage," 88–90.
33. M.B. Willey, "Engineering Characteristics of the Gulf Coast Continental Shelf," American Institute of Mining and Metallurgical Engineers, Technical Paper No. 2323; Willey, "Structures in the Sea," B-38-B-47.
34. National Petroleum Council, *Technology of Offshore Operations on the Gulf Continental Shelf* (Washington: 1953), II and VI.

CHAPTER 3

A Maturing System

The race to exploit offshore oil reserves in the Gulf of Mexico accelerated in 1954 after the resolution of the tidelands dispute and creation of a legal framework for offshore prospecting. But once oil companies had obtained their leases, the race became less a competition between companies than a renewed struggle to overcome the natural and economic constraints to moving offshore. The high costs of venturing out into deeper water fostered a community of interest within the oil industry in developing technologies to meet a new set of challenges. During the 1950s, the industry reduced the costs of operating offshore with technological innovations that achieved greater mobility in exploration, speed and capacity in transportation, structural design improvements in platforms, and large scale production aided by submarine pipelines.[1]

Success at prospecting at sea was not the result of a single technological breakthrough, but rather the result of a cluster of related innovations. Offshore operations moved gradually into deeper waters by accumulating knowledge as an industry. Oil companies increasingly contracted for specialized services to perform the tasks of drilling, platform design and installation, production, pipelaying and supply. Together, however, these services made up the interrelated components of a maturing industry with standardized methods of working offshore that could be applied in other waters of the world. Brown & Root assumed leadership in two of the most important parts of this new maturing system—platforms and pipelines.

Leaving the Cradle of Land

Before the offshore oil industry could move into deeper water, geophysical exploration techniques had to advance. Geologists, drillers, and surveyors learned to become sailors, divers, and navigators. The customization of the seismograph to single-ship transportation made marine operations rapid and continuous, offsetting the high costs of prospecting at sea. Between 1949 and

1954, seismic crews or "doodlebuggers" left the cradle of land as young offspring of the petroleum industry in its unending search for new oil.

Geological explorations of water-covered areas began in the Gulf of Mexico and South Atlantic in the 1930s. Geologists first employed gravity meters, or gravimeters, to isolate anomalies favorable for the accumulation of petroleum. In shallow water under favorable wind, water, and weather conditions, the land gravimeter could be read from tripods reaching the bottom. A few remote control underwater gravimeters and diving chambers appeared in the 1940s, and geologists began to probe the seabed with coring drills. But neither gravimeters nor coring methods were "a positive means of determining thickness of sediments overlying the basement rocks or the detailed geologic structure . . . nor the depths and characteristics of sediments beyond the ocean ooze."[2] The seismograph, which explored subsurface formations by recording the reflected and refracted energy waves set off by dynamite explosions, also appeared as early as 1936 in Galveston Bay and in the marsh and swamp areas of the Gulf. Seismometers were planted in holes drilled in the bay floor and surveying was carried out using conventional land instruments on platforms, since landmarks were always in sight.[3]

By the end of World War II, new developments in shooting and recording made the seismograph seaworthy and independent of landmarks. Seismic teams employed a fleet of small boats, as many as five or more, with separate units carrying recording, shooting, and radio-location surveying crews, the latter using hyperbolic positioning methods. The typical multiboat operation could shoot a new profile every three to four minutes and remain at sea for 10 days at a time, depending on the weather.[4] Still, even moderately rough waters and weather could bring operations to a halt. Small boats were unable to perform marine work in open water far offshore. "Seismograph exploration had to become fully seagoing, in both equipment and technique," wrote John Wilson of Geophysical Service Inc., a leader in marine exploration in the 1950s, "and a new concept of operation had to be formed."[5]

A new single-ship operation, developed in the early 1950s, combined all shooting, recording, and surveying operations in vessels as large as 160 feet long. The single ships contained six months of supplies and provisions and could withstand all but the most extreme weather. Using sonobuoys for surveying and pressure-sensitive detectors built into a special "detector streamer" assembly attached to a large reel on the stern of the ship, the single-unit could record profiles every one-half to two minutes continuously and cover up to sixty-five miles per day. This compared to 50 miles of continuous coverage per month on dry land. The cost of a water crew in 1955 greatly exceeded that for a land crew ($60,000–$100,000/month versus $15,000–$25,000/month), but the higher rate of production by the new marine exploration techniques, both small fleet and single-ship, yielded a much lower cost per profile obtained (approximately 75 percent of coastal land belt cost).[6]

Improvements in underwater geological reconnaissance stimulated a rapid growth in offshore geophysical activity over more extensive and deeper parts of the Gulf and supplied oil companies with enough accurate information to build confidence in bidding for offshore acreage. Better skill in verifying and interpreting seismic data, and the comparative uniformity of the offshore Gulf's sea bottom, helped identify quality salt domes offshore and decrease the finding costs of first class structures. Still, seismic records were not 100 percent reliable. Reflections of seismic impulses from the sea bottom itself sometimes led to erroneous interpretations of subsurface structures. As a California Company exploration specialist pointed out in 1955, "You never really know whether oil is present until you drill a well."[7]

Finding Sea Legs

The high costs of equipment, operations, and lease bonuses, plus the practical difficulties of drilling underwater meant that, to make offshore oil pay, innovations in drilling technology were essential. The flurry of lease auctions for lands off the coast of Louisiana and Texas set off by the resolution of the tidelands controversy opened a new era of marine drilling; which came to be marked by exceptional engineering achievements in mobile drilling rig design. "Although the tempo of offshore activity has increased on all fronts since the settlement of the tidelands dispute," Shell Oil engineer John Payne observed in 1954, "the evolution and development in mobile rig designs has been far ahead of advancements in other phases of the problem."[8]

By 1954, drilling contractors realized that mobility in offshore drilling equipment would be imperative for progress in exploratory drilling. Ninety percent of all offshore wells drilled in the Gulf of Mexico had used the floating tender-small platform method.[9] When war surplus vessels were no longer available for conversion into tenders, the expense of purchasing new ones skyrocketed. Rising steel prices and the need for larger boat designs to service platforms farther offshore increased the average investment to convert drilling tender from $325,000 in 1947 to $1.1 million in 1954.[10] Demurrage costs approached $4,500 per day, thus compelling an operator to "keep his rig busy when he discovers an oil field and develop it immediately and vigorously to make 100 percent use of his expensive facilities," noted Ben Belt, vice president of Gulf Oil Co.[11] As operators moved into rougher waters, however, they encountered mooring difficulties with floating tenders and more frequent platform evacuations, losing valuable drilling time.[12]

New mobile drilling rigs changed the cost structure of exploratory marine drilling and extended water depth capabilities. The initial costs of early mobile rigs equaled that for a YF tender and minimum platform, but the cost per well drilled, thanks to mobility, was much less.[13] Engineers set out to design drilling units to fulfill several criteria: no fixed structures, mobility under average weather conditions, short time horizons for rigging up and drilling, and the ability to withstand wind and wave forces and poor soil conditions. Under close guidance from lease-holding oil companies, many engineers contributed to advances in mobile drilling technique by taking the design initiative, assuming calculated risks, and learning from each other's mistakes. During spring 1954 and 1955

Offshore Operating Symposiums of the API's Southwestern District, Division of Production meetings, oil and drilling company representatives met to exchange engineering information and work through design problems. The main difficulty, as Mercer H. Parks of Humble Oil observed at the 1955 meeting, was that "the evolution of the mobile offshore drilling unit is not under engineering control but instead is under control of economic incentives and the challenge of the sea which has always reacted creatively on the spirits of men."[14] Three types of rigs—the submersible, jack-up, and floating—emerged to meet the challenge of attaining mobility in offshore drilling.

The first submersible rigs were modeled on John Hayward's *Breton Rig 20.* The submersible-pontoon barge design of the *Breton Rig 20,* however, could still only maintain stability in depths of no more than 22 feet. In 1954, Kerr-McGee developed a larger and more modernized version of the Hayward barge with a fixed-height deck, moveable pontoons to stabilize the rig during sinking, and storage space contained within the barge hull. The $2 million 140-by-72 foot *Kermac 44* submersible mobile drilling platform could drill in 40 feet of water and withstand high winds.[15] Other fixed-height deck submersibles developed between 1954 and 1956 exhibited new pontoon and hull design features to make rigs more stable while floating and sinking.[16]

These early submersibles, nevertheless, were susceptibile to soil scouring around the base and displacement by moderately severe storms. They could operate in a maximum of about 45 feet of water and their seaworthiness in harsh conditions remained in question. One notable demonstration of this problem was the 1955 disaster that befell the American Tideland's *Rig No. 101,* which capsized while being raised to shift to another location. The great advantage of the new rigs—their maneuverability—also turned out to be their greatest disadvantage. Without standard models or prototypes, mobile rig designers had to resort to trial and error in their search for more stable rig types. This explains why the early designs of mobile rigs varied so widely and underwent so many modifications.

Given the short time horizons for lease development, pressures for innovation in mobile drilling escalated. By the end of 1955, oil companies already had leased acreage beyond the 100 foot line, which had become the new depth threshold for mobile drilling rig designers. In 1956, Kerr-McGee introduced a new submersible rig that achieved a much greater degree of stability in deeper water. The 242-by-202 foot *Kermac 46* consisted of a platform mounted on top of bottle-shaped tanks that were flooded to submerge the unit in place, looking "for all the world like some prehistoric sea monster."[17] The bottles did not depend on soil for support during submerging—as did pontoon submersibles—and they gave the rig positive stability over a considerable angle of heel. Plus, the narrow bottlenecks on which the platform sat offered less resistance to waves. Other rigs imitated the column-stabilized design, and many pontoon submersibles eventually were converted to columns as well. Most regular submersibles of this kind cost around $3 million to build in the late 1950s, and they were not made to operate beyond 80 foot water depths.

Building them larger for deeper water was outrageously expensive. The number of regular submersibles in operation in the Gulf of Mexico reached 30 by 1958, but they were a dying breed thereafter.[18]

By the mid to late 1950s, the elevating-deck, or jack-up, rig emerged as the preferred design for drilling in 75- to 150-foot depths. Inspired by the famous Delong jacks first used by the U.S. Navy in the western Pacific to install and elevate docks, these rigs consisted of a barge platform which was towed to site and elevated out of the water by legs extended, or jacked, to the bottom. The 230-by-70 foot *De Long-McDermott No. 1,* built in 1954, was the first truly mobile jack-up unit. Glasscock Drilling Co.'s $3.5 million jack-up named *Mr. Gus I,* constructed the same year at the Bethlehem Steel Company shipyards in Beaumont, Texas, became the first mobile unit of any type capable of drilling in 100 feet of water under hurricane conditions. A testament to the collective efforts of the industry to meet the technological challenge of deep water, *Mr. Gus* was the product of years of research by Bethlehem Steel, Gulf Oil, Texas A & M University, the University of Michigan, Greer & McClelland of Houston, and the Stevens Institute of Hoboken, New Jersey. The rig had two connected platforms, one for the drilling rig and the other for supplies and living quarters, but after *Mr. Gus* capsized in a storm off Padre Island, Texas, in the spring of 1957, it was successfully salvaged and repaired as a single platform rig.[19]

Early disasters plagued jack-up rigs, such as the destruction of Royal Dutch Shell's *Qatar Rig No. 1* shortly after *Mr. Gus's* misfortune. Engineers solved the problem of excessive leg penetration in soft soils, which could tip rigs over, by attaching large-diameter spud cans or mats near the end of leg, features that were exhibited by The Offshore Co.'s *No. 51* and *No. 52.* Hydraulic pin or electric rack-and-pinion drives, in place of pneumatic boots or hydraulic slips, also enhanced leg-gripping. To address greater bending stresses on legs in deeper water, rig designers began using open-fabricated or truss-type legs that allowed for more strength without increasing wave forces. The Offshore Co.'s rig *No. 54* and Zapata Offshore Co.'s unique three-leg, triangular platform *Scorpion,* both built in 1956 by R.G. Le Tourneau Inc., employed the truss design and the new drive features. Jack-up rigs based on Le Tourneau's design became the most common type of mobile drilling unit for offshore operations, and they increased depth capabilities to 150 feet. About 30 jack-ups were in operation by 1960. New rigs were built at a fast pace in the 1960s and were eventually modified with legs slanted outward for stability in rougher weather and in depths up to 250 feet.[20]

Floating vessels or drillships also provided important advances during the 1950s. Drillships were not used for exploratory drilling, but for gathering geologic information. Although barges converted to core drilling units had operated off the coast of California since 1944, Brown & Root introduced the practice to the Gulf of Mexico in 1953. It converted the largest barge available for marine drilling, 165 feet long by 45 feet wide with the drilling rig welded to large beams cantilevered over the side. The barge drilled hundreds of test holes to determine factors affecting stationary drilling and production platforms.[21] In 1956, in association with McClelland Engineers, Inc., Brown & Root rigged out a larger, more sophisticated drilling barge, the *U-303,* with an over-the-side rig. The success of the *U-303* prompted Brown & Root/McClelland to build another floating unit, the *Champion,* for work in the Gulf of Mexico. These two barges performed stratigraphic boring and core drilling for the Texas Gulf Sulphur Company in 30-foot waters. In July 1957, the *Champion* was towed 3,000 miles to Passamaquoddy Bay in Maine to test the bay floor to determine foundation requirements for dams associated with the Passamaquoddy Tidal Power project.[22]

Innovations to mobile drilling into the 1960s were crucial to the early expansion of the offshore oil industry, especially for drilling wildcat wells. "The many unusual designs developed to solve various problems," wrote Dr. Richard Howe of Esso, "is one of the important engineering achievements of this century."[23] Drillships evolved to self-propelled vessels built from scratch with center-line derricks. Along with new semi-submersibles, drillships "carried exploratory drilling to all but the most remote regions of the world's continental shelf areas."[24] Dynamic positioning and other technological developments took marine drilling into waters of almost any depth. But advances in drilling had to await technological development in other areas before offshore oil operations could push the deep water frontier in a concerted way.

Filling Out the System

During 1955–1958, the offshore industry was still learning to work in water depths approaching 200 feet; although drilling in water up to 100 feet was standard. Orders for new mobile rigs bombarded shipyards all along the Gulf Coast.[25] By early 1956, twenty companies were drilling off the coast of eight Louisiana parishes between Lake Charles and New Orleans.[26] Firms also drilled off the coast of Texas counties Jefferson, Galveston, Nueces, and Kleberg; but Texas offshore exploration proved disappointing. Offshore Louisiana was where most discoveries were made. As 1957 began, 23 mobile units navigated the Gulf, with at least 11 more under construction. There were also 46 drilling tender operations and 29 self-contained platforms.[27] As oil was discovered, the industry developed new production systems to tap the offshore domain, with Brown & Root taking a leading role in platform and submarine pipeline construction.

High finding rates offshore and generous allowable production limits set by the State of Louisiana reassured oil companies that they would finally see an income stream to offset their "billion dollar adventure in applied science." Of 777 wells drilled in the Gulf by 1956, 410 were oil wells, 120 were gas wells, and 247 were dry holes. Of the total, 138 were wildcat or exploratory wells and 639 were field development wells. At least 26 percent of wildcat wells were producers, compared with the U.S. onshore average of 11 percent.[28] In the early 1950s, Louisiana's Department of Conservation had established larger daily production allowables from wells offshore relative to those onshore. Whereas a 10,000-foot-deep onshore well in Louisiana was allowed a daily limit of 132 barrels, a 10,000-foot-deep well offshore had an allowable of 242 barrels. The higher allowable encouraged greater spacing of development wells (up to 40 acres offshore compared to 30 acres on onshore salt dome fields), resulting in lower field development costs to compensate for higher individual well costs. In early 1955, Dean McGee, president of Kerr-McGee Oil Industries, reported that the "economics of offshore operation are attractive at current costs and are comparable now with some areas of the United States where exploration and development are being aggressively pursued."[29]

Despite favorable economic conditions, practical problems remained. While mobile rigs were preferable for wildcatting, fixed platforms remained the technique of choice for drilling in proved fields. By the mid-1950s, oil companies had determined that drilling several directional holes and multiwell completions from a single, self-contained platform was the most profitable approach in offshore fields. Other fixed structures were needed to protect wellheads and provide for oil storage and related tasks. More than 40 fields had been discovered since the late 1940s, at least seven of which promised more than one million barrels per year: South Pass Block 24, Bay Marchand Block 2, Main Pass Blocks 69 and 35, Eugene Island Block 126, Grand Isle Block 18, and West Delta Block 53.[30] Major field development required bigger templates, enlarged decks to house more drilling and production equipment, and improved pile design to take greater lateral loading. Even tender-type operations needed larger platforms in deeper waters. As Brown & Root engineer F.R. Hauber recalled, these new demands on platform construction "dictated the hard rule that as much prefabrication as was possible must be the hallmark about which the entire project must be conceived."[31]

The growing size of self-contained platforms in deeper water demanded changes in the techniques of fabricating and installing these structures. Because of the near impossibility of bracing pilings after they were driven, production platforms could not be built in place in deeper water. The entire structure had to be prefabricated on land, towed to site, and installed. Prefabrication also reduced the amount of expensive time that vessels and equipment spent offshore erecting structures. The 1949 hurricane had wiped out many of the early self-contained platforms, but by 1955 several large self-contained platforms, or "artificial islands" as the U.S. Coast Guard called them, again dotted the waters of the Gulf. In that year, J. Ray McDermott & Co., the original innovator in template or steel jacket structures, established the first fabrication yard specifically geared toward the construction of offshore platforms, the Bayou Boeuf facility just east of Morgan City, Louisiana. McDermott also developed an 800-ton capacity marine crane, hailed as the "largest single mechanical lift ever created," to perform the herculean task of installing platforms and setting massive deck sections on top of them.[32]

"Now that the first big rush to get a number of platforms in operation is over," wrote one observer in 1957, "designers and fabricators are taking a critical look at their handiwork."[33] Early 1950s platforms were rectangular frame structures consisting of vertical "H" columns interlaced by smaller horizontal and diagonal pipes. Many small steel tubular piles driven into the seabed generally supported the platforms. As marine designers processed the data from research projects on wave and hurricane behavior, they realized that the close spacing of jacket columns hindered the free passage of storm waves and that the small diameter piles were too weak to withstand large lateral loads. Platform designers increased piling diameters from about 10 inches to 30 inches, reduced their number, and widened their spacing. The bearing capacity of pilings thereby increased from about 80 tons to 300 tons, and their soil penetration could reach 300 feet if needed. Pile columns were battered (sloped out at the bottom), giving the jacket a slender pyramid shape. This design also reduced wave impedence by limiting the area exposed to the most wave action.[34]

Designing these $2 million structures seemed relatively easy compared to building and installing them. The longer spans between pile columns on the new jackets required trusses rather than beams to support the deck. Consequently, deck sections had to be prefabricated as integral units set intact on the structure. Launch barges large enough to transport templates and deck sections had to be built, as did more derrick barges such as McDermott's to install the massive structures. The larger and heavier piles demanded more powerful steam hammers to drive them, which in turn called for higher capacity steam boilers and hose. The fabrication of giant tubular sections created whole new areas of construction dedicated to rolling pipe and columns. "As the quest for petroleum and gas proceeds into deeper waters and different climate zones," noted F.H. Hauber, "the size of structures becomes larger and heavier, and as before, the materials methods, facilities and equipment are forced to keep pace." The new scale of operations raised daunting challenges, but at the same time offered new opportunties for specialized services that contributed to the growth of offshore oil as a full-fledged industry unto itself.

Along with J. Ray McDermott, Brown & Root was at the forefront of this growth. These two companies emerged as the leading specialists in template and platform engineering and construction. After World War II, Brown & Root set up a separate company, called Brown & Root Marine Operators, Inc., to tackle construction jobs in the Gulf waters. Initially, the driving forces behind Marine Operators were "Ox" Hinman, who had close ties to companies such as Humble Oil and Tennessee Gas; R.A. Turrentine, a former naval officer who directed the company's purchase of surplus government cargo vessels for conversion into the first construction barges, the *Herman B* and the *L.T. Bolin;* and Hal Lindsay, who took over the management of Marine Operators after Hinman's departure. The company prospered through the efforts of such men as Frank Motley, a master of "applied engineering" in the formative years offshore; F.R. (Ferd) Hauber, Brown & Root's leading marine structural designer; and Benny Lawrence, an ingenious naval architect who was largely responsible for the development of Brown & Root's early pipelaying barges.

Drafting room at Brown & Root's Clinton Drive headquarters

These men and their assistants worked out of the big drafting room at Brown & Root's Clinton Drive headquarters, which housed the company's engineering department. By the mid-1950s, according to long-time Brown & Root engineer Rick Rochelle, they were the ones who "kept Greens Bayou busy."[35]

Brown & Root offered services for the complete job of engineering, fabricating, and erecting drilling platforms, both self-contained and tender types. Contracts also included transportation to the jobsite and installation. Brown & Root developed platform designs that usually called for 8 piles on a tender-type platform and 10 to 24 piles to support a self-contained platform. "In all cases," reported the *Engineering News-Record* on Brown & Root's method, "the template, piles and deck structure are unitized at the site into a single structure, by welding, to resist the overturning forces of wind and waves and to carry the vertical loads of the drilling operations."[36] Brown & Root designed all its platforms to withstand a 25-year storm producing 125 mph winds and 49-foot waves.[37]

Installing these structures required bigger equipment to lift and lower the template or jacket, drive the piles, and set the heavy deck sections. Brown & Root upgraded the lifting capacity of its two derrick barges, the *L.T. Bolin* and the *Herman B.* The *Herman B* was outfitted with a revolving and traveling gantry crane with a lifting capacity of 75 tons, and a 250-ton, heavy-lifting, hammerhead crane. The *L.T. Bolin* acquired a hammerhead crane with a lifting capacity of 300 tons. In 1956, the company launched its most modern derrick barge, the 300 × 90 × 19 foot *H.A. Lindsay,* equipped with a 250-ton revolving crane.[38] After a jacket was fabricated on shore and towed by cargo barge to site, two of these giant derrick barges would lift it off the barge and set it in the water in a horizontal position. One derrick barge then rotated the jacket into the vertical position, flooded the legs, and eased the structure down to the bottom.[39] Piles were driven by large pile-drivers through the template tubes and cut off at grade. Finally, the prefabricated deck sections were lifted by the cranes off another cargo barge, set in place on the jacket, and welded to the supporting piles and tops of the template tubes.[40]

In 1956, Brown & Root constructed one of the largest platforms to date and installed it at a record depth of 112 feet for the CATC Group (Continental, Atlantic Richfield, Tidewater, Cities Service) in a producing field in Louisiana's Grand Isle Block 43. After analyzing wave, wind, and soil conditions at this depth and location, Brown & Root engineers designed a jacket to house sixteen 33-inch piles with a slope, or batter, from the top of two feet per twelve feet. The deck measured 100 × 146 feet, allowing 12 wells to be drilled from two positions by a derrick having a 40 × 45 foot base. Constructed in subassemblies to permit three lifts for final position placement, the main deck also supported pipe racks, pumps, mud and storage facilities, living quarters, and all other auxiliary equipment and supplies. The total height of the structure was 472 feet, from 150 feet below the mud line, to a deck 55 feet above water, to the derrick crown 155 feet above the deck. The final structure contained 3.2 million pounds of steel, 200,000 pounds of other material, and 50,000 pounds of paint. Total cost was $1.2 million.[41]

Constructed in the winter of 1955–1956 and installed in the spring, the CATC platform withstood the 110 mph winds and 20-foot waves of Hurricane Flossie and provided a nice advertisement for Brown & Root work. George Brown later explained, "We had studied tides, currents, waves, and winds, and we knew that it wasn't the winds that hurt, even hurricane force winds; it

The *L.T. Bolin* lifting a deck section onto the CATC Group's jacket in Louisiana's Grand Isle Block 43

was the waves. The waves can batter the hell out of a platform. So we built our platforms high off the water so the waves couldn't get to them. Our platforms were higher than the others, and we rode out the storms."[42]

Brown & Root specialized in platforms and marine pipelines, and the company decided not to get involved in mobile drilling rigs or support services which played an increasingly important role in offshore work. As offshore drilling and field development increased in scale and distance into the Gulf of Mexico, the transportation of personnel and material, not to mention the oil and gas, became a critical cost and logistical factor in offshore operations. Transportation to a drilling site could run a company up to $1,000 per day, per well. Mobility was the key to drilling and producing oil in a marine environment. Mobile drilling rigs, drilling tenders, and construction barges set the trend for the industry. "Full-fledged miniature navies are maintained by the industry," reported one trade journal, "miniature if the expenditure of a million dollars for transportation facilities and services can be compressed into the meaning of this description."[43]

Drilling platforms had to be supplied with labor, fuel, drilling mud, cement, casing and drill pipe, fresh water for drilling, crew provisions, and specialty equipment. Oil companies contracted for larger and faster crew and cargo boats and, increasingly, for the use of airborne travel to transport personnel farther offshore. Supply boats evolved from tugboats and barges to war surplus LCTs and eventually to purpose-built vessels 125 to 165 feet in length.[44] Crew boats were initially converted from their military equivalent, the "PT" boat.[45] As crews were ferried greater distances into unpredictable waters, larger boats were constructed (up to 85 feet) with fast-planing hulls. Crew boats could be dangerous, as twenty-one Brown & Root employees found out in July 1958 when a 51-foot crewboat taking them to a pipe-laying barge sank 45 miles out in high seas that were running six to eight foot waves. All were rescued without serious injuries after remaining in the water for an hour and a half.[46]

By the mid-1950s, the introduction of helicopters provided greater economies of speed and safety, especially during rough weather. They paid "a sizeable dividend in transporting contract service companies and other part-time specialists."[47] Helicopters saved crews from the sickness-inducing 8- to 12-hour rides that often left them in no shape to work for an entire day after they arrived on board. Heliports became a ubiquitous feature on offshore platforms, mobile rigs, and tenders.[48] Contractors such as Petroleum Helicopters, Hawk Helicopters, General Air Transport, and Rotor Aids served offshore needs with better and larger choppers. They were an example of how independent offshore transport and supply companies sprouted up in the 1950s to provide specialized transport services to the offshore industry.

Transporting offshore oil to market presented the most formidable challenge to the industry. As *World Oil* magazine put it in 1957, "The major problem confronting production men offshore is not well equipment or workovers. The big question is 'where will we put it after we get it?'"[49] Each operator in each producing area faced a unique set of considerations in making decisions about production handling and transportation: depth of water, distance from land, prevailing weather, rate of production, reserve estimates, location of surrounding wells and fields, and crude market values. Early field development operations with moderate levels of production employed some combination of offshore storage and barging. Often, the oil was separated on the well platform, stored in a tank battery, and then transferred to a barge moored off the platform for shipment to shore. If production merited it, a separate tank battery platform with a separating unit was built and connected to the well platform(s) by a flow line. Some submersible tank-battery barges and underwater storage systems were placed in shallower and more tranquil waters of the Gulf. However, barging was very expensive, adding 40 cents a barrel or more to production costs, depending on the distances involved. Storage tank costs increased in proportion to their size and became increasingly costly and impractical with higher levels of production. "The ultimate answer," wrote *World Oil* in 1957, "is pipelines if and when production warrants it."[50]

Pipelines became a trademark Brown & Root service to the offshore industry. Frank Motley, who was in charge of the firm's first offshore crew, recognized that pipelines would be the umbilical cords for production operations offshore. Brown & Root had laid lines in Galveston Bay as well as the first oil pipeline in the Gulf of Mexico to connect the early Creole platform to shore, but laying pipeline in deep, open waters introduced new and difficult problems. Like all other aspects of the emerging offshore industry, standard methods and expertise for laying submarine pipelines were attained by gradually moving out into deeper waters of the Gulf of Mexico.

The first barges used to lay pipe were small and crude. Typically, several barges were hooked up end to end and equipped with side booms to stalk the pipe. Work proceeded slowly, and laying operations routinely damaged or lost pipe and equipment. In its descent from the barge to the sea floor, the pipestring often buckled due to excessive stresses on the pipe and welds. To control bending stresses on the pipe, oil drums were lashed to pipe sections for buoyancy, but this method involved tremendous physical exertion and time. In the early years, pipelines were "laid more by sheer muscle power than by efficient specially designed equipment."[51] Until the early 1950s, offshore pipelining projects still were restricted to bays, rivers, and marsh areas of the shallow water Gulf.[52]

In the early 1950s, Brown & Root took the lead in advancing the development of marine pipelining. The need to improve on the "flotation system" fostered innovation. Frank Motley came up with the idea of attaching a ramp to the side of Brown & Root's *Herman B* to allow the pipeline to angle more gently down to the ocean floor. Motley claimed little special insight; he later explained that his idea had been provoked by his "experience with losing pipe so many times." Motley said, "It wasn't that we were so smart. It was just obvious that a ramp was needed to allow the pipeline to slope to the bottom, and the side of the barge was the only place to put it."[53]

At first, Motley's crew laid pipe off the *Herman B* in welded sections 120 feet long, but they subsequently developed an assembly line approach in which the pipe went through four stations on the deck of the barge, where it was continuously prepared, welded, weighted, and laid. The barge also was equipped with an eight-point mooring system. The Gulf of Mexico received the first truly offshore pipeline in 1954 when the *Herman B* laid a 10-inch diameter, concrete-coated line 10 miles out from the Louisiana shore to the Cameron field for gathering gas. The line was laid in water depths ranging from 14 to 30 feet at an average of 2,500 feet per day. Because there were no crew quarters on board, the work day was short. But the crew and the system worked quickly.[54]

The new system earned plaudits from the industry and encouraged Brown & Root's offshore division to move aggressively into pipeline laying as one of its main specialties. In 1954, Brown & Root won a $7 million contract from the Magnolia Petroleum Company to lay the longest undersea pipeline. When completed in 1955, the "Eugene Island Field Flowline System" followed a 48-mile route, constructed of 12-inch pipe, from the coast of St. Mary's Parish, Louisiana, to the Magnolia-Conoco-Newmont well platforms in Blocks 45 and 126 of the Eugene Island Area. Brown & Root used the *L.T. Bolin,* which was also outfitted with a side ramp, along with the *Herman B* to lay the pipe in depths up to 40 feet. The pipeline was designed with a capacity to carry 50,000 barrels of crude per day and the ability to handle two-phase oil and gas flow.[55]

The *M-211* pipelaying barge with stinger

Such projects convinced the company to develop specialized vessels and equipment to take on pipelining in deeper waters. Through a joint venture with the Southern Production Company in 1956, Brown & Root developed a new pipe-coating process for weighing down pipelines and protecting them from corrosion and abrasion. The process applied a hot mix of rubberized asphalt mastic, weight material, and fiberglass onto preheated pipe as it was laid. Since no curing time was required, in contrast to concrete-wrapped pipe, the process also speeded up pipelaying.[56] In 1956, at a new facility in Lafitte, Louisiana, Brown & Root engineers converted a number of barges into full-time pipelayers, named the *M-211,* the *M-140,* and *Sample No. 2.* In less than two months in 1957, those barges laid a 5.5-mile line connecting five offshore platforms in Chevron's gas field off the coast of Kleberg County, Texas, to Standard Oil Co.'s condensate plant on Padre Island. That same year, Brown & Root also began laying 60 miles of 26-, 20-, and 12-inch pipe in the East Cameron Area to connect the offshore gas fields of the CATC Group and the MCN Group to the Tennessee Gas Transmission Company's compressor station near Kinder, Louisiana.[57]

On the Tennessee Gas project, Brown & Root unveiled a new pipelaying innovation. Brown & Root's newest barge, the 300-foot long *M-211,* possessed a novel device called a "stinger," which was a cradle consisting of two parallel 30-inch pipes with controlled buoyancy used to ease the pipeline to the bottom in deeper waters and thereby minimize bending stresses. The stinger was the first in a series of pipelaying innovations pioneered by Brown & Root in the 1960s that would take marine pipelining into deeper waters.

Transitional Period

The year 1958 marked the end of what might be called the first major phase of offshore oil development in the Gulf of Mexico. Exploratory drilling in the Gulf dramatically fell off its torrid pace of the mid-1950s. Whereas 100 percent of drilling rigs available were at work in 1957, by January of the following year only 37 percent of those rigs were working.[58] New development wells declined from 282 in 1957 to 172 in 1958. "The rapid rise and correspondingly rapid decline in offshore drilling operations in the Gulf of Mexico," wrote the President of the American Association of Oilwell Drilling Contractors in 1959, "is one of the most surprising phenomena which has occurred in the oil business in many years."[59]

A combination of economic and natural constraints brought about the deceleration in drilling. A recession and oversupply of U.S. crude oil discouraged oil companies from embarking on new exploration. Excess supply also lowered the allowable production rate for offshore Louisiana wells at a 10,000-foot depth from 242 to 173 barrels per day, driving up operating and unit production costs, and driving down profits. Capital outlays of $1 million per well were not unusual. Lower finding rates in deeper water contributed to increased drilling costs. In water deeper than 60 feet, 40 percent of the wells drilled were dry, compared to a dry-well rate of one-third for those drilled in less than 60 feet of water. During the period of rapid expansion, companies had skimmed off the cream by drilling into the most promising salt dome structures. Oil finds were now thinning out.[60]

Hurricanes and sustained bad weather in the Gulf of Mexico forced companies to re-evaluate their risk minimization calculations and structural designs. Hurricanes Flossie, Audrey, and Bertha rocked the Gulf in 1957. Even though advance warnings and preparations saved offshore installations from catastrophe, Hurricane Audrey in July 1957 nonetheless inflicted damages estimated at between $15,000 to $30,000 per rig and destroyed Cameron, Louisiana, an offshore support center near the Louisiana-Texas border.[61] The disruptive weather continued for months afterward. *World Petroleum* journalist James Calvert described a common scene from Morgan City, Louisiana, during the harsh winter of 1958: "Battered supply vessels, their steel bulwarks hammered and crushed—crew boats crusted with salt spray with their seasick and injured human cargoes—seagoing oil barges, pipeline lay barges, giant mobile drilling rigs seeking shelter from the relentless pounding of the sea in one of the worst winters within the recollection of even the oldest natives of the Gulf Coast region." Engineers stepped back to reassess the challenges of deepwater drilling and structural design.[62]

Still, the late 1950s was a transition for the industry rather than an extended slowdown. Instead of drilling for new reserves, companies focused on cultivating the skills to perform all facets of offshore petroleum development in the 50- to 150-foot depth range. Offshore Gulf of Mexico oil production rose steadily from 16 million barrels in 1957, to 25 million in 1958, and hit 36 million in 1959. Gas production climbed from an average of 82 billion cubic feet per year during 1955–1957, to 128 billion in 1958, to 200 billion cubic feet in 1959.[63] By 1958, around 200 miles of underwater pipeline had been laid in the Gulf, more than one-half of that in 1957 alone, thanks largely to the efforts of the Brown & Root marine unit. Notable improvements made to fixed structures included radio communication systems, rig electrification, pre-packaged modular units to maximize space usage, and underwater cathodic protection systems to safeguard big offshore structures from corrosion.[64]

The expansion of offshore drilling in the Gulf may have declined, but exploration and development in other parts of the world, namely Venezuela, the Persian Gulf, California, and eventually Alaska and the North Sea, increased. The offshore experience in the Gulf of Mexico had given companies optimism for the future and the confidence to ply other waters. Already drilling in 200 feet of water by 1959, the industry now talked of drilling in 600 to 1,500 feet. Conoco vice president Ira H. Cram, speaking in 1958, exclaimed that "offshore appears to be today the most promising area for the discovery of major fields of crude oil and natural gas at ordinary or unusual depths."[65] The offshore oil industry continued to innovate and expand, and Brown & Root Marine continued to be one of its pioneers.

Notes

1. This chapter draws from interviews with Edward Tallichet, Edward Blaschke, and W.R. Rochelle.
2. W.B. Agocs, in *Geophysics* (April 1944), quoted in Curtis A. Johnson and John W. Wilson, "Marine Exploration Comes of Age," *World Petroleum* (March 1954): 76.
3. John W. Wilson, "Single Ship Takes the Place of a Fleet," *World Oil* (June 1954): 163.
4. Johnson and Wilson, "Marine Exploration is Coming of Age," 77; James W. Calvert, "You Never Really Know Until You Drill," *World Petroleum* (March 1955): 64–65.
5. Wilson, "Single Ship," 164.
6. Wilson, "Single Ship," 164; Johnson and Wilson, "Marine Exploration," 77; William H. Newton, "Doodlebuggers Afloat," *World Petroleum* (February 1954): 35.
7. "Special Offshore Report," *World Oil* (May 1957): 125; Calvert, "You Never Really Know," 65.
8. Payne, "Mobile Units for Offshore Drilling," Paper presented to Offshore Operating Symposium, Houston, Texas, March 1954, 257.
9. W.D. Stine, "Tenders Will Get Wide Use in Off-shore Rig Programs," *Drilling* (August 1954): 75–131; "New Methods Aid Progress of Offshore Drilling," *World Petroleum* (March 1954): 62–65.
10. Dean A. McGee, "Economics of Offshore Drilling in the Gulf of Mexico," *Offshore Drilling* (February 1955): 16.
11. "Tests $3 million Offshore Lease," *World Petroleum* (August 1954): 32.
12. James W. Calvert, "Offshore Drilling Techniques," *World Petroleum* (August 1957): 42.
13. McGee, "Economics of Drilling in the Gulf of Mexico," 16.
14. Discussion of Richard J. Howe, "Some Factors in the Engineering Design of Offshore Mobile Drilling Units," Paper presented at the API Southwestern District, Division of Production, New Orleans, March 1955.
15. "Mobile Units for Offshore Drilling," *World Petroleum* (November 1954): 78–80; "Here's the Newest 'Ship' in the Petroleum 'Navy'," *World Oil* (December 1954): 140–141; Paul Wolff, "Kerr-McGee Adds Deep Water Submersible Drilling Barge," *Drilling* (June 1954): 107–108.
16. Howe, "Offshore Mobile Drilling Units," 38–41; "Fixed-Height Deck Type Mobile Platforms," *Drilling* (August 1955): 84–91; "Operation of Mobile Drilling Units Offshore," *Offshore Drilling* (March 1955): 11–30; and "Economics Designed the 'John Hayward'," *Offshore Drilling* (June 1955): 12, 20.
17. James W. Calvert, "Gulf Offshore Activity Booming," *World Petroleum* (January 1957): 49.

18. Richard J. Howe, "The History and Current Status of Offshore Mobile Drilling Units," *Ocean Industry* (July 1969): 42–42; Robert H. Macy, "Trends in Offshore Drilling Rigs," Paper presented at the Louisiana Engineering Society Annual Meeting, January 11, 1963, 6–7.
19. "Twin-Hull "'Mister Gus' Built by Bethlehem," *Offshore Drilling* (December 1954): 5–6; "'Mr. Gus'," *Drilling* (December 1954): 54–55; James W. Calvert, "Mr. Gus Goes to Sea," *World Petroleum* (December 1954): 34–37; Calvert, "The Mobile Rig Disasters," *World Petroleum* (June 1957): 30–33; "Special Offshore Report," *World Oil* (May 1957): 138.
20. "Zapata Offshore Company Purchases Huge Barge Rig," *Drilling* (February 1955): 91, 127; "The Scorpion Goes to Sea," *Drilling* (April 1956): 121; James W. Calvert, "Seagoing Skyscrapers," *World Petroleum* (March 1956): 74–77; Howe, "Offshore Mobile Drilling Units," 42–43.
21. "There is Where the Oil Is," *Brownbilt* (Fall 1972): 12.
22. "New Era of Offshore Expansion," *Offshore* (November 1957): 53–54, 76.
23. Howe, "Offshore Mobile Drilling Units," 38.
24. L.C.S. Kobus, "Early Drillships Evolved 'Logically'," *Offshore* (September 1979): 94.
25. "Shipyard Summary," *Offshore Drilling* (February 1957): 24–25.
26. "Optimism Offshore," *World Oil* (February 1, 1956): 62.
27. James W. Calvert, "Gulf Offshore Activity Booming," *World Petroleum* (January 1957): 48.
28. Ben C. Belt, "Louisiana and Texas Offshore Prospects," *Drilling* (March 1956): 119.
29. McGee, "Economics of Offshore Drilling in the Gulf of Mexico," 30–32.
30. "Special Offshore Report," *World Oil* (May 1957): 120.
31. F.R. Hauber, "Drilling and Production Structures for Oil and Gas on the Continental Shelf," API Paper, January 1966.
32. "Derrick Barge is Engineered for Heavy Duty Gulf Work," *Drilling* (December 1955): 93.
33. J. Roland Carr, "New Designs in Offshore Drilling Rigs," *Engineering News-Record* (October 3, 1957): 54.
34. Hauber, "Drilling and Production Structures," 154–155; Griff C. Lee, J. Ray McDermott & Co., "Offshore Structures: Past, Present, Future and Design Considerations," *Offshore* (June 5, 1968); R.W. McDonald, "Analysis of Offshore Structure Design," *Offshore* (February 1958): 35–40; Richard A. Geyer, Humble Oil and Refining Co., A.H. Glenn, A.H. Glenn and Associates, and Robert O. Reid, Texas A & M College, "Design Criteria for Fixed-Piling Structures and Mobile Units," Paper Presented at the API Offshore Operating Symposium, Southwestern District, Division of Production; National Petroleum Council, *Impact of New Technology on the U.S. Petroleum Industry* (NPC, 1967): 205–209; "New Platform Has Slanted Piling for Deepest Gulf Drilling," *World Oil* (July 1956): 116–117.
35. Interview, Ed Tallichet, September 13, 1995; Interview, Ed Blaschke, October 12, 1995; Interview, W.P. Rochelle, January 16, 1996.

36. Carr, "New Designs in Offshore Drilling Rigs," 56.
37. *Ibid.*
38. Carr, "New Designs in Offshore Drilling Rigs," 56–57.
39. See Griff Lee, "Offshore Platform Construction Extended to 400-foot Water Depths," *Journal of Petroleum Technology* (August 1963): 386.
40. Carr, 56.
41. Brown & Root, "Marine Operations," Marine Division pamphlet in company possession; "CATC Group Builds Huge Platform to Explore Deepest Water Yet Tested," *Offshore Drilling* (March 1956): 16–17.
42. Brown & Root, "Marine Operations;" Joseph Pratt and Christopher Castaneda, *Builders* (Texas A&M University Press, forthcoming).
43. "Transportation Major Cost in Off-Shore Rig Operation," *Drilling* (June 1954): 110.
44. Robert H. Macy, "Trends in Offshore Boat Designs," *Offshore Drilling* (July 1957): 19–21.
45. J.C. Craig, "Auxiliary Floating Equipment From War Surplus Conversions to Drawing Board Dreams," *Offshore Drilling* (July 1955): 14, 33–34.
46. "Twenty-one Marine Workers Saved after Crewboat Sinks," *Offshore* (July 1958): 26.
47. "Personnel Transfer is Major Problem in Off-Shore Safety," *Drilling* (June 1954): 121–122; "Helicopters in Offshore Operation," *Offshore Drilling* (July 1955): 22.
48. Richard J. Howe, "Economics in the Use of Helicopters in Offshore Drilling and Production Operations," Paper Presented at the API Southern District, Division of Production meeting, San Antonio, Texas, March 1956.
49. "Special Offshore Report," *World Oil* (May 1957): 143.
50. Isaac L. Ault, "Production Handling and Transportation of Petroleum Hydrocarbons—Crude Oil," Paper presented to the API Offshore Operating Symposium, Southwestern District, Production Division, Houston, Texas, March 1954; National Petroleum Council, *Impact of New Technology,* 215; "Special Offshore Report," *World Oil* (May 1957): 143, 147.
51. Delbert R. Ward, Brown & Root, Inc., "Submarine Pipeline Construction Techniques," API Paper (1966), 153.
52. "A Pioneer Among Pioneers," *Brownbilt* (Fall 1972): 48.
53. "A Pioneer Among Pioneers," 21.
54. "'There is Where the Oil is,'" *Brownbilt* (Fall 1972): 11.
55. "Brown & Root Gets Pipeline Contract," *Offshore Drilling* (February 1955): 24; R.H. Illingworth, W.L. Montgomery, Clyde Aldrige, and C.V. Temple, "Proposed Offshore Production and Gathering Facilities," *Offshore Drilling* (May 1955): 29–30; and "Longest Undersea Pipeline," *World Petroleum* (January 1956): 92.
56. "New Pipe Coating Process Boon to Submerged Oil, Gas Lines!" *Offshore Drilling* (December 1956): 16–17, 38.
57. "TGT Gets Offshore Pipeline," *Offshore* (October 1957): 45–46, 66.

58. Robert O. Frederick, "Marine Drilling: The Future Remains Bright," *Drilling* (December 1959): 55–56.
59. Joe Zeppa, "What is the Outlook for Drilling in the Gulf of Mexico," *Drilling* (July 1959): 59.
60. Zeppa, "Outlook for Drilling," 59–60; Bouwe Dykstra, "Costs, Allowable Rate Hinders Offshore Work," *Drilling* (August 1959): 65, 114.
61. "Tide Forecasts Offshore Industry Precautionary Measure," *Offshore Drilling* (October 1957): 72–74; "Forethought Minimized Hurricane Damage to Offshore Installations," *Offshore Drilling* (August 1957): 15–18, 25.
62. James W. Calvert, "Pipeline Gathering Systems Essential to Profitable Offshore Operations," *World Petroleum* (April 1958): 44–47.
63. Richard Gramling, *Oil on the Edge: Offshore Development, Gridlock, Conflict* (Albany: State University New York Press, 1996), 71.
64. Calvert, "Pipeline Gathering Systems," 44–47.
65. Calvert, "Pipeline Gathering Systems," 47.

CHAPTER 4

New Sophistication

The 1960s was an exhilarating period of expansion for Brown & Root Marine. The company improved its capabilities for analyzing the behavior of marine structures and laying pipeline in deeper waters. New engineering teams brought higher mathematics and computer-assisted innovations to structural design and pipelining. These innovations, in turn, led to the design and building of new equipment and vessels. Taken as a whole, the work of Brown & Root Marine in the decade before the energy crisis of 1973–74 marked a dramatic advance in scale and complexity from its previous offshore work.[1]

These advances were made in an era when the price of oil remained securely in the $2 to $3 per barrel range. Even at those low prices, the offshore industry moved forward briskly, using new technology and economies of scale to lower the price of recovering oil and gas from under the seas. As the industry attempted to keep pace with buoyant demand for oil and gas, it established offshore oil and gas as an important element in the energy mix. This was especially true of natural gas in the Gulf of Mexico, where the construction of a massive pipeline gathering system connected the large new reserves of offshore gas into the onshore gas transmission system, bringing an important new source of energy to market.

In the 1960s, Brown & Root Marine was the leader in the laying of this submarine natural gas pipeline system, and it remained at the forefront in the construction of submarine oil pipelines which hastened the development of new offshore oil and gas reserves. Building platforms and laying pipelines in progressively deeper waters required staggering investments in new equipment. Purpose-built lay barges with more sophisticated pipe-handling equipment expanded the frontier of marine pipelining. Bigger derrick barges enabled the lifting and setting of increasingly larger jackets, and they allowed for operations in rougher seas. Even before the energy crisis dramatically raised the price of oil and gas, Brown & Root and the offshore industry as a whole had already developed much of the new technology that would be adapted and applied to harsher offshore environments once higher prices justified such initiatives. "I don't give nearly enough credit for all the outstanding things that were being done on a continous basis," admitted John Mackin, former head of Brown & Root's Marine Industries Deparment in the 1960s, "because it just got to be commonplace."

The New Brown & Root Marine

Most of Brown & Root's early marine engineers practiced a hands-on approach to their work, which stressed solving practical problems in the field. This was true of Brown & Root as a whole, which remained primarily a construction company with a growing engineering corps. As the company diversified, it improved its ability to conduct preliminary feasibility studies, prepare detailed proposals for prospective projects, complete detailed engineering plans, and manage the resulting work. From the engineering staff inherited from Brown Shipbuilding at the end of World War II, Brown & Root's engineering department grew to more than 500 by the mid-1950s and 700 by the mid-1960s. At that time, the company's brochures listed the following engineering groups: air conditioning, chemical, civil, electrical, materials handling, mechanical, power, port facilities, oil and gas processing, overseas projects, and marine. In practice, these groups overlapped, and marine projects drew on the engineering expertise of the company.

In the 1950s, Brown & Root's marine engineers and construction specialists built a solid client base and rapport with the offshore fraternity of producers, drillers, and related subcontractors operating from the Gulf Coast. Brown & Root's top platform designer F.R. (Ferd) Hauber "was doing a good job and people out in the Gulf of Mexico loved him." The original group of marine engineers had grown up solving pressing practical problems while meeting construction deadlines out in the Gulf, and they preferred working in this established realm. The design work itself often was treated more as a project-related task—how can we build a sturdy, durable structure on a particular job?—rather than as an abstract, theoretical task—how can we understand the general design parameters for offshore structures? But in the late 1950s, a new generation of engineers, trained in sophisticated mathematics and design analysis techniques, entered the company. As they introduced these new analytical techniques to the more complex design problems encountered in deeper waters of the Gulf and different environments around the world, they pushed Brown & Root Marine into new geographical and technological frontiers.

The new marine engineers coalesced into a group of recent engineering graduates organized by H.W. Reeves as a part of a special new marine engineering group. Reeves had joined Brown & Root in 1946 and had worked on major projects such as the Lake Ponchartrain Causeway. The group, referred to within the firm as Reeves's "chicks" or the "elite squad," included Dick Wilson, John Mackin, Tim Pease, Ber Pieper, Bob Gibson, and Delbert

Greens Bayou fabrication yard during the 1960s

Johnson. Most were graduates of Rice University, although Mackin was an "Aggie" from Texas A&M.

Pieper noted that the new group's mission was to "bring a little more sophisticated design capability into the offshore design arena." Ferd Hauber and Frank Motley had pioneered early marine structure designs and construction concepts using "a good crystal ball approach."[2] By the mid-1950s, they were still "designing the whole structure on about two size 'A' drawings with overturning moments, simulated wave forces and a lot of guesses and be damned sure it was heavy enough." Many assumptions about wave forces, lateral loads, joint designs, and the interactions between the pile and surrounding soil needed validation as construction moved further offshore. According to A.B. Crossman, one of the pioneers in computer-aided design of offshore structures at Brown & Root, the approach of the new engineers was "different from the way that the old marine engineering department analyzed platforms. They were doing a graphical truss analysis that was basically not correct. It was something that had worked from trial and error, but it was not at all rigorously accurate." By the late 1950s, some individuals working on offshore projects at Brown & Root realized that "we ought to have more calculations, that we ought to be more precise."

The new engineering group moved platform analysis "past a hip pocket art form." These engineers were the products of the research that the offshore petroleum industry had sponsored beginning in the early 1950s to generate more reliable methods of structural analysis.[3] Tim Pease, for example, had written his master's thesis at Rice on the effect of dynamic forces on "T" joints. He "found out right away that it didn't take very much load on the small member to overstress the wall on the big member," something that jacket designers had not fully understood. Pease applied this knowledge to developing special fins to reinforce the joints on the aluminum jackets Brown & Root built for Lake Maracaibo. "We were lucky in that group," recalled Pease, "because Reeves always let us do innovative things and get involved in things that were normally not done by recent graduates."

Hal Lindsay, head of Brown & Root's Marine Operators, began giving projects that demanded innovative design thinking to the new marine engineers. "When it was something different," Pease remembered, "he'd always come and get us involved." The real innovative work began in Venezuela. Many of the younger engineers such as Pease, Wilson, and Johnson were assigned to projects in Lake Maracaibo. A rivalry of sorts developed between the old marine group, who generally stayed with work in the Gulf of Mexico, and the new guys, who began to handle ambitious new overseas projects in South America and, increasingly, the Middle East. Rochelle recalled that "Ferd would take care of his old Gulf of Mexico customers and these other guys would go out and go for the wild and weird and sell the higher mathematics and all of this stuff. That was not without a little rancor." Symbolic of their new status, members of the new marine group even moved out of the main drafting room into a new one behind George and Herman Brown's offices, a place that came to be known as "the Palace." "It wouldn't be a palace by anybody's wildest dreams today," Rochelle said, "but anyhow, it was called 'the Palace.'" Despite tensions, the two groups coexisted working in their own realms until, eventually in the late-1960s, all marine structural design was incorporated into a single Marine Industries Division.

One of the most significant innovations in the early years of the new marine engineers was the development of computer programs to help analyze and design offshore structures. As the oil industry pushed into deeper water—building platforms in 200 feet of water by 1962 and 300 feet by the late 1960s—design considerations were becoming much more complicated. The traditional problems of wave load, hurricanes, and soil composition were more complex in deeper water. The growing size and weight of the jackets focused additional attention on stresses encountered in the fabrication and installation of platforms. New methods of "launching" larger and heavier jackets introduced new stresses on the jacket, and these had to be studied and accounted for in fabricating the structure.

Before the mid-1960s, the application of computer technology to most engineering design problems was not yet practical, since the computational algorithms had not been programmed and the hardware lacked sufficient capacity. As computers became larger and faster, however, they became practical tools in engineering design. The last half of the 1960s witnessed the application of computer technology to all types of engineering problems. Brown & Root Marine was one of the leaders in using computers to assist in designing offshore structures. In the words of Stanley Hruska and Albert Koehler—two of Brown & Root's pioneers in this area—"with the ever-increasing complexity of engineering problems surrounding the development of natural resources in the oceans, the marine engineer has also found that use of the computer is a must."[4]

Initially, in 1964, Brown & Root engineers worked with FRAN, a program developed by IBM for the aircraft industry to analyze large, indeterminate structures. The output included joint deflections and rotations, member forces and moments, support reactions, and joint equilibrium checks.[5] This program proved cumbersome, requiring considerable programming to adapt it to the design of marine structures. Hruska and Koehler spent "at least two years before we developed our first edition of 'DAMS,'" an acronym for Design and Analysis of Marine Structures. They developed a preprocessor which made FRAN more user-friendly for marine engineers by eliminating "some of this drudgery of an engineer having to put this thing together to run." At the time, Brown & Root still shared time on mainframe computers owned by other Houston-based companies. As this work continued, however, the company installed its own computer at its engineering offices on Clinton Drive in Houston.

Using DAMS revealed the need for further enhancements: "We had to go into developing the postprocessor so that you could take and apply a code check, the AISC code check, using the output you got from FRAN or the analysis program. . . . So, it was sort of like a never-ending development process." By 1970, DAMS had been joined by DYAN, a dynamic analysis program for predicting structural response, and FLAP (Flotation and Launching Analysis Program), which was developed under Jay Weidler to study and predict the motion characteristics of platforms as they were being launched from a barge. Launching eventually became the standard approach to installing large platforms. As the first offshore engineering company to offer clients such capabilities, Brown & Root "took a leadership role in the business." DAMS and its offsprings helped solidify Brown & Root's reputation as an innovator in offshore engineering. Hruska noted that "We were out there in front of the clients, and they recognized us."[6]

The strengths of computer-assisted design included tremendous savings in time, more rigorous analysis, and ultimately the creation of a "three dimensional blueprint" to replace the two-dimensional frame analysis process. While McDermott and other oil companies still used this two-dimensional process, Brown & Root had moved well into 3D. In the mid-1960s, "Brown & Root and Shell were among the few to use 3D," claimed John Mackin. Clients could "choose a total conceptual package, or lease the programs and analyze their own concepts." Using Brown & Root's software, a client could "load a space frame with any design wave, tide and current combination desired. . . . The output permits the engineer to see the axial loads, shears, moments and stresses in all members under any combination of loads." These programs "resulted in more accurate analysis of a structure under all the desired loading conditions," yet this could be done "at a savings in design man-hours."[7]

DAMS and its successors became essential to marine engineers. By applying computer analysis to the design of marine structures, the new engineers in Brown & Root's marine department contributed greatly to one of the key technological breakthroughs that transformed the offshore industry in the mid-1960s. In the process, they announced that Brown & Root Marine was to remain a leader in the new era of marine engineering. Computer-assisted design, said Rick Rochelle, "was one of the things that helped us to make the jump into the North Sea, because nobody else had the mathematical background."

Pipelines

Brown & Root's leadership in platform analysis during the 1960s was complemented by the strides it took in pipelining technology. Thanks to the ingenuity of people like Frank Motley, the company had developed improve-

Computer room at Clinton Drive

ments to pipelaying such as the pipe ramp and stinger device. But as pipelines were laid into deeper water, Brown & Root's engineers and management realized that the company could not maintain a leading position in this realm through minor adaptations to conventional practices. They would need to rethink the challenge of laying pipe in deeper water, which required a commitment to in-house development work.

The first step involved increasing the speed and scale of pipelining. In 1957, the pipeline division created a base for more extensive pipeline operations with the founding of the Belle Chasse pipe yard across the Mississippi River from New Orleans. The following year, Brown & Root built the *L.E. Minor,* the industry's first purpose-built lay barge. Designed by F.R. Hauber, the *Minor* was built by Levingston Shipbuilding in Orange, Texas, at a cost of $2.35 million. It incorporated the state-of-the-art in pipelaying technology.

Named after Ed Minor, then the vice president of Brown & Root's pipeline division, the giant barge (by the standards of the day) was 350 feet long by 60 feet wide, with a 22-1/2 foot draft. It could lay 40-inch pipe weighing 25,000 pounds per joint. It had air-conditioned crew quarters for 88 men, and storage area large enough to hold three miles of pipe insured against transportation delays caused by rough seas. A custom-designed spoonbill bow enhanced stability and permitted normal pipelaying in seas with four- to five-foot waves. The barge could work around the clock and through turbulent weather, laying up to a mile of line per day. The size alone of the *Minor* represented a step forward in the pace and scale of pipelaying capacity. The significance of this new vessel for the offshore industry was reflected in the wide coverage given to its launching in trade publications.[8]

As one of the earliest examples of what was later classified as the second generation of lay barges, the *Minor* was akin to a giant floating pipelaying factory, complete with assembly line methods. Pipe and practically all machinery were placed below deck, giving the vessel a virtually flat silhouette, which made it more stable in rough seas. An 80-ton Manitowoc traveling gantry crane handled all lifting chores, including placing pipe between a set of rollers on a special stalking machine. The machine moved pipe into position on the welding ramp, where hydraulically operated lineup shoes placed the pipe in position for the stringer bead to be made. Five welding stations protected from the elements allowed for welding under most weather conditions. Once welded, the pipe moved automatically through an X-ray station, a doping station, and a concrete station for making the final field joint in the weight coating. Cement and dope were stored below deck and delivered directly to the laying ramp through doors in the barge's side. The 12-foot-wide work ramp extended along the starboard side, with arrangements for moving pipes along the ramp and out onto the stinger. Five pipe-handling davits spaced at sixty-foot intervals above the work ramp had the capacity to lower and raise pipe to the bottom without the assistance of a crane if needed to avoid rugged weather. A control tower above the work ramp gave the anchor man an unobstructed view of all work stations. From here, he operated the system of winches used to move the barge once a joint of pipe was completed. Eight 10,000-pound anchors, four forward and four aft, held the barge in place. By taking up slack on the forward anchor cables and letting it out on the rear ones, the anchor man moved the barge out from under each pipe joint. Then the stalking machine dropped another joint of pipe into the lineup rollers on the work ramp, and the process began again, as workers took shifts to allow for around-the-clock operations.[9]

Brown & Root's marine pipeline business expanded rapidly. The company built a fleet of pipelaying barges with roughly the same capabilities as the *L.E. Minor.* In the early 1960s, one of Brown & Root's new barges laid 20-inch pipe through 7-foot seas by brute strength, reporting that "Progress . . . was excellent, sometimes running over 200 joints per 24 hours."[10] Although advances in later lay barges that were developed in response to the harsher demands of the North Sea would make such statistics seem almost modest. But in the context of the Gulf of Mexico in the 1960s, these first purpose-built lay barges represented a dramatic step forward in speed, power, and dependability. Led by Brown & Root, those companies willing and able to develop and build these specialized vessels became the leaders in marine pipelaying worldwide. In the process, they altered the economics of the business by providing cost savings and more predictable service.

The key challenge to laying pipe in deeper water was to control stress on the pipe during installation. Submarine pipelines had to be welded and then lowered to the bottom of the sea. As pipe size and water depth increased and seas became rougher, the installation of pipe confronted greater problems. "We were having to analyze them for laying stresses," explained Rick Rochelle, "and we were getting into such things as bottom stability and currents that nobody ever paid any attention to." Engineers knew that a basic requirement in pipelaying was "to provide continuous control of the pipeline to prevent overstressing during installation."[11] But for pipelaying to progress into deeper, rougher waters, they had to develop new techniques.

From the mid-1950s forward, almost all pipelines in the Gulf of Mexico had been laid from barges. Brown & Root had introduced the use of pontoons to reduce the stress on pipe by easing it downward to the ocean floor. In shallow water with relatively small diameter pipe, a basic pontoon system extended from the work ramp of the pipelaying barge worked efficiently: the welded sections of pipe simply rolled on down the ramp and out onto the pontoon, where they continued to roll between barrels or other sorts of buoyancy devices down to their final resting place. As water depths increased, the industry moved to a controlled buoyancy pontoon, or "stinger." Attached off the stern of the lay barge as an extension of the launching ramp, the stinger consisted of two continuous parallel pipes joined by cross members carrying rubberized or polyurethane rollers. Cross-connected tanks within each pipe allowed for adjustments of the ballast in the stinger to assure that the combined weight of the pipeline and the stinger was zero, thus reducing stress on the pipe as it made its journey to the bottom. Alterations in the length and angle of the stinger based on calculations of the stresses on the pipe allowed for continuous support for the pipeline. The stinger's strength also helped resist stresses due to water and barge movement.[12] The stinger had to become longer to account for increasing water depths, and the modern lay barges were equipped with 300–400 foot-long stingers. The basic "straight stinger" arrangement proved effective in laying standard-sized pipe in more than 200 feet of water in the mid-1960s.

By this time, however, the effectiveness of traditional stingers was limited. Even in waters of 200–300 feet, under certain circumstances, a stinger as long as 500 to 700 feet might be required to provide the proper angle of inclination for laying pipe. Rick Rochelle remembered that beyond 200 feet of water, the pontoon became "so big, so long," and that "even in the Gulf of Mexico, a 500-foot pontoon was hellaciously hard to handle." A more significant limitation of a greatly extended stinger in deep water was the overbend on pipe as it left the rear of the barge and the sag-bend as it left the stinger near the ocean floor. In Rochelle's words, "from the end of the pontoon to the sea floor was a simple span. And if you got that simple span too long, you buckled the pipe." Brown & Root's marine engineers observed that "the stinger has its self-imposed limits of length due to handling problems." Although this limit varied by pipe size, weight, and coating, bottom currents, and other variables, they concluded that "a generalized depth of 300 feet can be constructed as a limit." According to a Brown & Root estimate in 1967, using traditional stingers, the offshore industry had the capacity to install 8-inch pipe at a maximum depth of 400 feet in calm water, with this depth decreasing for larger pipe to a range of 225 feet for 18-inch pipe and 140 feet for 40-inch pipe, which was, at the time, very large submarine pipe.[13] In 1966–1967, J. Ray McDermott used an extra-long stinger to lay 90 miles of 22-inch pipe in 300 feet depths of the Persian Gulf, but with great difficulty.

By the mid-1960s it was becoming evident to Brown & Root's upper management that spending more money on existing systems and a "brute force" approach would not be enough to hold a lead position in the increasingly difficult pipeline projects anticipated in the future. "Somewhere along the line, the Gulf of Mexico started picking up a bit and we needed better methods of laying pipelines," remembered Rochelle. As Brown & Root managers cast about for possible answers to this dilemma, Project Mohole was brought to a close. The loss of this contract provided a new opportunity for pipeline management.

L.E. Minor selected approximately a dozen veterans of Mohole to set up a research and development organization called the Marine Technology Department to develop techniques and equipment for the future. Minor had concluded that "We need to have an 'ivory tower group' to look completely beyond anything that our engineering has even thought of." The people chosen for the new organization—Rick Rochelle, Joe Lochridge, Clyde Nolan, and Adi Desai, among others—had been exposed to almost every contemporary naval and civilian research project pertaining to deep water. They were also practical designers. They brought such disciplines as drilling and pipe-handling equipment design, advanced tubular analysis, specialized naval architecture for deep ocean drilling, subsea acoustical navigation, and dynamic positioning systems into Brown & Root's pipeline world. Their charge from L.E. Minor, according to Rochelle, was "Take all of that deep ocean drilling equipment experience and turn it horizontal." The introduction of this new R&D group brought about some internal personnel problems. Altering methods, equipment, and customer designs naturally provoked resistance from some quarters. In time, however, these problems were resolved to everyone's benefit.

The charge was to look to the future, but present as well as future problems loomed. Brown & Root vice president Hugh Gordon informed the group that "when we lose a million dollars a day on that barge sitting out there [in the North Sea], your guys' salaries and your plane fares and all of that are peanuts. Just do it." Offshore operators were planning for relatively large diameter pipelines in 275 to 350 feet of water, and a high premium was placed on new techniques. Brown & Root responded, and by the early 1970s, the company—along with others in the industry—had transformed pipelaying technology for deep water in several critical ways.

In 1965, while laying pipe in 235 to 290 feet of water for Shell's Marlin System to carry crude oil up to 40 miles to shore from platforms in Louisiana's West Delta block, Brown & Root cooperated with Shell in developing what came to be called the "catenary system." Laying of a portion of this pipeline in 235 feet of water required a 728-foot stinger. To avoid further lengthening of the pontoon, the companies experimented with a "patented constant tension device," which involved a shorter stinger. Instead of attempting to support the pipe all the way to the ocean floor, this new "reverse catenary under tension" approach allowed the pipe to support its own weight as it moved toward the bottom. The successful use of this innovation on the Shell project, and on another job laying a 26-inch natural gas pipeline in about 300 feet of water for Southern Natural Gas Company, convinced Brown & Root engineers Joe Lochridge and Adi Desai that this was a practical solution for laying pipe in deeper water.[14]

Shell Pipe Line had patented a design for an articulated stinger that was without rigid joints and completely flexible. After examining model tests that Shell had run, Brown & Root's engineers were "a little scared" of this design, which they saw as too flexible. They modified the design by adding "some rigid members between the bottom parts of the pontoon so that it could be adjusted. It wasn't free—it was a rigid pontoon. It was curved. You could adjust it through jacks. You could make it flatter, longer, change the angle of it, do whatever you wanted." In the offshore "fraternity" of the time, all parties understood that most technological progress came from the efforts of many different people in different companies; there was too much work to do to get involved in an extended legal conflict over the pedigree of a much needed innovation that had quickly proved its worth.

Adopting an adjustable and shorter stinger depended on a more sophisticated understanding of the stresses on pipe as it was laid. "This, again," Rochelle said, "involved throwing the book away and saying, what do we really need?

The *L.E. Minor* pipelaying barge

Let's quit doing it like it has always been done and see what we really need." The new design required a more powerful and precise tensioner to maintain a constant tension on the pipe.

From the earliest days of lay barges, some sort of safety brake had been needed to control the pipe as it rolled down the ramp and the stinger. The tensioner had to be far advanced beyond traditional brakes to be effective. In Lake Maracaibo in the late 1950s, barge work ramps had been rigged with rows of giant truck tires above and below the pipe for stopping the pipe on its journey down the rollers. As stingers lengthened, "pipe would want to run downhill on them whenever they moved forward. So they took this old brake tensioner and put some motors on it. This was kind of how the conversion to the . . . first power tensioning machine [occurred]." The first generation of purpose-built lay barges used cables and winches as brakes, but in 1965 Brown & Root began using a "newly patented hold-back shoe," which was a "hydraulic device operating through 10 sets of dual rubber tires placed on the top and bottom of the pipe."[15] Clyde Nolan at Brown & Root developed the more sophisticated tensioners required in the North Sea and other environments. The modern tensioner advanced the articulated stinger by giving pipelayers a tool to use in maintaining proper pipe tension. The "pull" from the lay barge

Wiring up the *M-211*'s 728-foot stinger to lay pipe for Shell's Marlin system

compensated for the vertical weight of a long, heavy string of pipe, relieving stress on the unsupported portion of the pipe from the end of the stinger to the ocean floor.

Experimentation, computer analysis, and model testing helped determine the proper tension under different conditions. The stress on pipes laid with this new equipment, said Rochelle "was extremely hard to analyze because the programs didn't exist. Incidentally, we were willing to write most of our programs." With time and testing, Brown & Root and the industry as a whole learned how to make effective use of the articulated pontoon with tensioner under different conditions. This combination transformed the laying of pipe in water depths greater than 300 feet, allowing the offshore industry to continue moving out into deeper, harsher waters.

One significant breakthrough in pipelaying technology came in 1969, when Brown & Root used a computerized microwave survey system to control a lay barge during pipelaying operations. Lay barges traditionally had used preplaced buoys to mark their desired course; but tides and currents could move these buoys. As Brown & Root sought to lay larger pipe for longer distances, more precision was needed to reduce the possiblity of buckling and the danger of laying pipe across another line. A straighter line also resulted in the use of less pipe and less time. To achieve such results, the company used two microwave ranging sets at stations onshore, a specially developed computer program to convert the microwave range data into grid coordinates, and a digital plotter to display both the preplotted path and the actual path taken during laying operations. Deviations from the planned course could be observed and corrected through simple right or left change by winch operators. The beauty of this innovation was that it used advanced microwave and computer technology to produce simple, direct instructions for operators.[16]

The microwave-survey system was first used to lay 27 miles of giant 42-inch concrete-coated pipe off Kharg Island in the Persian Gulf. This project revealed the challenges often involved in applying new technologies to varied marine environments. "As with almost anything new in the marine business," said Rochelle, "if you wanted it done, you did it yourself." The microwave system used triangulation by line of sight and required active transponders high enough to receive and send signals that would not be interrupted by the earth's curvature. Consequently, one transponder had to be located on the top of a coastal mountain in Iran. No one else would do the job, so Jerry Jones and Jack Harris, who built the system, took one technician, an Iranian guide, and all the equipment in a Jeep up to a village in the mountain. They then rented mules and hired a guard to go the rest of the way. The mules balked toward the top, so they had to carry the equipment themselves. When climbing, the technician lost his footing and almost slid over a precipice to what would have been certain death. The group left the equipment, including batteries, set up on the mountain top with the lone Iranian guard armed with an old rifle. The equipment was still there at the end of the project.

The microwave survey system added technical precision to the art of pipelaying, while larger and better bury barges improved efficiency. Brown & Root invested in a fleet of large bury barges. They were 250 to 300 feet long, and they employed Brown & Root's patented combination of high pressure jets and suction dredge. By the late 1960s, the demands of laying miles of pipeline in the rugged North Sea compelled the company to build bury barges and lay barges with hulls specifically strengthened for the tough conditions encountered there. The company also acquired barges equipped to both lay and bury pipe. These investments symbolized Brown & Root's commitment to keeping pace with the needs of a growing offshore industry.

By 1971, Brown & Root had laid approximately 6,000 miles of submarine pipelines. The firm was acknowledged as the leader in this area of the offshore industry, boasting a fleet of specialized vessels and claiming an impressive list of pipeline "firsts." Its major pipeline projects as of 1971 show that Brown & Root had laid more than 75 percent of the pipelines in the Gulf of Mexico. The Persian Gulf accounted for about 10 percent of the total, with Venezuela

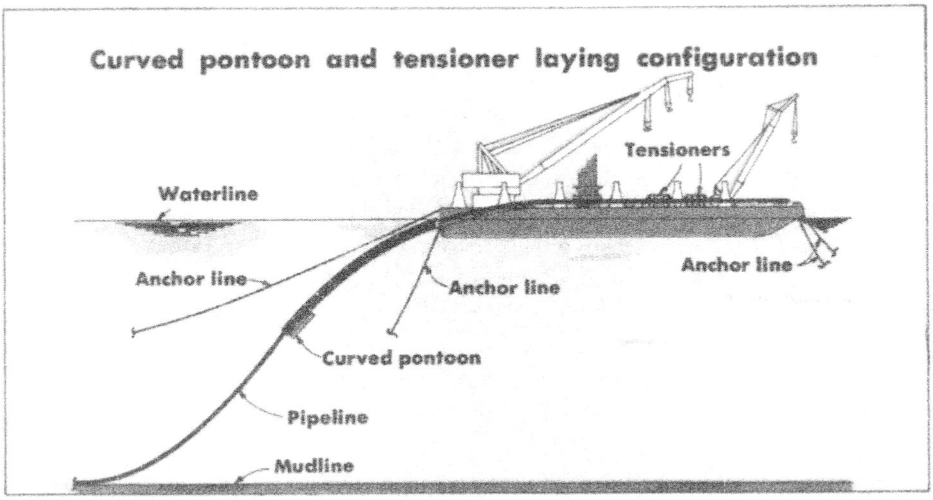

Schematic diagram of pipe tensioner and curved pontoon. Photo courtesy of *Offshore* magazine

at only 3 percent, the emerging North Sea at about 5 percent, and Alaska at about 2 percent. (These figures would change in the years after the energy crisis.) Shell Oil led the list of clients, hiring almost 25 percent of all the pipelining performed by Brown & Root up to 1971. Humble and Gulf Oil also show up prominently on the list, but the figures that stand out are projects for companies active primarily in the natural gas business: Transco, Tennesee Gas, Texas Eastern, United Gas, Southern Union, and Trunkline. These and other natural gas companies accounted for almost 2,000 miles of the pipeline laid by Brown & Root in this era. Gas projects for major oil companies boosted the total of natural gas pipeline to perhaps 3,000 miles, or half of the pipe laid by the company. These gas lines tended to be larger-diameter pipe than oil lines. Brown & Root regularly laid pipe larger than 20 inches in diameter, with numerous listings of 40- and 48-inch diameter pipes laid in the Middle East.[17]

Such figures point toward one of Brown & Root's key accomplishments in the 1960s, the laying of much of the pipeline needed to integrate gas fields in the Gulf of Mexico into the nation's gas transmission system. Giant, Houston-based gas transmission companies—notably Texas Eastern, Tenneco, and Transco—transported natural gas after World War II from the fields of the Southwest to the cities of the Northeast. In the 1940s and 1950s, Brown & Root had laid pipe and built compressor stations for these transmission systems. In the 1960s and 1970s, Brown & Root continued to work for them offshore Texas and Louisiana where large deposits of natural gas had been discovered. Pipelines were the vital link that could make offshore gas an important new energy source for a nation starved for domestic energy. The company laid thousands of miles and millions of dollars worth of pipelines to gather this gas from offshore fields and feed it into onshore pipeline systems.

Many gas companies tapped into Louisiana's important new gas producing area. Brown & Root completed major projects offshore Louisiana involving more than one hundred miles of pipelines for Transco, Texas Eastern, Southern Natural Gas, and Trunkline, for whom it built in 1969 the largest offshore pipe system yet constructed as a single project. In the early 1970s, the company joined with J. Ray McDermott to lay the Stingray System, a 228-mile gas gathering system capable of delivering more than one billion cubic feet of natural gas per day from the Gulf of Mexico to a treatment facility owned by Sun Oil at Holly Beach, Louisiana, near the Texas-Louisiana border. Such large diameter pipe projects kept the company's fleet of lay and bury barges busy in the Gulf.

Building the Boom

Brown & Root Marine's growing engineering expertise in platforms and pipelines helped expand offshore construction during the years before the first energy crisis. The 1960s experienced a torrent of oil production around the world, with offshore oil leading the way. From 1960 to 1970, "free world" oil production more than doubled, from 18 million to almost 38 million barrels per day. During the same period, offshore production almost quadrupled. With a daily production of 7.6 million barrels in 1970, offshore oil accounted for more than 20 percent of the free world's total production.[18] As large reserves on land in the United States became harder to find, much of this expansion occurred in the Gulf of Mexico, which was recognized "as one of if not the major undeveloped basin in North America."[19] Development of offshore reserves in the rest of the world also surged, with the Persian Gulf, Lake Maracaibo, and Africa leading the way. Brown & Root Marine participated in this expansion, becoming an international presence in the development of offshore oil and gas.

Technological innovations facilitated this expansion by allowing for production in deeper and harsher waters, economies of scale, and aggressive international expansion to bring new sources of offshore oil to market. In the 1960s and early 1970s, all aspects of the basic system had to be improved if offshore oil and gas were to expand as major sources of energy at a time of skyrocketing demand. The challenge for the offshore industry was to develop large, new sources of offshore oil and gas at roughly the same price ($2 to $3 per barrel) that had prevailed since the 1950s. The added costs of recovering oil and gas from deeper waters or locations geographically distant from major markets that lacked highly developed infrastructure, would have to be borne by the industry. Therefore, technological change would be crucial to cutting costs and keeping the new oil competitive in world markets.

Brown & Root Marine met this challenge with new technology and massive investments in new equipment. The company backed up its commitment to innovations in marine technology by building up a marine fleet that allowed for the practical application of those innovations. Brown & Root invested in new and more sophisticated lay barges, derrick barges, diving chambers, and associated equipment. An internal estimate made in 1971 placed the company's total investment in the 1960s in "marine equipment" at about $90 million. This expansion was required to meet a "five-fold increase in the volume of business in its worldwide activities in marine construction." According to the company's planners, the "offshore construction market" (which they defined as 'activities normal to the marine-related operations of Brown & Root, McDermott and several other worldwide construction firms') grew from about $100 million in the early 1960s to a peak of $660 million in 1969. In these years, "Brown & Root apparently accounted for over one-third of the estimated world market." By 1970, vessels valued at about $425 million served this market, with Brown & Root's fleet accounting for more than $125 million of this total.[20] The fleet eventually numbered over 100 vessels.[21]

The new vessels were not conversions as in earlier years, but "purpose-built" vessels designed by Brown & Root's marine engineers—Hauber, Lawrence, Lochridge, and others—with the latest technology and the size and power needed to support the complex needs of the offshore industry. Brown & Root also remained committed to a constant upgrade plan to ensure that those vessels did not become obsolete as technologies advanced.

During the 1960s, Brown & Root helped revolutionize marine design by developing sophisticated computer programs to produce three-dimensional perspectives capable of analyzing data more rigorously than had been possible earlier. In the familiar surroundings of the Gulf of Mexico, the company improved existing pipeline and platform technology while designing and building an impressive fleet of vessels for offshore work. Such work reduced costs, thereby extending production into deeper waters, which had been considered uneconomic at prevailing prices. Technological innovation thus made possible an oil and gas boom in the Gulf of Mexico even before the price hikes of the 1970s. Throughout the 1960s, the oil and gas companies reached deeper offshore in the Gulf of Mexico and around the world in search of other offshore provinces. In possession of new marine construction vessels, Brown & Root and its major competitors reached with them.

Notes

1. This chapter draws from interviews with John Mackin, W.R. Rochelle, Ber Pieper, A.B. Crossman, Tim Pease, S.J. Hruska, J.C. Lochridge, and A.R. Desai.
2. F.R. Hauber, "Drilling and Production Structures for Oil and Gas on the Continental Shelf," OECON Proceedings, 1966.
3. Griff C. Lee, J. Ray McDermott & Co., "Offshore Structures, Past, Present, Future, and Design Consideration." Proceedings of OECON II, 1968.
4. Stanley Hruska and Albert Koehler, "Computer Aided Design of Marine Structures," Proceedings of OECON IV (1969), 169–183.

5. "DAMS . . . The Newest Technique in Marine Engineering," *Brownbilt* (Fall 1968): 10–12.
6. "Engineering," *Brownbilt* (Spring 1970): 8.
7. "Engineering," *Brownbilt* (Spring 1969): 10.
8. Larry Resen, "New Sea-Going Pipe Layer," *Oil & Gas Journal* (March 17, 1958): 180–181; Donald DePugh, "Brown & Root, Inc., Launches L.E. Minor for Offshore Pipe Line Construction Work," *Pipe Line Industry* (June 1958): 33–34; "Largest Pipelaying Barge 'L.E. Minor'," *Offshore Drilling* (July 1958): 25.
9. "New Sea-Going Pipe Layer," 180–181.
10. Article cited in A.R. Desai and J.R. Shaw, "Marine Pipelaying Methods Defy Odds," *Offshore* (September 1979): 106.
11. D.R. Ward, "Laying Large Diameter Offshore Pipelines," *Offshore* (June 2, 1967): 53.
12. L.E. Minor, "Improving Deep Sea Pipeline Techniques," *Offshore* (June 1966): 55–56.
13. D.R. Ward, "Laying Large Diameter Offshore Pipeline," 56.
14. H.M. Wilkinson and J.P. Foster, "How Shell Laid the World's Deepest Pipeline in the Gulf of Mexico," *Oil & Gas Journal* (March 28, 1966): 187–192; "Deepwater Delivery," *Brownbilt* (1st Quarter, 1966): 18–19; "Five Giants for the Gulf," *Shell News* (July 1965).
15. "Federal Gas to Flow Soon Off Texas," *Oil & Gas Journal* (August 23, 1965): 56–57.
16. "Microwave Survey System Used for the First Time," *Brownbilt* (Winter 1969–70): 10–13.
17. "BAR Pipelines up to May 20, 1971," Document in Brown & Root Archives.
18. These figures are cited in "The Position of the Brown & Root Companies in the Offshore Construction Industry, 1971–1974," Study Prepared under Direction of Hugh Gordon, May 1971. Copy in Brown & Root Archives.
19. "Gulf of Mexico is Still Booming," *Offshore* (June 20, 1967): 61.
20. "The Position of Brown & Root in the Offshore Construction Industry," 1–15. The companies listed in Brown & Root's analysis of the offshore construction market were as follows: Brown & Root, J. Ray McDermott, Ingram Contractors, Houston Contracting, Saipem, Heerema Engineering, Santa Fe-Pomeroy, E.T.P.M., Fluor Ocean Services, Teledyne Movable Offshore, Micoperi, Raymond International, Murphy Pacific Marine Salvage, and Williams-McWilliams.
21. "Equipment for Marine Services," Brown & Root Marine Brochure (n.d.).

CHAPTER 5

Wading into Deepwater

The offshore industry in the Gulf of Mexico continued to grow during the 1960s, extending depth capabilities and redefining the meaning of "deepwater." Deepwater had always meant the water just past the deepest existing platform.[1] One engineer pointed out that deepwater was a "variable parameter, gradually increasing with the age and progress of the offshore industry."[2] By the early 1970s, when companies geared up to take on the central North Sea, anything beyond 400 feet was considered deepwater.

Just past the edge of the continental shelf loomed the truly deepwater of the continental slope and abyssal plain. Today, the edge of the shelf is taken to be the 200 meter (656 feet) level. Beyond the shelf, the continental slope drops off rapidly to 3,000 feet. By the early 1970s, technological developments made operating beyond the shelf conceivable for the first time. Marine engineers prepared for the challenge of deepwater by improving jacket designs, upgrading fabrication and installation techniques and equipment, and developing theories about the dynamic response of tall offshore structures to wave excitation. Before those theories could be put into practice, however, the industry needed an economic incentive to move out past the shelf: high oil prices and prolific finds.

The oil embargo of 1973 by the Organization of Petroleum Exporting Countries (OPEC) provided the necessary impetus. High oil prices and the urgency of exploiting alternative sources of oil stimulated offshore development in the North Sea and other non-OPEC locations and revitalized the U.S. Gulf of Mexico. Pressures from environmental groups had slowed deepwater development offshore California, so the Gulf emerged as the main arena for deepwater exploration in the United States. Companies made a few significant plays on continental slope tracts offered for the first time by the U.S. government. The euphoria that characterized the industry's approach to the North Sea infected the Gulf coast. In 1975, Brown & Root geared up for the deepwater market by establishing a new fabrication yard at Harbor Island near Corpus Christi.

The period 1976-1984 witnessed stunning achievements in deepwater development, many of them made by Brown & Root Marine. In 1979, the company built and installed for Chevron the first single-piece jacket, the Garden Banks structure, in more than 600 feet of water. Brown & Root topped that achievement in 1983 by fabricating and installing Exxon's Lena guyed tower platform in 1,000 feet of water. The Lena was undoubtedly one of the most innovative projects of this period, a compliant platform that spawned technological advances in structural design, fabrication, launching, and installation.

When the oil boom in the Gulf of Mexico subsided after the peak in oil prices in the early 1980s and virtually ended in 1986, deepwater development slowed. Hundreds of leases were returned to the U.S. government, drilling fell off sharply, and oil service businesses in the Gulf went bankrupt or restructured. Deepwater Gulf of Mexico was all but dead until the nineties. As Jay Weidler admitted, it "was a lot of fun from a technical standpoint," but "it didn't produce the continuity of revenues that were necessary to have a business." Even so, deepwater development beyond the shelf paved the way for spectacular renewal of deepwater activity in the mid-1990s.

Steady Progress during the 1960s

During the 1960s, as the offshore industry expanded into other areas of the world, the platform business remained attractive in the Gulf of Mexico. Brown & Root and J. Ray McDermott remained the two dominant companies in the fabrication and installation of platforms. When Brown & Root's Marine Operators reviewed their work on platforms in April 1966, they found that they had completed 315 platforms with a total value of almost $180 million. Their three major clients—Shell Oil, Gulf Oil, and the CATC group—had accounted for about $70 million of this total. The rest was spread among almost every company active in the Gulf of Mexico.[3]

Brown & Root dueled McDermott throughout these years for the boasting rights to constructing the platform in the deepest water. Brown & Root laid claim to this distinction in 1955, and it did so again in 1962 with the completion of a platform for Gulf Oil in 209 feet of water. In 1967, the company moved out to 285 feet. In 1973, its platform for Sun Oil in 362 feet of water was the second deepest platform then standing, but it was the farthest from shore at a distance of 125 miles. Then in 1974, Brown & Root set another record by building and installing a platform for Tenneco in 375 feet of water.

The largest structural component of the steel platform was the jacket. Designing and handling larger jackets were the primary considerations involved in moving into deeper water. As water depths increased, jackets evolved in response to design factors such as environmental loading, foundation conditions, and the stresses placed on tubular joints. Structures had to satisfy the

Brown & Root derrick barges setting a heavy deck section

requirements of functionality, strength, practical construction, and reasonable cost. For a given offshore site, designers had to decide upon the method of installation that would be needed and then select a layout that would fulfill the operational requirements of size, load-carrying capacity, and arrangement efficiency. By the early 1970s, basic platform designs and installation techniques had tended to stabilize up to the 400-foot depth level. Engineering and construction techniques were not radically different within this depth range. Moving into much deeper water—600 feet and beyond—introduced fundamentally new problems. For the first time, structural designs had to consider the dynamic interaction between those structures and waves. Fabrication and installation methods also had to be revised to handle larger structures.

A brief review of standard designs and techniques for waters extending out to 400 feet (the working definition for deepwater in the Gulf by the early

1970s) is necessary to demonstrate the departures that deepwater technology would make in the late 1970s. Oil companies employed a variety of structures depending on water depth, reservoir characteristics, and production rates. The minimum structure for a single well in 60 feet of water or less was a 10 × 10 foot deck and caisson linked by a flowline to a nearby manifold or production platform. For small numbers of wells at depths out to 200 feet, a simple three- or four-leg well-protector platform was used that could be worked over by a jack-up unit. Operators installed production platforms in conjunction with caissons or well protectors to produce oil in fields with scattered wells and small reservoirs. In deeper waters, self-contained drilling and production platforms were the norm. Drilling and production equipment had to be combined on a single structure because the cost of a separate production platform was prohibitive at greater depths. There was some overlap, however, in 50- to 200-foot water depths for self-contained drilling/production platforms and well-protector caissons, depending on reservoir size, depth, and shape.

The most common structure in the Gulf of Mexico in the 1970s was the minimum self-contained platform. It was desirable for developing multiple wells with relatively low production. The minimum self-contained unit usually was built with four legs and two decks accommodating six to twelve wells, typically in 100- to 200-foot water depths. Although they were less expensive (at around $5 million by the mid-1970s) than larger platforms, the drawback of the mininum self-contained units was their inability to carry on drilling and production at the same time. For larger and more prolific fields, where early production was a significant cost factor, operators preferred a self-contained drilling and production platform, which allowed for simultaneous drilling and production. Based on the new standardized layouts approved and recommended by the API, the self-contained platform had eight to twelve legs and twice the deck space as minimum units; it housed up to twenty-four conductors. These structures typically cost $10–12 million.[4]

Fabrication techniques for Gulf of Mexico structures also had become standardized by the early 1970s. Designing a platform for 400 feet of water was not much different from designing one for 200 feet, except for the greater size and weight of the components. Few yards, however, were large enough to assemble a jacket designed for 400 foot depths, which required a larger base. Even though a deepwater jacket was built on its side, the height of its lower panels still exceeded the reach of most fabrication-yard gantry cranes. Brown & Root's Greens Bayou yard and J. Ray McDermott's Morgan City facility were two of the few yards that could handle the bigger structures. Large jackets were skidded onto a launch barge—whose sizes had to increase with larger jackets—and towed in a horizontal position to the site.[5]

During the 1960s, installation procedures were modified for deepwater jackets. By the mid-1960s, even the largest platforms generally were set in place using derrick barges to lift the jacket onto and off of barges. Standard platforms in the Gulf of Mexico were installed by derrick barges with 250-

ton revolving cranes that lifted the jackets and placed them into position. The barges would then drive the piles and set the deck sections and equipment. Increases in jacket weight presented limits to the lifting capacity of barges. In response, Brown & Root and others developed a "dual lift" system that used two derrick barges working in tandem to handle the jackets. Brown & Root and its major competitors invested heavily in new very large derrick barges to perform this work and other heavy lifting needed offshore. Nevertheless, a new method would be needed to install larger platforms in deeper water.

As jackets grew to be too heavy for lifting, marine construction firms devised ways of launching them from barges. Launches were performed by cargo barges equipped with launching-ways matched to a set of runners or skids on the jacket. At the installation site, the barge was tilted and the jacket was winched off into the water. A derrick barge would then attach a hook to the horizontally floating jacket. Next, the jacket's legs would be flooded, and the hook of the derrick barge would hold the top of the jacket while the structure rotated to the vertical. Launching eliminated the need for a second derrick barge to lift the jacket off the barge, but it had its disadvantages. Jackets had to be stiffened by affixing more trusses or braces to resist launching stresses, which involved adding costly extra steel to the jacket.[6]

Launching by this method still required large derrick barges with massive cranes to position the ponderous jackets. Increasing jacket sizes, though, ultimately exceeded the capacity of launch barges and created unsafe installation conditions. Waves could cause a barge to heave on the surface, while the submerged jacket would take longer to respond, a situation that inflicted excessive dynamic loads on the barge. During the mid-1970s, experiments were made in setting steel jackets in the North Sea through controlled up-ending and levelling without a derrick barge hook. "Self-floating" jackets also made a brief appearance. Such methods of installation were short-lived, however, since they required complex control systems and extra steel for buoyancy. The development of launch barges with extended capacities and new methods of installing deepwater jackets characterized the new era of deepwater.[7]

Launching and installation by hook-assisted derrick barge remained the preferred method for structures installed out to 400-foot depths. In 1974, Brown & Root launched the Gulf of Mexico's largest steel jacket structure for Tenneco in the West Cameron field. Brown & Root fabricated the jacket in sections, using eleven cranes to place the sections together. Members with diameters up to 57 inches were used in its construction. The 400-foot jacket, designed to rest in 375 feet of water, measured 143 by 234 feet at its base and weighed 3,700 tons, a huge structure for its time. It was fabricated at Greens Bayou, skidded onto a barge, and towed out of the Houston Ship Channel to the location. Brown & Root's giant *Foster Parker* derrick barge lifted and positioned the jacket, drove the 48-inch diameter piles, and set the deck sections by means of 400- and 500-ton lifts. Additional pilings and mud mats were used to add stability. Counting the jacket, piles, and deck, Brown & Root had firmly attached 8,000 tons of metal to the floor of the Gulf of

Top section of Tenneco's steel jacket for the West Cameron field

Mexico, providing facilities for up to twenty-four wells at a location about 125 miles out in the Gulf.[8]

Even this huge Tenneco platform did not remain the Gulf's largest for long. It would soon be dwarfed by new platforms that revolutionized the design, fabrication, and installation of deepwater structures and extended record water depths by quantum leaps.

The Industry Prepares to Move Deeper

Two developments enabled the move into truly deepwater. One was the sense of common purpose within the industry in exploring new technological frontiers. By the early 1970s, Brown & Root and others had begun to share what they had learned about offshore structures, establishing standardized practices and uniting to confront technological challenges. The other key development was the sustained period of high oil prices caused by the OPEC oil embargo in 1973. Higher prices encouraged oil companies to invest in technologies to exploit new sources of oil and gas from greater ocean depths.

During the 1960s, the industry had significantly enhanced its understanding of offshore structures in the Gulf of Mexico. Three "one hundred year" hurricanes hit the Gulf—Hilda in 1964, Betsy in 1965, and Camille in 1969—destroying or damaging many platforms that were designed to resist only a 25-year storm. Camille was the most powerful, producing a measured wave height of 75 feet. Engineers learned much about meteorology, oceanography, soil movements, and hydrodynamics associated with these storms, which enabled them to make design improvements in tubular joints, foundations, and platform instrumentation programs. Higher quality steels and better welding practices improved the strength of joints, reinforcing their resistance to large waves. Computer programs revolutionized design techniques and the analysis of load-bearing capacities on marine structures. "In a little over seven years," wrote F.P. Dunn, a Shell Oil official, "engineers went from electrical/mechanical desk calculators to finite element analyses of complex tubular joints and complete three-dimensional computer models of 400-foot structures."[9]

The calamities of the 1960s rallied oil companies around a stronger sense of technological purpose. In 1969, the first annual Offshore Technology Conference was held in Houston as a forum for exchanging technical information and recognizing accomplishments. "It loosened the secrecy surrounding companies' research efforts," said Dunn, "making it much easier to release important technical results."[10] That same year, the API published its first API Recommended Practice (RP) document for the design, fabrication, and installation of offshore platforms. "The RP's were written by the more knowledgeable industry engineers who integrated their experience and their companies' research efforts into consensus documents, providing guidelines for building reliable, cost-effective platforms."[11] These guidelines further reduced industry secrecy and were later incorporated into the federal government's regulatory program.

Just as the offshore system in the Gulf of Mexico was entering a new stage of refinement and development, it received a mighty boost. The 1973 OPEC oil embargo jolted the posted price of crude up from around $3 per barrel to more than $10 per barrel. The oil industry already had been chasing rising demand around the world. The supply crunch now magnified excess demand and created a worldwide "energy crisis." As fuel rationing was implemented in the United States and automobiles lined up at service stations, the oil industry mobilized to make new investments. Ten-dollar oil opened up a new realm for offshore oil exploration and development, particularly in the North Sea. The restriction of Middle Eastern oil, on which the United States had grown dependent, also provoked a frantic search for new domestic oil sources in the United States and a reassessment of U.S. offshore reserves.

The higher prices that crude commanded allowed companies to look beyond the edge of the continental shelf for new sources. Exploration already had progressed to depths past 1,000 feet. Brown & Root's Project Mohole had produced innovations in mobile drilling, most notably a practical semisubmersible drilling rig design, first commercially developed by Shell Oil and Bluewater Drilling in 1962. The semisubmersible consisted of a braced assembly of large cylinders that were connected to a drilling deck. The cylinders could be ballasted down far enough to minimize pitching and rolling, allowing the rigs to operate in rougher and deeper water. Mohole also contributed to other technological advances in deepwater drilling, such as dynamic positioning systems, well-head sonar re-entry systems, electric BOP controls, and buoyant marine risers. The Mohole-inspired evolution of semisubmersibles (led by companies such as Sedco, Zapata, and Odeco), dynamically positioned drillships (dominated by Global Marine, The Offshore Company, and Fluor Drilling), and associated technologies greatly extended the range of offshore drilling.[12]

By the mid-1960s, California had become the world's primary site for deepwater activity. In 1965, the drillship CUSS I, a veteran of prior Mohole work, opened the new era of deepwater drilling in 632 feet of water off Southern California for Humble Oil (soon to become Exxon U.S.A.). Deepwater activity during the remainder of the decade moved to tracts leased in 1966 and 1968 in the Santa Barbara channel, where semisubmersibles drilled test wells in progressively deeper waters, mostly for Exxon U.S.A. Drilling in that area had reached 1,400-foot depths by 1971.[13]

Continued deepwater development off California, however, was constrained by political pressures. Citizen protests against offshore operations in California, sparked by an offshore blowout in 1969, which spilled 50,000 to 70,000 barrels of oil into the Santa Barbara channel, led to the eventual suspension of lease sales off both U.S. coasts. Later that year, Congress passed the National Environmental Policy Act, which required an Environmental Impact Statement and slowed deepwater development in the channel.[14] "The west coast was becoming an increasingly difficult place to attempt to do business because of the heightened interest in environmental matters at the time," Jay Weidler noted.

Consequently, the focus of deepwater exploration shifted to the Gulf of Mexico and Brazil. Brazil's oil sector was dominated by the state-owned company, Petrobrás, and international oil companies were largely excluded from participation. Thus, many major U.S.-based companies interested in offshore concentrated their efforts on finding new sources in the Gulf of Mexico. By 1974, 800 producing platforms were at work there, with at least another 200 installed for pumping or gathering. Beginning in 1970, however, crude production had started to decline from a daily average production peak of 1.4 million barrels per day. The largest and most productive structures in shallower waters had begun to play out. Although natural gas production from federal waters was increasing, gas output from state waters was declining.[15] In the crisis atmosphere of 1973-1974, U.S. government and oil industry officials reckoned that deepwater plays in the Gulf of Mexico were vital to the nation's hopes for maintaining energy independence.

Technological progress and soaring oil prices bolstered confidence that deepwater development could pay in the Gulf of Mexico. Advances in bright spot seismic technology and gravity/magnetics data had enhanced the industry's ability to evaluate geological structures. Soil boring and sampling techniques had improved dramatically as well. Early survey results from the Gulf were encouraging. In March 1974, the U.S. government opened up 212 tracts (940,000 acres) in the Gulf for lease, including 42 deepwater tracts (199,000 acres) on the Continental Slope offered for the first time.[16] The offshore industry spent a record amount of money in bonuses ($2.16 billion) in the March sale. Only 13 of the 42 deepwater tracts were bid on, but the 11 bids that were accepted pulled in a whopping $321 million. Shell and Amoco spent most of that amount on tracts located in waters of 1,000 feet or more. The new deepwater leases created an unprecedented demand for deepwater rigs and fueled a burst of deepwater drilling rig construction. Rates for semi-submersibles and dynamically positioned drilling rigs doubled. "It's just wild," said Exxon's deputy exploration manager about the craze. "The thought of moving into deeper water doesn't scare us," affirmed another official in 1974. "We are ready now."[18]

Beyond the Shelf

The central problem faced by marine engineers attempting to design jackets for waters at the edge of the continental shelf was the dynamic interaction between waves and the structure. In water beyond about 400 feet, a platform's natural vibration frequency tends to approach the frequency of cyclical ocean waves. McDermott engineer Griff Lee put it simply: "You see, when we have a platform in, say, 50 feet of water that's 100 feet wide, it's pretty low and stocky. It's not going to vibrate much. You put that same platform in 500 feet of water, it becomes more like a flag pole. And you run into the problem that the dynamic response really becomes significant." Platforms in 300 feet of water have a natural period of about two seconds, which is much smaller than the period of the largest storm wave or the periods of fatigue-inducing sea states. The structure is stiff enough so that the structural response to wave action is essentially static. The structural period increases with depth, greatly amplifying the force of waves on the structure. "Take, for instance, a wave that comes along every 10 seconds, and you have a deepwater, rather floppy structure that has a 10-second period, it's like a pendulum," Lee explained. "If you ever got to the point that it was moving in resonance with the waves, it would be hit by the waves with maximum force just at the time it was already . . . leaned over as far as it could go. And you have a resonant condition that the structure will not be able to stand."[19]

Engineers knew that in waters 400 to 1,000 feet, the stress in individual jacket members caused by the amplified wave forces had to be lowered. As water depths increased, the susceptibility of structures to stress amplification made stiffness of the structure as a whole, rather than strength of the members, the controlling aspect of design. Stiffening a structure reduced its natural vibration period. One way of adding stiffness was increasing the leg batter, or slope, which was one of the main determinants of its overall stiffness. Increasing the stiffness, therefore, meant increasing the batter and thus the base area of the structure. Other ways of imparting stiffness included adding steel to the legs and piles, increasing the area of brace pipe, and lowering the mass of the structure.

As the size, weight, and base areas of jackets increased with depth past about 600-foot depths, new methods of fabrication and installation had to be developed. Offshore construction firms began to realize they would need bigger fabrication yards and new equipment for transporting and installing the new steel jackets.

Brown & Root decided to acquire a new fabrication yard, because new deepwater jackets would be too large for Greens Bayou and the Houston Ship Channel. Company officials searched for a site that could serve as a major assembly point, with open access to the Gulf but within a reasonable distance to the main yard at Greens Bayou. They found such a site at Harbor Island, Texas, located about one-half mile from the open Gulf waters on the main ship channel leading into Corpus Christi Bay.

Harbor Island was open, protected from heavy waves by Mustang Island and San Jose Island, and close to the intracoastal canal system that provided a direct transport link to Houston. Brown & Root purchased 300 acres on Harbor Island in 1975 and developed a fabrication yard that opened the next year. Employing 500 skilled workers by 1980, the yard was equipped with 20 cranes, including the first two Manitowoc-6000 cranes ever built, each capable of 500-ton lifts. It was prepared to load out the largest imaginable structures with two 600-foot skids, one 900-foot skid, and one 1,400-foot skid. To handle single-section jackets up to 1,000 feet in length that would emerge from the yard, Brown & Root developed the world's largest launch barge, the

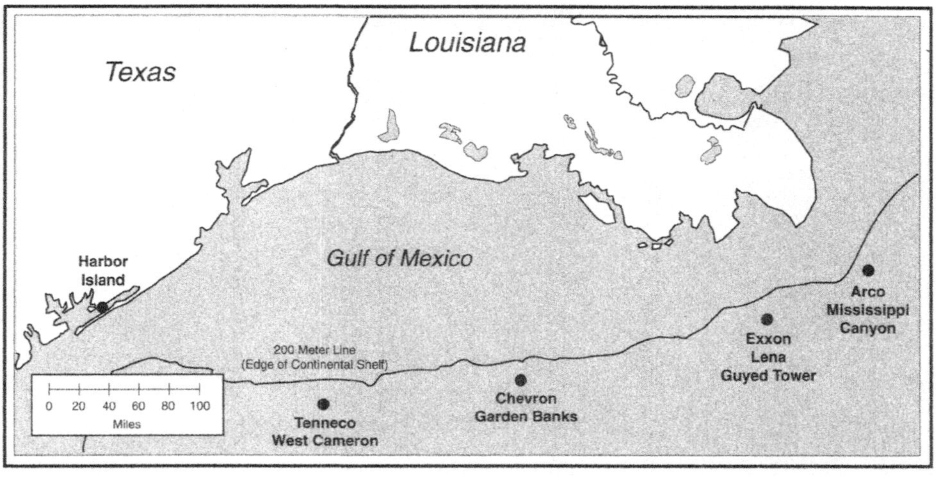

Deepwater Gulf of Mexico

BAR-376, which was 580 feet long by 160 feet wide. *Brownbilt* magazine described Harbor Island as a place "where giants lie in the sun fanned by gentle Gulf breezes while they await the day they can stand on their own and go to work."[20]

The first giant brought to life at Harbor Island was Chevron's Garden Banks structure. Brown & Root built the 708-foot, 12,000-ton steel jacket for Chevron's gas development 140 miles south of Cameron, Louisiana. A Chevron official described the Garden Banks as "the beginning of a new era of deep-water, one-piece construction."[21] Launched in August 1979 by Brown & Root's *BAR-376,* the Chevron Garden Banks was not only the farthest offshore of any U.S. platform, it was the world's largest single-piece jacket. The base of the structure measured 180 feet by 300 feet.[22]

While it was an impressive sight to citizens of Corpus Christi who observed the curious edifice erected as a backdrop to the city's skyline, the Garden Banks did not involve anything innovative other than its unprecedented size. The $42.5 million jacket resembled others that supported drilling and production in the Gulf of Mexico, though it made them look like scale models by comparison. "This one here was just kind of taking the expertise and technology we had and just extending it," said Stan Hruska, "extending it into deeper water. Although it was challenging, it wasn't mind-boggling like some other projects were." A year later in 1980, Brown & Root installed a similar

Fabrication of Chevron's Garden Banks jacket at Harbor Island

single-piece jacket for ARCO at a depth of 651 feet in the Mississippi Canyon area off Louisiana.[23]

The Garden Banks and the ARCO platforms may have been the deepest single-piece structures, but they were not the largest jackets installed during this period. Exxon and Shell both had placed jackets in deeper water past the benchmark depth of 800 feet by sectionalizing the jacket components. In 1976, after the drilling off California had resumed, Exxon installed its Hondo platform in 850 feet of water in the Santa Barbara channel. Installation posed a dilemma to the structure design. Because the controlling design criteria on the west coast was seismic rather than wave-loading, a heavy, self-floating jacket was ruled out. A self-floating structure would leave too much mass in the installed structure requiring an impractical number of piles to resist an earthquake. A standard pile-supported jacket was preferred, but such a large structure was not installable with the equipment available. Exxon sectionalized the jacket into two pieces. Kaiser Shipbuilding constructed the $67 million Hondo in one piece in Oakland, and then separated it into two for transport and launching. At the site, the two pieces were launched and mated horizontally in protected water by welded connections from inside the legs. The mated jacket was then uprighted and set on the bottom. Sectionalizing bypassed yard and equipment limitations for fabricating and installing larger structures.[24]

In 1977–1978, Shell Oil modified the sectional launching concept to install a giant 1,040 foot structure in the company's Cognac prospect located in the Mississippi Canyon. Horizontal mating at the surface was not feasible in the Gulf, where the structure would be exposed to unpredictable weather, including hurricanes. So Shell built the Cognac in three pieces, upended each one, and assembled them vertically, or "stacked" them in place. The Cognac's design was basically a deepwater application of the API drilling/production platform for 62 wells.[25]

The Cognac's installation was the most sophisticated ever attempted of an offshore structure. It established records for the deepest water (1,025 feet), the most wells, and the heaviest steel platform (59,000 tons). The two derrick barges used to handle the sections had to employ heave-compensating lowering hoists to minimize wave action in setting the 380 × 400 foot base section.[26] Once set on the floor, the base was leveled with hydraulically actuated mud mats and secured with skirt piles. Hydraulic underwater hammers, three times as powerful as their predecessors, drove the 84-inch diameter, 600-foot-long main piles. The other two sections were then stacked on top of the base. An estimated 14,000 person-hours were logged in saturation diving time at 900 foot depths during installation.[27] The complex installation was expensive, costing Shell $275 million, which led one of the platform's designers to admit that "matching up large pieces working in a marine environment was very, very expensive."[28]

Although the Cognac may have set records, it did not set a trend. Two years later, Union Oil proved that a single-piece jacket could be installed by conventional methods at similar depths in the Gulf, and for much less money.

The main challenges to designing and installing Union's platform, located in 955 feet of water in the East Breaks area, were reducing the weight of the jacket and simplifying the design. Brown & Root did a preliminary conceptual design for the jacket, but McDermott won the turnkey contract for the Union platform. Only 500 tons were added to Brown & Root's conceptual design in the detailed design. McDermott used computer design programs to analyze dynamic launch, tow, and in-place loads. With help from Brown & Root design engineers, the McDermott team calculated that installation loads affected the size of only about 40 percent of the structural members. These analyses permitted designers to reduce the amount of steel called for by their original scaled-up model. "We hated to put any more steel than necessary in a jacket, which will stand in the Gulf for 40 years, just to accomodate installation loads that would last only weeks or minutes," said McDermott's supervising engineer.[29]

The lighter construction of the jacket permitted it to be installed as a single piece. It weighed approximately 26,000 tons, less than half of the 59,000-ton Cognac structure. In 1981, McDermott commissioned the largest "super" launch barge ever built (650 × 170 feet) to launch the jacket. Once in the water, it was righted and lowered into place by lifting and controlled flooding. Union cleverly named its platform Cerveza, the Spanish word for beer, because at $90 million it came in on a beer budget compared to the exorbitant price of Shell's Cognac. McDermott engineer Griff Lee remarked that the "low cost and rapid schedule of this platform represents a new direction to extend the economic range of conventional fixed platforms."[30] The following year, Union installed a companion platform, the Cerveza Ligera (or Light Beer), with fewer wells in 925 feet of water.[31]

The Cerveza platform achieved a more economical design for structures at 900-foot depths. Yet it did not extend the technology to deeper water. The Cerveza stood near the maximum depth limit for conventional platforms. In waters much beyond 1,000 feet, standard jacket designs became less practical. The base dimension required for providing the desired dynamic behavior at these depths was too large. Even new fabrication yards, such as Brown & Root's Harbor Island, could make heavy lifts up to only 250 feet, the height of the Cerveza base. Dynamic considerations also were more complicated in depths greater than 1,000 feet. The natural period of a conventional structure in such depths (about 4–6 seconds) approached that of the operating sea state. The high cycle/low stress loading caused by day-to-day waves introduced fatigue as a critical design factor. Although the forces induced by such waves are individually small, the cumulative effect of millions of small waves whose fundamental frequency matches that of the jacket shortens fatigue lifetimes. Modifications to stiffen steel structures and reduce the natural period would require dramatic, though not necessarily prohibitive, increases in steel tonnage and costs.[32] Engineers returned to their drawing boards to fashion an alternative that would provide cost efficiencies and structural soundness in water exceeding 1,000 feet.

Exxon's Lena: The Guyed Tower Concept

During the early phase of deepwater development in the Gulf of Mexico, engineers initially thought about limiting the dynamic response of jackets in deeper water mainly in terms of reducing the natural period of the structures. In the mid-1970s, they started asking if the same effect could be achieved by making the fundamental period of the structure *larger* than the wave energy period, while keeping all other natural periods well below the wave period. Was it possible, in other words, to design a structure that oscillated so slowly that waves moved past before it had time to respond and reach its maximum condition? These inquiries led to new designs for "compliant" structures, ones that could yield to waves in a controlled manner.

Two types of compliant designs for deepwater applications emerged as the most promising in the late 1970s. One was the buoyant tethered or "tension-leg" platform, such as the one Brown & Root engineered for Conoco's Hutton field in the North Sea. Another concept consisted of a slender steel tower held upright by a radial array of anchor cables or guy lines, which would allow the tower to tilt from the vertical with a high natural period. The guy lines would support the tower while it moved in response to environmental forces, eliminating the need for a large foundation on the sea floor. Therefore, a guyed tower would be lighter than a conventional jacket at the same depth. A guyed tower in 1,000 feet of water would have a fundamental period of about thirty seconds, whereas a steel jacket of similar dimensions would have a period of about five seconds and would resonate with many more waves over time.[33]

The guyed tower idea was not new. It was patented in the 19th century and considered by one oil company in 1950 as a way of extending offshore operations out to 100 feet.[34] The Exxon Production Research Company first evaluated the guyed tower in 1965 and continued to research it over the years. In 1975, as deepwater Gulf of Mexico really began to heat up, Exxon and twelve other companies built a one-fifth scale test model of a tower designed for 1,500 feet of water and placed it in 300 feet in the Gulf of Mexico near Grand Isle, Louisiana. Test data over a 3-1/2 year period confirmed the soundness of the concept and encouraged Exxon to pursue full-scale development.[35]

In mid-1978, Exxon decided to design a guyed tower for a 1,000-foot water depth at its Lena Prospect in the Mississippi Canyon. Exxon performed its own engineering and project management, but in 1981 awarded to Brown & Root the contracts for fabricating and installing the Lena platform. Brown & Root was a logical choice for Exxon. With the Cerveza platform already on the skids at McDermott's yard in Morgan City, Brown & Root's Harbor Island was really the only yard along the coast that was equipped to tackle the construction of the audacious tower.[36] Brown & Root's own cost research on the guyed tower versus the tension-leg platform suggested that the guyed tower was the solution of choice in the 1,000- to 2,000-foot range. Beyond that point, the tension-leg platform became more cost effective.[37] At an estimated $420 million, the guyed tower cost more than a conventional steel jacket would, installed in a similar depth. Taking a longer term view, however, Exxon proceeded with the guyed tower to develop the technology "in anticipation of the later need for guyed towers in water depths well in excess of 1,000 feet."[38]

The Lena was one of the most sensational and innovative offshore undertakings of the period. "When I had worked on the Forties and some of the other projects, it was sort of like, well, this has to be a career project, one of those once-in-a-lifetime projects," said Stan Hruska, Brown & Root's deputy project manager on Lena. "Well then, all of a sudden, here comes Lena. I mean, if you want to identify a once-in-a-lifetime or career project, that's it, Lena." Lena spawned innovations that included compliant pile foundations, guying system components, tower buoyancy and dynamic response designs, side-launching, and the first dynamically positioned derrick barge. Lasting four years, 1979–1983, from design through installation, the Lena became legendary in the industry. In 1984, the Offshore Technology Conference in Houston devoted an all-day session to Lena. "Of course, after that," said Hruska, "everybody wanted to hear about that project."

Although Exxon handled Lena's design, Brown & Root made a significant engineering contribution, even before the company won the fabrication and installation contracts, by finding a pile solution for the tower. One of the questions in the conceptual design for Lena, as Brown & Root engineers pointed out, was "how to obtain sufficient compliancy to enable the structure to oscillate with the waves without overstressing the foundation."[39] Brown & Root designed a pile system that transferred most of the vertical load onto the piles and minimized the overturning loads carried by them. A pile arrangement was selected based on the premise that a greater number of piles and the spacing between them increased the structure's load-carrying capacity, but decreased its compliancy. Locating the main piles in the exterior tower legs, as in a conventional foundation, was impractical because it would have created resistance to the tower rotating at the mudline. Instead, eight conventional ungrouted piles would be closely spaced in a circle in the center of the tower, all of them driven to deep penetration. These long clustered piles would carry the vertical loads and act as springs to give the tower more flexibility.[40] Explained Hruska, "At the bottom it's almost like a pin joint." The guy lines would take out the lateral loads, thereby reducing the stress on the foundation. The pile foundation was "flexible enough to accomodate tower tilt allowed by the guying system, yet strong enough to resist base torsion and vertical forces."[41]

Organizing and scheduling the Lena project to be completed within four years was challenging. The project had three design functions: tower design, production facility design, and installation design. Engineering for all these functions tallied hundreds of thousands of staff hours. Brown & Root spent 200,000 hours on installation engineering alone. Specialized material procurement and component prefabrication involving many contractors and consultants added complexity to the task. Exxon mobilized a special project management team to oversee all aspects of the project. Advance planning and organization were vital to the intimidating task of building and installing the monster. This was Brown & Root's job. From 1981, when construction began, to 1983, when the Lena started producing, Brown & Root, in cooperation with Exxon, developed innovative techniques and equipment needed to complete the job.

The sheer size of the tower, the tallest single-piece marine structure ever built, presented challenges to fabrication. Site preparation at Harbor Island began in the summer of 1981 and construction finished just over two years later. Brown & Root had to build new concrete skidways and pile caps strong enough for the concentrated loads that would be encountered during load-out. The construction site was divided into nine areas for erecting each of the nine symmetrical sections of the tower, six box sections and three framing panels. The 38 × 120 foot box sections were fabricated on their sides, like conventional jackets. When completed, they were rolled up to a vertical position. As many as twelve cranes were needed to roll up the boxes past their pivot point, and a special holdback winching system was devised to keep the boxes stable during roll-up operations.[42] The tower also contained some unconventional components, such as twelve buoyancy tanks and twenty guy line bending shoes, that required special fabrication and very heavy materials. Many closely grouped, thick-walled, and large-diameter members were welded near the top to distribute pile-to-jacket and guy line-to-jacket loads.[43] "It was very, very congested in a lot of areas," said Hruska, "expecially up in the top where you

Box roll up of Exxon's Lena tower at Harbor Island

had all the guy lines coming in." When assembled, the giant, 27,000-ton statue of tubular steel was awe-inspiring. "We were at Harbor Island on a cold winter day," Hruska said. "The wind blowing like hell from the north. All you had to do was get on the back side of that tower and man, you were blocked from the wind. You can imagine how much steel it had in there."

Loading the tower out in May 1983 nearly equalled the accomplishment of its fabrication. Hruska noted that "It was probably the most sophisticated loadout that I had ever been involved with." Because of its length, the tower had to be loaded transversely, or crosswise, onto the barge. Brown & Root convened a team of naval architects and engineers to plan for the industry's first transverse loadout. The tower had been built on six skidways, the four in the center matching up with those on the *BAR-376* super launch barge. Once transferred onto the barge, the two ends of the tower jacket would hang over sides, with around 300 feet of overhang on one side and 200 feet on the other. When planning the transfer of the tower from the skidways to the barge, however, engineers encountered a problem in taking the load off the two secondary skidways at the ends of the tower. Said Hruska, "As you bring it onto the barge, you have to take the load off, and you can't just drop it off at the end when you come off." The stress on the tower would have been too extreme.

Brown & Root engineers solved the problem by adding a loadout beam, split in tapered halves, to each outside skidway. The lower half was anchored to the concrete skidway and the upper half was welded to the tower. As the tower jacket moved onto the barge, the upper wedges moved down the lower wedges, deflecting the ends until separation occurred and the full load was shifted to the four interior skids. Brown & Root devised a special levelling and surveying system to read, shoot, and correct the position of the tower as it was brought onto the barge. This system was correlated to ballasting the barge as the continuous 29-hour loadout progressed.[44] "It was a unique loadout," Hruska marveled, "it really was."

Each phase of the project seemed to offer an even greater challenge than the last. Once the tower arrived on site, the next challenge was to carry out the first side launch of a major offshore structure. Exxon decided to go with a side launch instead of a conventional end launch, which would have placed too much stress on the tower. The side-launch technique was quicker and saved nearly 3,000 tons of steel that otherwise would have been required to sustain the cantilevered loads produced by an end launch, but it also introduced a design complication. The tower was likely to skew when sliding off the barge and thus become damaged through an imperfect separation. Scale-model and computer-simulated tests demonstrated this tendency, which was due to nonsimultaneous release of both ends, misalignment between the centers of gravity on the barge and tower, or a variation in friction between the four skidbeams.[45] "Every time we would do it, one end or the other would come off first," Hruska remembered. "We were all debating, well, how in the world are we going to eliminate skew?"

Further tests revealed that Brown & Root needed a way to ballast the barge to a 7-degree heel angle, hold the tower in place during the heel, and then simultaneously release the ends. Accomplishing this meant modifying the barge and designing a holdback/release system. The *BAR-376* received new starboard tilt beams and internal modifications. Designing a way to hold back the 27,000-ton tower "required some fairly substantial structural systems." The engineers borrowed a technology from the U.S. space program. In launching the space shuttle, NASA used an explodable release mechanism called a "frangible nut assembly." This mechanism consisted of a large-diameter bolt, an explosive nut, and a remote electrical firing system. Frangible nuts assured a simultaneous release and were adopted as the key component of the holdback/release mechanism for the Lena launch. On the skidways, hydraulic jacks held back the tower. The jacks were connected to a release subassembly of six 24-inch diameter pipes, each with a flanged joint and a frangible nut through the center. All twelve frangible nuts were set to fire simultaneously. Twelve back-ups were provided in case of failure.[46] Everything was thoroughly inspected, tested, and checked, up to the very moment of release. "This was a one-time shot," said Hruska. "You didn't get a chance to go in there and shoot it more than one time."

Tow out of the Lena tower

On June 17, 1983, after the 36-hour prelaunch procedure, barge personnel were evacuated, the barge was heeled, and the frangible nuts were fired. Stan Hruska and others observed the launch from a workboat a couple thousand feet away. "We watched and it seemed like when that thing went off, nothing happened. It just took time for that thing to start moving. And then all of a sudden, you could see it start moving. And boy, it hit the water. That was an experience there to see it come off like that." The launch was a success. Equipped with twelve buoyancy tanks in the upper part of the structure, the tower self-uprighted to about an 85-degree angle. Divers descended to a control panel and began ballasting the lower buoyancy tanks. Tugboats then towed the tower to Brown & Root's derrick barge, *Ocean Builder,* which hooked it up to its winches, and lowered it into position.[47]

Setting the tower required precise positioning and orientation. Four of the twenty guy lines had been preinstalled and fastened to the *Ocean Builder* before the launch so that the lines could be immediately tied into the tower, securing it against a potential hurricane. The guy lines needed to enter the jacket within three degrees of the optimum entry angle to prevent wear on the guy line fairleads. This area of the Gulf of Mexico, however, was susceptible to a strong loop current that threatened to pitch the barge and frustrate installation. To reduce this possibility, Brown & Root deployed an unconventional system of twelve preset moorings to keep the lines taut and anchor the derrick barge at the tower site directly over the guy line arrangement. Aided by the use of a baseline acoustic positioning system located on the *Atlas I* derrick barge, the *Ocean Builder* managed to set the tower jacket on bottom within five feet of the desired location and within 1 degree of the desired orientation. Thereupon, the *Ocean Builder* drove the piling and began to install the well conductors.[48]

Installing the guying system—the distinguishing feature of the Lena—proved to be one of the most difficult and unique aspects of the project. The system consisted of 20 guy lines that attached near the top of the structure and radiated outward about 3,000 feet away from the tower to anchor piles driven into the seabed. Each line was composed of three segments of 5-inch galvanized steel structural strand and a 200-ton articulated clump weight joining the catenary segment to the anchor line. The clump weights stiffened the tower and limited motions during short-period sea states. During storms, they would lift off the bottom to give the tower more compliance and increase the sway period further away from the period of the storm waves. "All that equipment was specially designed," Hruska said. "We had different groups looking at handling the anchor pile. You had another group looking at handling the clump weight. Another group of engineers we had looking at handling and installing the guy lines. . . . And, of course, we were working with Exxon, with their team. They had an engineer assigned to each of these areas of responsibility."

To hook up the guy lines, Brown & Root converted the *Atlas I* into the first dynamically positioned derrick barge. Dynamic positioning of the barge was necessary for controlled mobility at the crowded site and for accurate placement of the system's components. The barge could not use mooring lines that would interfere with the guy lines. The $17 million conversion was a project in itself. Brown & Root worked with Honeywell to design an automatic control system for generating thruster commands that governed barge motions. Position measurements were achieved by baseline acoustic communication with a grid of transponders arranged symmetrically on the sea floor around the tower site.[49] The *Atlas I* installed each line by first driving the 6-foot diameter, 115-foot long anchor pile with an underwater hydraulic hammer. Next, the anchor line and clump weight were lowered. The catenary segment was then unspooled and connected to the pendant segment, which was joined to a pull line and brought into the tower through a specially designed fairlead and bending shoe.[50]

After the guying system was completed in September 1983, the *Atlas I* finished installing the 58 well conductors and set the prefabricated deck modules and production facilities. Exxon laid its own crude and gas lines from the tower to its South Pass 89-A platform. Lena's first well was spudded in November and production began soon thereafter.[51] When completed, the entire Lena structure, including the decks and drilling derrick, weighed approximately 47,000 tons and measured 1,300 feet in height—50 feet taller than New York's Empire State Building. R.S. Rugeley, Exxon Co. U.S.A.'s guyed tower project

Launching the Lena tower

manager, called the Lena a "culmination of twelve years of innovative technology development."[52]

Lena was a one-of-a-kind project. To complete it, Exxon and Brown & Root developed new technology, expertise, and analytical tools for deepwater structures. Achievements in fabrication, load-out, launching, and installation established Lena as a case study in innovation. Lena, however, was not followed by other guyed towers, which are less desirable in water beyond 2,000 feet. Increased tower stiffness is required for those depths, and steel guying systems become less efficient as more of the guy line's strength goes to supporting its own weight.[53] Tension-leg platforms, compliant towers (without guylines), and subsea systems emerged as more practical and cost-effective solutions. "It's sort of like what we do in the Space Program," explained Stan Hruska: "You wouldn't do it again because now you've taken that and advanced the technology into other solutions."

The End of an Era

The regime of high crude oil prices during the 1970s and early 1980s embargo stimulated the expansion and revitalization of the offshore system in the Gulf of Mexico. During this boom period, Brown & Root added to its impressive resumé of engineering and construction projects in the Gulf that date back to the 1930s.

That resumé promised to grow as another spurt of big projects was anticipated in the 1980s. The price of oil appeared to be rising steadily into the indefinite future. The Iranian revolution and oil embargo in 1979 (the "second shock") created a windfall of high crude prices and sustained the expansion of offshore oil in the Gulf of Mexico. By the early 1980s, oil prices climbed to as high as $44 per barrel. Prices of $100 per barrel were projected for 1990. Deepwater projects were on the leading edge of this expansion. Gulf of Mexico operators devoted an increasingly larger share of their exploration and development budgets to deepwater activities. In 1982, Brown & Root designed, built, and installed a large steel jacket structure for Zapata, called the Tequila, in 658 feet of water. Brown & Root engineers also drew up preliminary and basic designs for other deepwater fixed platforms. In 1983, the U.S. government introduced "areawide" leasing in the Gulf of Mexico, and oil companies snatched up nearly 2.5 million acres of leases in water depths beyond 600 feet. Platform and mobile rig costs skyrocketed. New marine yards opened to take advantage of the backlog.[54] The offshore industry appeared healthier than it ever had been. By 1984, offshore production around the world surpassed 15 million barrels per day, around 28 percent of total world production.[55]

Oil prices stopped rising in 1982 and began to decline gradually. The bubble burst in 1985. Oil prices plunged to $10 per barrel as crude supplies overshot demand. A combination of energy conservation measures, global economic recession, and the decision by Saudi Arabia to step up production created the surplus. The price plunge spelled disaster for Gulf operators. The application of advanced technology for deepwater operations had depended on the economics of high oil prices. Drilling fell off sharply and hundreds of leases were returned to the U.S. government. Contracts evaporated, followed by massive layoffs and bankruptcies. "Ten-dollar oil made the oil industry one time," as Ber Pieper put it, "and the second time, it damned near broke it."

Marine construction at Brown & Root suffered under the contraction. The fabrication yard at Harbor Island idled away its potential. "With the exception of the guyed tower," said Jay Weidler, "nothing else was ever built, or very little was ever built, that that yard was intended for. So, we geared up for this wonderful market that never really appeared and never sustained anyone in our end of the business." Smaller structures were fabricated there, including many of the jackets and platforms for Mexico's Bay of Campeche development, but nothing else that approached the magnitude of the Lena or even the Garden Banks.

The era of deepwater in the Gulf of Mexico during the 1970s and 1980s produced landmark technological achievements in marine design and construction. Through its work on Garden Banks and the Lena, Brown & Root maintained technological leadership in the industry and helped extend offshore operations to depths scarcely imaginable a decade earlier. It was a bittersweet experience for the company. The deepwater era ended abruptly and with it, Brown & Root's commitment to the next generation of deepwater technology. "It was challenging, interesting, fascinating," remembered Weidler. "We did maintain a tremendous capability to handle deepwater, little realizing how quickly it would be passed by and never used."

Notes

1. This chapter draws from interviews with J.B. Weidler, G.C. Lee, S.J. Hruska, and W.B. Pieper.
2. Griff C. Lee, "'Deep' Thoughts on Conventional Concepts," *Offshore* (April 1978): 90.
3. "Resume of Marine Operators Work Up To April 20, 1966," Copy in Brown & Root Archives.
4. Leonard LeBlanc, "Platform Economics: A Costly Game," *Offshore* (December 1978): 46–50.
5. Griff C. Lee, "Offshore Platform Construction Extended to 400-foot Water Depths," *Journal of Petroleum Technology* (April 1963): 385; Lee, "'Deep' Thoughts on Conventional Concepts," 92.
6. Lee, "Offshore Platform Construction Extended to 400-foot Water Depths," 385–386.
7. Lee, "Offshore Platform Construction Extended to 400-foot Water Depths," 285–386; Lee, "'Deep' Thoughts on Conventional Concepts," 96.
8. "The Search for Domestic Energy," *Brownbilt* (Winter 1974–1975): 14–21.
9. F.P. Dunn, "Deepwater Production: 1950–2000," OTC 7627, Offshore Technology Conference, Houston, TX May 2–5, 1994, 923.

10. Dunn, "Deepwater Production: 1950–2000," 924.
11. Dunn, "Deepwater Production: 1950–2000," 924.
12. Susan Thobe, "Deepwater Exploration Calls for Floating Drilling Rigs," *Offshore* (June 5, 1975): 60.
13. "Drilling Technology Keeps Pace with Deep Water," *Offshore* (June 5, 1975): 47–49; "Industry Has Drilled 60 Wells in Waters Exceeding 600 Feet Since 1965," *Offshore* (June 5, 1975): 50–51.
14. Robert E. Kallman and Eugene D. Wheeler, *Coastal Crude in a Sea of Conflict* (San Luis Obispo: Blake Printery and Publishing Co., 1984): 47–80.
15. Jim Carmichael, "U.S. Gulf of Mexico is Still a Large Producer after 25 Years of Activity," *Offshore* (June 20, 1974): 93–94.
16. Ron Londenberg, "Deep Water Leases May be Offered in Gulf by U.S. Government," *Offshore* (March 1973): 31–34; "Deepwater Tracts will be Offered on Continental Slope for the First Time," *Offshore* (October 1973): 52–53.
17. Quoted in Daniel Yergin, *The Prize: The Epic Quest for Oil, Money, & Power* (New York: Touchstone, 1992), 664.
18. Jim Carmichael, "March Offering of Louisiana Blocks Becomes Landmark Sale," *Offshore* (May 1974): 71–79, quote from 79.
19. Lee, "'Deep' Thoughts on Conventional Concepts," 95.
20. "In the Land of Giants," *Brownbilt* (Summer 1981): 21–25; quote on 21.
21. Quoted in "The Birth of a Giant," *Brownbilt* (Fall 1979): 26.
22. "Single-Piece Jacket Being Installed in Gulf," *Offshore* (October 5, 1979): 11; Leonard LeBlanc, "Long Jackets Advance Deepwater Operations," *Offshore* (November 1979): 46–48.
23. "ARCO Platform Gulf's Second Largest," *Brownbilt* (Winter 1981): 14–18.
24. Griff C. Lee, "Design and Construction of Deep Water Jacket Platforms," Paper presented to the Third International Congress on the Behaviour of Offshore Structures (BOSS), Massachusetts Institute of Technology, Cambridge, MA, August 2–5, 1982.
25. Lee, "'Deep' Thoughts on Conventional Concepts," 96.
26. Lee, "Design and Construction of Deep Water Jacket Platforms."
27. A.O.P. Casbarian, "Cognac: Unique Diving Skills Aid Project," *Offshore* (August 1979): 52.
28. Quoted in "Platform Comes in on Beer Budget," *Engineering News-Record* (August 27, 1981): 50.
29. Quoted in "Platform Comes in on Beer Budget," 51.
30. Lee, "Design and Construction of Deep Water Jackets."
31. Lee, "Design and Construction of Deep Water Jackets."
32. Andrea Mangiavacchi, Phillip A. Abbott, Shady Y. Hanna, Rudolf Suhendra, Brown & Root, Inc., "Design Criteria of a Pile Founded Guyed Tower," OTC 3882, Offshore Technology Conference, Houston, TX, May 5–8, 1980, 275–276.

33. Fred S. Ellers, "Advanced Offshore Oil Platforms," *Scientific American* Vol. 246, No. 2 (April 1982): 48.
34. Lee, "Design and Construction of Deep Water Jacket Platforms."
35. D.E. Boening and E.R. Howell, Exxon Co. U.S.A., "Lena Guyed Tower Project Overview," OTC 4649, Offshore Technology Conference, Houston, TX, May 7–9, 1984.
36. "Newsletter," *Offshore* (October 5, 1979): 5.
37. "Newsletter," *Offshore* (December 1979): 5.
38. Boening and Howell, "Lena Guyed Tower Project Overview," 9.
39. Mangiavacchi, *et al,* "Design Criteria of a Pile Founded Guyed Tower," 276.
40. Mangiavacchi, *et al,* "Design Criteria of a Pile Founded Guyed Tower," 276–277.
41. M.S. Glasscock and J.W. Turner, Exxon Co. U.S.A., and L.D. Finn and P.J. Pike, Exxon Production Research Co., "Design of the Lena Guyed Tower," OTC 4650, Offshore Technology Conference, Houston, TX, May 7–9, 1984, 19.
42. R.E. Parnell Jr. and R.J. Houghton, Exxon Co. U.S.A., and J. D. McClellan, Brown & Root, "Fabrication of the Lena Guyed Tower Jacket," OTC 4651, Offshore Technology Conference, Houston, TX, May 7–9, 1984, 29–31.
43. Parnell, *et al,* "Fabrication of the Lena Guyed Tower Jacket," 32.
44. Parnell, *et al,* "Fabrication of the Lena Guyed Tower Jacket," 31–32.
45. J.K. Flood and P.Q. Pichini, Exxon Co. U.S.A.; M.A. Danaczko, Exxon Production Research Co.; and W.L. Greiner, Brown & Root, Inc., "Side Launch of the Lena Guyed Tower Jacket," OTC 4652, Offshore Technology Conference, Houston, TX, May 7–9, 1984, 39–40.
46. Flood, *et al.,* "Side Launch of the Lena Guyed Tower Jacket," 40–42.
47. E.C. Smetak, J. Lombardi, and H.J. Roussel, Exxon Co. U.S.A., and T.C. Wozniak, Brown & Root Inc., "Jacket, Deck, and Pipeline Installation—Lena Guyed Tower," OTC 4683, Offshore Technology Conference, Houston, TX, May 7–9, 1984, 319.
48. Smetak, *et al,* "Jacket, Deck, and Pipeline Installation—Lena Guyed Tower," 320.
49. L.D. Ziems, Exxon Co. U.S.A., W.C. Kan, Exxon Production Research Co., E.S.B. Stidston, J.J. McMullen Assocs. Inc., and M.L. Neudorfer, Honeywell Inc., "Dynamically Positioned Derrick Barge and Position-Measuring Equipment for Lena Guyed Tower Installation," OTC 4685, Offshore Technology Conference, Houston, TX, May 7–9, 1984, 339–342.
50. C.P. Brown and L.M. Fontenette, Exxon Co. U.S.A., and P.G.S. Dove, Brown & Root Inc., "Installation of Guying System—Lena Guyed Tower," OTC 4682, Offshore Technology Conference, Houston, TX, May 7–9, 1984, 309–313.
51. Boening and Howell, "Lena Guyed Tower Project Overview," 13.
52. Quoted in "Exxon's Guyed Tower at Home Now in Gulf," *Brownbilt* (November 2, 1982): 7.

53. Boening and Howell, "Lena Guyed Tower Project Overview," 2.
54. Rick Hagar, "Deepwater Drilling Hits Record Pace in Gulf of Mexico," *Oil & Gas Journal* (October 8, 1984): 25–28.
55. "Worldwide Offshore Daily Average Oil Production," *Offshore* (May 1985): 114.

Part 2
The Challenge of New and Extreme Environments: Depth, Earthquakes, Ice, and Fire

Chapter 6
Brown & Root Marine Goes Abroad

After World War II, Brown & Root gained considerable experience in international work. The firm had obtained U.S. government projects for erecting townsites and military installations on the island of Guam, constructing NATO airbases in France, and building U.S. air and naval bases in Spain.[1] Brown & Root also constructed dams, power plants, and raw materials processing complexes in developing countries. By the 1960s, approximately one-third of the company's work was outside the United States. The company also moved aggressively into offshore work around the world. From the late 1950s through the first energy crisis in 1973, Brown & Root Marine's offshore achievements helped develop an international offshore oil and gas industry and the transport systems that linked supply and demand.[2]

The surging worldwide demand for oil drove the international search for oil and gas. The construction of the U.S. interstate highway system and the suburbanization of American metropolitan areas made motor vehicles, and the gasoline to fuel them, ever more important staples to American lifestyles. The reconstruction of European economies, and their wholesale conversion from coal to oil-powered electric generation, created another source of demand. The industrializing strategies of Japan and some developing countries raised their thirst for oil. The extension of U.S. political influence in the postwar period over all these regions, including the oil-rich Middle East, facilitated the expansion of the U.S. offshore oil industry into foreign waters.

The high success ratio of finding oil in the Gulf of Mexico in the early 1950s encouraged petroleum geologists to examine other marine areas exhibiting similar kinds of sedimentary layering. They had found rich oil deposits in the upper Mesozoic, Tertiary, and Quaternary stratigraphy of the coastal U.S. Gulf plain and adjacent continental shelf. "This fact," geologists noted, "leads to speculation regarding other marine areas in the world where geosynclinal conditions of deposition have been continuous from the Mesozoic to the Recent."[3] Explorations converged on shelf areas around the world; the most promising basins were thought to be located in the Gulf of Mexico, the Caribbean, Mediterranean Sea, Persian Gulf, west coast of Africa, South China Sea, and the North Sea.

Oil companies scrambled for concessions, and drilling rigs set off for new marine environments around the world. Although exploration and development concentrated on protected and shallow waters similar to those in the Gulf of Mexico, each producing area presented different challenges to the innovative spirit of offshore engineers. Two regions were especially important: Venezuela's Lake Maracaibo and the Middle East's Persian Gulf, where enormous oil deposits justified path-breaking departures in the scale and cost of offshore projects.

Lake Maracaibo

Venezuela's Lake Maracaibo witnessed the biggest boom in offshore oil in the 1950s. Brown & Root became involved in all aspects of oil development in the lake, from driving concrete piles, to moving derricks, to laying pipelines, to building and installing gas compression and injection stations. According to Rick Rochelle, "Maracaibo was where things really began to reach dimensions beyond backyard stuff."

Venezuela had emerged as a world-class oil producer in the mid-1920s with Shell Oil's development of the La Rosa field bordering Lake Maracaibo. Shell's discovery brought an oil frenzy to the country; by 1929, Venezuela produced 137 million barrels annually, second only to the United States in total output. Other oil companies moved in, including Standard of New Jersey (now Exxon) through its Creole Petroleum Corporation affiliate, and Gulf Oil through the Mene Grande Petroleum Company. These three companies established themselves as the dominant oil producers in Venezuela and in the Lake Maracaibo region for many years to come. Oil production in Venezuela turned out to be prolific and profitable. For a time in the 1950s and 1960s, for example, Creole Petroleum generated about half of Jersey Standard's total world production and income.[4]

In the 1920s, Creole began to drill shoreline wells in the Bolivar coastal field in the northeast part of Lake Maracaibo. Operations consisted of small platforms serviced by equipment housed on tender vessels or onshore. Platforms and derricks were erected first on piles made from wood, but the marine termites or shipworms, known as the teredo navalis, which infested the lake made quick meals out of the wooden piles. Even heavily creosoted wood could not fend off the voracious teredo. Creole then turned to precast reinforced square concrete piles, which were bacteria-resistant and long-lasting. Small wave and wind forces in the protected lake permitted the use of such piles without the kind of lateral support needed for structures in harsher marine environments such as the Gulf of Mexico. In 1928, the Raymond Concrete Pile Company set up a yard to fabricate precast piles, and over the next two years Creole installed 160 platforms with all-concrete foundations, reaching out into 60 feet of water by 1934. As the water depth increased, even the small waves and wind in the lake required stronger piles, and installation costs rose.[5] The decline in oil production and drilling in Maracaibo during the depression years, however, slowed the installation of concrete piles and offshore platforms.

In the early 1950s, offshore Maracaibo boomed again. The 1948 revision of the oil companies' concession contracts had given the Venezuelan government a fifty-fifty split in oil profits in exchange for forty-year concession renewals, encouraging exploration. Shell Oil and Creole Petroleum took the lead in developing large acreage blocks in the lake. By 1960, almost 4,000 wells could be counted on the lake's surface, with maximum production exceeding 1.4 million barrels per day. This oil flowed to 176 gathering stations, mounted on piling platforms and connected to a network of pipelines running to shore. The work of building and operating this offshore complex attracted many of the companies previously active in the Gulf of Mexico.[6]

From Brown & Root's perspective, "Lake Maracaibo was obviously good pickings because it was getting deeper and deeper and deeper." Brown & Root entered the Maracaibo scene in 1952 under a contract with Creole to construct what would be the first of four integrated gas compressor stations—also known as pressure maintenance plants or secondary recovery operations. Lacking

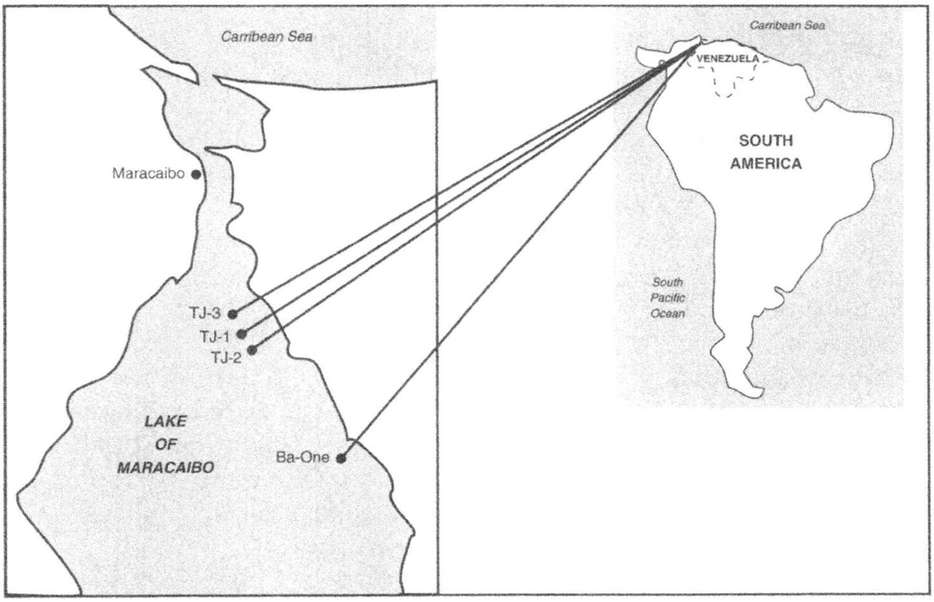

Lake Maracaibo

markets for the gas produced from its Maracaibo wells, Creole and other companies had long flared natural gas associated with oil. By the 1950s, this practice was neither technically nor politically attractive. The reinjection of gas could maintain reservoir pressure, thereby increasing the rate of crude recovery and extending the life of the reservoir.[7] The compression and reinjection of low-pressure gas might also conserve the gas for the future, when growing markets might justify the construction of expensive natural gas pipelines.

Having constructed numerous gas compressor systems for Texas Eastern and other gas transmission companies, Brown & Root was a logical choice for Creole. Brown & Root was also a leader in applying gas turbine power to the centrifugal compressor. Industrial use of the gas turbine, an outgrowth of military jet engine development during World War II, was still new, but it offered great increases in engine power.[8]

Designed under the management of Brown & Root engineer Bill Rice, Creole's conservation system called for "the largest concentration of turbine-driven centrifugal compressors anywhere in the world, particularly offshore." Built in place in 1954, seven miles offshore in 60 feet of water, the Tía Juana Plant No. 1 rested on 350 concrete piles, 28 inches square and 165 feet long. The plant itself covered the area of a football field, centered on a web-like network of gas gathering and injection pipelines, and connected to flow stations and production wells in the Tía Juana field. Ten gas turbines were installed, each rated at 6,000 horsepower (hp) and coupled to a centrifugal compressor. The ten units were connected as a combination of parallel units arranged in a series with individual speed adjustment by a master controller sensing inlet pressure. The plant also was outfitted with a unique "Tracer Gas System" that injected carbon monoxide along with natural gas into the oil reservoir. By measuring the minute quantities of tracer gas that showed up in producing wells, the operator could determine the sweep pattern of the natural gas in the system. With the capacity to compress 170 million standard cubic feet of gas per day (MMSCFD) from 10 pounds per square inch of gas (psig) to 1,950 psig, the Tía Juana 1 was the first of its kind.[9]

After the pioneering success of Tía Juana 1, Creole awarded Brown & Root the contract for the design and construction of Tía Juana No. 2 in 1955 and Tía Juana No. 3 in 1957. The $20 million Tía Juana 2 was similar to its predecessor, only larger, with twelve gas turbine centrifugal compressor units running at 72,000 hp and capable of drawing 330 MMSCFD. Tía Juana 3, completed in 1958 by Brown & Root, was larger still, incorporating fourteen turbine compressors with 84,000 hp and a compressor capacity of 377 MMSCFD. In 1960, Brown & Root completed the system for Creole with the Bachaquero No. 1, another fourteen turbine compressor facility, located 25 miles south of the three Tía Juana plants in the Bachaquero field.

Concrete-pile platforms supported all of these gas compressor stations in water ranging from 65 to 100 feet. Pipelines integrated the plants, permitting alternate delivery of gas from one area to more than one compressor plant.

The expansion of the last two plants in the mid-1960s to conform with reservoir mechanics and oilfield development gave the $93 million gas compression system a total capacity of 420,000 hp from 50 turbines. The four semiautomatic plants, designed for minimum staffing or unattended operation, could compress 1.47 billion square cubic feet of gas per day and inject gas into any one of twelve reservoirs served by the system.[10]

The construction and installation of these plants was a remarkable feat for Brown & Root, as was designing and laying the world's largest submarine gathering system to connect them. Plans called for 800,000 feet of pipeline to integrate the plants with production wells, injection wells, and 40 flow stations containing scrubbers and separators. Each flow station received production from 20 to 40 producing oil wells surrounding them. In 1956, Brown & Root laid its first Maracaibo pipeline, 20 inches in diameter, for Shell Oil using two barges.[11] But hooking up the conservation system was no ordinary pipelaying operation. The biggest challenge involved laying fourteen miles of 40-inch diameter pipe, also known as the "Super Inch," the largest pipeline yet laid underwater. To give the line negative buoyancy and to protect it from the corrosive Maracaibo waters and damage from anchors, other pipelines, and cables, the Super Inch had to be wrapped with steel mesh, fiberglass, and six inches of concrete coating. Each 40-foot joint weighed approximately twenty tons. Submerging the massive line from barge level to seventy feet below onto the highly uneven floor of the lake without buckling section joints or cracking the concrete coating made the job "one of the toughest ever attempted."[12] Brown & Root set up yards on both the east and west sides of Lake Maracaibo to prepare the pipes and assemble the three million square feet of steel mesh and 23,000 cubic yards of concrete required for the project.

In 1958, Brown & Root's *M-211* pipelaying barge was loaded with 2 million pounds of material and towed to Lake Maracaibo to tackle the job. The *M-211*'s 40-ton capacity davits equipped crew members to handle the massive pipe sections of the Super Inch, and its new stinger device enabled the barge to ease the line to the bottom in deeper waters. Upon completion, the gas gathering-compression-injection system consisted of 18 lines and nearly 800,000 feet of pipe ranging from 6 to 40 inches in diameter.[13]

While specializing in the construction of gas compressor systems in Venezuela, Brown & Root also took on other innovative marine construction assignments. Basil Maxwell, one of the original Brown & Root marine construction specialists, served as the company's Lake Maracaibo superintendent during this time. In 1958, he and Brown & Root engineer Dick Wilson devised a method of moving tender-supported drilling rigs from platform to platform in the lake. Brown & Root had used the derrick barge *L.T. Bolin* to set flow stations for the Creole gas system. Equipped with a 250-ton capacity hammerhead crane, the barge could move drilling rigs by means of a single lift carrying the substructure, derrick, draw-works, engines, and the rest of the drilling equipment—a feat that had never been tried before. Such structures had offset centers of gravity, so to make level lifts, the barge crane had to be equipped with special lift adapters to adjust its sling lengths. The *L.T. Bolin* first moved rigs for Shell Oil, and then other companies began leasing the barge to make their lifts. "Naturally, everybody wanted us to keep trying to get more on, so we ended up with more," Maxwell said. "When we went to work for Mene Grande, they wanted the whole cheese—the motors, the rotary, and the blowout preventers which hang underneath." After a mishap on the first rig move, when the derrick peeled off the hammerhead and sunk to the bottom, the *L.T. Bolin* executed many successful single-unit rig moves, sometimes two in one day. Between December 1958 and July 1959, the Bolin moved 45 rigs.[14] "I had chills and fever every time we picked up one," confessed Maxwell, "but sometimes we'd move them ten miles or fifteen miles across that lake."

Venezuela became a new spawning ground for marine technology and "the most fabulous overwater oil producing area in the world."[15] The success of companies like Shell, Mene Grande, and Creole in developing near-shore fields in the lake encouraged the government to open up new tracts in the central lake. The Gulf of Paria, a shallow gulf separating the island of Trinidad from the Venezuelan coast, also attracted the attention of bidders. In addition to the "old" concessionaires (Texaco, Phillips, Atlantic, and the big three), many new participants (Signal, Pure, Sun, Union, San Jacinto, Kerr-McGee, Sohio, Tennessee Gas, and others) snatched up leases. Because the lease acreage offered was so large (mostly in blocks of 10,000 hectares or 24,700 acres), with prices ranging as high as $2,000 an acre in the lake, bids were typically made by groups, permitting many independents to participate on a share basis. The 1957 lease sales brought Venezuela $113 million for new Lake Maracaibo acreage and $120 million for the Gulf of Paria offering. For 1958, Creole and Shell budgeted $168 million and $200 million, respectively, for oil operations in Lake Maracaibo. Rigs, barges, drilling equipment, and men rapidly descended upon the lake, and new platforms and derricks sprouted up across the water horizon.[16]

Because the waters of Lake Maracaibo were calm, operators could build platforms as little as eight feet above the water. Most drilling in the lake and in the Gulf of Paria was conducted using the fixed platform-tender type method, which was logistically easier in these waters than in the Gulf of Mexico. A few mobile units traveled to the Gulf of Paria in 1958, but the first mobile drilling rig in Lake Maracaibo did not appear until 1961.[17]

Since the costs to install concrete pile structures increased with water depths, the newer steel jacket and pile technology was introduced into the lake from the Gulf of Mexico. Although steel lowered construction costs and drastically shortened construction schedules, the brackish waters of the lake corroded steel structures. Traditional remedies proved ineffective. Cathodic protection in the form of sacrificial anodes did not work because of the low amounts of electrolyte in the waters; impressed current also did not prove efficient.[18]

Designers thus began to search for an alternative to both concrete and steel. Aluminum jackets were introduced as a logical cure-all to the expensive, time-consuming-to-install concrete structures, and the hard-to-maintain steel variety. Aluminum offered an alternative to steel as a corrosion-resistant metal for jackets. J. Ray McDermott & Company fabricated and shipped the first aluminum jackets to Venezuela in September 1957. These lightweight jackets were installed easily and quickly, and McDermott manufactured more than 40 of them for Lake Maracaibo over the next few years. "They really had a bonanza," recalled Dick Wilson. Aluminum soon became a popular material for all kinds of marine equipment used in the lake.[19]

Brown & Root briefly experimented with making aluminum jackets. Working with ALCOA in 1958, Wilson managed the development, fabrication, and installation of three aluminum jacket structures for Lake Maracaibo. A discouraging pile problem soon convinced Brown & Root to seek a different answer to the problems posed by Lake Maracaibo. The aluminum jackets were supported by steel piles driven through the legs. Unfortunately, the interaction of the steel with the aluminum made the jacket behave as a giant sacrificial anode which overcame the lack of electrolyte through its sheer magnitude. After trying various ways of insulating the steel piles, Brown & Root abandoned aluminum jackets in favor of a new approach using prestressed concrete cylinder piles.

This new technology illustrates how Brown & Root's diversified construction actitivies could contribute to its offshore activities. During the early 1950s, the Raymond Concrete Pile Co. had developed a process for making standardized, prestressed, and unusually strong concrete cylinder piles, ranging in diameter from 36 to 54 inches, at its yard in Madisonville, Louisiana. The piles achieved success in the construction of the 24-mile causeway across Lake Pontchartrain on the north side of New Orleans. In 1955, Brown & Root and T.L. James Company designed the causeway and established an integrated plant, Prestressed Concrete Products, Inc., with Raymond's license and engineering assistance, for the assemblyline production of 185-ton prestressed deck slabs, caps, and 54-inch diameter piling. The assemblyline method reduced both the cost and installation time involved in using concrete piles. Brown & Root completed the causeway in fifteen months, four months ahead of schedule. "The concept of doing all of this from the water and prefabricating steel forms that you could use again a thousand times and prestress all within the form, all this was absolutely new stuff," said Rick Rochelle. Described by the *Engineering News-Record* as "a bold venture, requiring unusual foresight, ingenuity and resourcefulness," the Lake Pontchartrain Causeway demonstrated the merits of Raymond's new prestressed cylinder pile in marine environments.[20]

Brown & Root decided to try these new cylinder piles in Lake Maracaibo. Raymond Concrete was well-established in Venezuela, driving square concrete piles and building offshore loading piers. Brown & Root in 1957 entered a joint venture with the company to engineer and build loading piers for Creole Petroleum. As construction progressed, the joint venture obtained a contract with Shell Oil to develop cylinder piles. According to David K. Smith, one of the original engineers and later manager of the Raymond-Brown & Root joint venture, Shell felt that "with the advent of the cylinder pile, which had been very successful up in Lake Pontchartrain, this same type of piling should work well in Lake Maracaibo."

Raymond-Brown & Root set up a $5 million fabricating plant in Shell's Bachaquero yard and brought down personnel, equipment, and barges from the Lake Pontchartrain job. Included among this fleet was Raymond's main pile-driving barge, the *S-10,* a "very special rig" equipped with very heavy, tall leads to guide the hammer and pile with underwater pile extensions down to minus 40 feet below the water level. "The old way of trying to set a pile and then put your hammer on top of it, you had so many different difficult

Installing concrete cylinder piles in Lake Maracaibo

soil conditions that when you put the hammer on it, the whole pile would start to run on you and then you lost control of the pile" said Smith. "But with the *S-10,* you had complete control at all times of that pile and hammer."

The new cylinder piles initially faced stiff competition from McDermott's aluminum jacket platforms. Many companies initially opted for the aluminum jackets because they could be installed in two to three days, compared to four to six weeks to erect structures with cylinder piles. Early problems with the quality of concrete further hampered Raymond-Brown & Root, as did a major accident in 1958, when 18 men died in the collapse of a Shell flow station for which the joint venture had driven the pile foundation. Such problems slowed the introduction of the cylinder pile technology, as potential buyers took a wait-and-see attitude.

Raymond-Brown & Root regrouped. Engineers improved their methods of designing and installing the piles, and they placed additional piling on structures to meet the new design criteria. They carried out more extensive soil boring and actual pile load testing on all heavily loaded structures with the help of McClellan Engineering. They established stricter quality controls, located higher strength coarse and fine aggregate used in making their concrete, and erected a new precast, poststressed cylinder pile plant at Potrerito on the western shore of the lake. The installation of a concrete plug in the bottom of the bearing piles enabled inspectors to look down inside the piling after driving to check for any leaks in the joints or any damage to the pile from hard driving. Refined grouting techniques and the use of larger pile driving hammers (30,000 and 60,000 lb. rams with adjustable strokes) enabled the prestressed piles to be driven deeper into the lake bottom to support heavier loads. With these improvements and the rig-moving capability of the *L.T. Bolin,* Brown & Root reduced the time required to install a derrick platform, ready for drilling, to less than seven days.

The joint venture's improved cylinder piles became very popular. "They are easy to handle and drive," wrote *Offshore* magazine in 1958, "and their section moduli permit them to carry large loads without need of expensive lateral bracing."[21] In 1958, Mene Grande ordered 50 new cylinder pile platforms, which Raymond-Brown & Root produced at a rate of one per week. Later, the joint venture also developed larger diameter piles, 54-inch and 66-inch, allowing for the construction of deeper-water derrick bases with a smaller number of piles and single-well platforms, called "unipiles," for mobile drilling barges.[22]

The Raymond piles became emblematic Lake Maracaibo marine structures. As McDermott's aluminum jackets corroded, they were often replaced with Raymond cylinder piles. David Smith remembers having "the great pleasure . . . of loading out all of McDermott's equipment on a marine barge and waving good-bye as it went to Nigeria." During the late 1950s, Dutch entrepreneur Pieter Heerema developed his own prestressed concrete pile, poured in one section rather than assembled. By the early 1960s, most of the lake's drilling platforms, derrick bases, flow stations, and gas conservation plants were constructed on concrete piles made by either Raymond-Brown & Root or Heerema.[23]

These piles provided the needed durability for offshore structures. Other oil companies followed Creole's lead in adding gas injection systems to their operations, and Brown & Root designed and fabricated nearly all of these plants. By the late 1960s, Brown & Root had engineered nine out of the ten plants operating in the lake. But the new generation of gas injection plants built in 100 feet of water in the central lake beginning in the mid-1960s ushered in a "new and drastically different design era."[24] Economic considerations dictated the choice to use reciprocating rather than centrifugal equipment on the new plants, introducing yet a new set of formidable marine engineering challenges to Brown & Root designers.

The first of these plants was called the Unigas I. Brown & Root designed, built, and installed it during 1965–1966 for a four-partner venture operated by Shell in association with Mene Grande, Conoco, and Creole. The companies desired a single system to serve reservoirs underlying concessions held by each of the four partners. The plan called for a projected 200 million cubic feet per day of gas flow with injection pressures generated by power from seven 5,500 hp compressors. Shell and its partners preferred not to use centrifugal compressor units that "staged" compression in series from one turbine-compressor to the next. In a series operation, the entire system would be shut down if one of the stages failed. Shell desired a "parallel" design, using integral gas engine-reciprocating compressor units, each operating independently. "In the parallel arrangement, any one or more of the compressors can maintain injection pressure with the number of compressors in operation determining the capacity of the system."[25]

Vibration was the major design problem caused by the shift to reciprocating compressors, which created imbalances that "resulted in the most complex and severe machinery-induced vibrational loads ever encountered on a marine structure."[26] The speed of the machine in reciprocating compression was relatively close to the natural frequency of the piling, raising the likelihood of harmful resonance and vibration transfer from the plant to the piling. The long piling needed to support the structure in 100-foot water depths magnified this dynamic design problem. To conduct the analytical and testing work, Brown & Root "had quite an array of talent . . . brought into the company to assist." Specialists on pile design and dynamic loading came from the University of Illinois. Engineers from Holland conducted vibrating load and harmonics tests simulating actual full-scale loading. They computed the response of the entire platform and piling system to expected vibrational loads. Data was also obtained from soil boring tests at another Brown & Root project, the Livingston Dam, which was being constructed in preconsolidated clays very similar to those under Lake Maracaibo.[27]

The design worked out from these tests called for seven 5,500 hp gas engine-compressor packages fitted with special balancing cylinders. These cylinders worked off the unit's crankshaft creating an inertial out-of-phase force that cancelled out the ones generated by the reciprocating compressors. "The compressors, for the first time, were balanced to a degree never heard of before," said Harris Smith, a Brown & Root engineer involved with all the Venezuela gas conservation plants after Tía Juana I. Suction and discharge pulsation filters and interstage piping were tailored to eliminate vibrations caused by gas pulsation. The truss deck and plant were set on one hundred prestressed concrete piles, most of them driven on a batter inside a small area after being spotted by marine divers on the bottom. Soon after the startup of the $15.6 million plant in 1966, tests proved that vibration amplitudes on the structure were far less than expected. It "ran like a Swiss watch," Smith said.[28]

The Unigas I established a new standard in offshore plant design using reciprocating compressors. In the late 1960s, Brown & Root extended this technology in the design and installation of four other compression facilities in Lake Maracaibo. The first three of these were the Lagogas 2 for Shell Venezuela, the Lamargas for Phillips Petroleum, and the Lama for Venezuela-Sun Oil Company in association with Texaco and Atlantic Richfield.[29] Perhaps Brown & Root's most significant work in Venezuela came in 1968–1969 with the fourth plant, the Menegas I, designed for the Mene Grande Petroleum Company to repressurize two Bachaquero reservoirs on the eastern side of Lake Maracaibo.

Shell's Unigas I gas conservation plant

What set the Menegas I apart was its reliance on prefabrication. Brown & Root completely prefabricated and tested the $11 million plant at Greens Bayou in Houston. At the time, Menegas I was the largest structure built at one location for shipment and reassembly at a distant site. The 45-MMSCFD unit was broken down into nine modules, consisting of five gas-engine reciprocating compressors, an alternator module, an air compressor module, a process module, and a switchgear-control room module. After the pieces arrived in Lake Maracaibo by barge from Houston, the 400-ton fixed crane on Brown & Root's *Atlas* construction barge lifted each module onto a concrete pile and beam platform erected in 100 feet of water. Prefabrication of Menegas I saved the project an estimated nine to twelve months in on-site construction and assembly time,[30] which helped prove the benefits of prefabrication to Brown & Root as it undertook projects at other locations.

Venezuela's Lake Maracaibo launched the offshore oil industry into new waters beyond the Gulf of Mexico. The lake's particular environmental conditions stimulated a series of creative adaptations in the design of marine structures. The results were readily apparent in Lake Maracaibo's growth into one of the world's greatest offshore oil-producing regions. By 1970, offshore production from Venezuela averaged an astounding 2.7 million barrels per day.[31] Brown & Root Marine had been at the forefront of engineering and

Setting a prefabricated module of the Menegas I

construction in the lake, and the company's leadership in the early development of both the Gulf of Mexico and Lake Maracaibo prepared the company to tackle new challenges in offshore environments around the world.

International Terminals and Pipelines

Under the new management of vice president L.E. Minor, Brown & Root's pipeline division aggressively pursued offshore projects overseas. As more international projects came along, Hal Lindsay, head of Brown & Root's Marine Operators, began giving those that demanded innovative design thinking to the new marine design group under H.W. Reeves. Many of the young engineers in this group received their first assignments in Venezuela.

The design and construction of early "supertanker" terminals became a specialty of the Reeves group. The increase in the amount of petroleum transported on the high seas during the 1950s had stimulated the construction of ever-larger tankers that could move oil faster and cheaper than smaller ones. New and bigger facilities were thus needed to load, unload, and berth the new floating giants that tied together world supply and demand. Brown & Root designed many of these facilities and laid the large-diameter pipelines that linked them to shore.

Brown & Root's new marine group anticipated the need for new marine terminals to accommodate ever-larger oceangoing oil tankers. "No longer can the docking facilities for tankers be developed by using only age-old rules of thumb as the design basis," wrote Reeves in 1958. The design of a modern terminal to handle supertankers had to be approached with the best scientific data and engineering know-how available. Brown & Root also started using 3D structural analysis in terminal design. "From a business standpoint," said John Mackin, "that was one of the most profound things we did in that group during that period."

Developing proper docking arrangements required thorough research on harbor conditions and the characteristics of the tankers intended to use the facilities. The size and displacement of the new tankers ruled out several options. Full-fledged harbors often were not deep enough. Constructing large jetties in deep water to serve the tankers was expensive and technically unfeasible. Modifiying most existing terminals by dredging also appeared to be prohibitively costly. New deepwater terminals connected by submarine pipelines to the shore, however, afforded "one of the better solutions to the problem."[32]

The first terminal of this kind designed by Brown & Root—and indeed the company's first foreign marine project outside of Venezuela—was a marine oil handling structure in Rio de Janeiro's Guanabara Bay for Petrobras, the Brazilian state oil company. Brazil wanted to develop a greater capacity for handling imported oil from the Middle East and from offshore fields being developed off the country's northeast coast.[33] In 1959, Reeves designed and supervised construction of the $30 million Guanabara terminal and pipeline facility with the aid of Ber Pieper, Tim Pease, John Mackin, and others, who analyzed data from field investigations, ocean current measurements, soil tests, and estimated tanker-approach velocities. Brown & Root engineers oversaw the construction of the terminal in 60 feet of water. Brown & Root lay barges laid two 26-inch submarine pipelines to connect the terminal to the Duque de Caxias Refinery on the mainland. The island wharf terminal structure consisted of two banks of hydraulic marine oil loading devices and two square breasting platforms connected to two mooring dolphins. The terminal could simultaneously berth two tankers varying in size from 2,000 dead weight tonage (dwt) to 105,000 dwt.[34]

The Guanabara Terminal in Rio de Janeiro launched Brown & Root into the business of designing terminals and loading facilities in South America. Subsequent projects in Brazil included a breakwater tanker and mooring facility at the port of Santos in Sao Paulo and a terminal offshore Ilheus, Brazil, about 200 miles south of Salvador, Bahia.[35] In 1961, Brown & Root crews traveled to the "bottom of the world" for a project off the Isla Grande de Tierra del Fuego, Argentina. Under the engineering supervision of Delbert Johnson, Brown & Root surveyed and laid a 20-inch underwater pipeline from shore out into the 50-foot water depth of San Sebastian Bay. The line joined a 35,000 dwt tanker sea berth anchorage and loading buoy, designed by the Brown & Root engineers for Tennessee Gas's Argentine subsidiary. A cold and constant wind in this extreme southern latitude generated almost unbearable working conditions. "We did a survey and tried driving sheet pile in that wind," Ber Pieper recalled. "The wind would take those pieces of sheet pile, and hold them straight out." High, 28-foot tides also frustrated operations. "When the tide was out, you'd go out and survey and locate all the big rocks," said Pieper. "When the tide came in, you'd just come in and lay off until the tide would go back out and keep going." The pipelaying barges would lay lines until the tide went out, leaving the barge at rest on the bottom. Laying would resume after the tide returned and the barge became buoyant again. Braving the extreme weather, Brownbuilders, as Brown & Root employees were called, also installed a pumping station, onshore gathering lines, and a water flood injection system to complete the project, the southernmost installation of its kind in the world.

By the early 1960s, the real action in tanker terminals and submarine pipelines was in the Middle East and Mediterranean. The growth in the world tanker fleet in the late 1950s had paralleled the spectacular increase in oil production by Middle Eastern countries, much of which was shipped from that region to Western Europe and Japan. Middle Eastern oil exports rose from 507,000 barrels per day in 1946 to 4.69 million in 1960 to more than 16 million by 1972, as Western Europe's share of Middle Eastern exports rose from 25 percent to 58 percent. Brown & Root saw a ripe opportunity in the design and construction of deepwater tanker terminals and the underwater loading lines to transport oil to them from onshore fields.[36]

Offshore prospects in the Middle East heightened Brown & Root's interest in the region. Surrounded by the world's greatest oil fields and possessing a

Persian Gulf Region

continental shelf shallow enough for drilling, the Persian Gulf attracted keen interest from world oil companies. Opportunities abounded for the offshore industry. Shell Oil, which was already producing oil from a 1952 offshore concession granted by the ruler of Qatar, obtained access to Kuwait's offshore territory. Iran granted a large offshore concession to Indiana Standard's Iranian subsidiary, the Iran Pan American Oil Co. (IPAC). The sheikdom of Abu Dhabi awarded a concession to joint venture between British Petroleum (BP) and France's Cie. Française des Petroles (CFP).[37] As new fields were developed in the early 1960s, the Persian Gulf emerged as "the hottest offshore oil patch in the world."[38]

Brown & Root entered the Persian Gulf in 1961 when Iraq's Basrah Petroleum Company commissioned the *L.E. Minor* pipelaying barge to lay two 32-inch submarine pipelines from the Iraqi mainland 25 miles to a new supertanker loading dock in the Arabian Gulf. Having seen very little action since it was built in 1958, the *L.E. Minor* left Houston to much fanfare in the spring of 1961, towed by two tug-boats on its 100-day journey across the Atlantic, through the Mediterranean Sea and Suez Canal, and up the Persian Gulf to Iraq. Some 150 Brownbuilder welders, crewmen, and technicians followed the barges to the Persian Gulf, where they remained for several years on other marine pipeline jobs.[39]

The *L.E. Minor*'s voyage to the Persian Gulf marked a key moment in the history of Brown & Root Marine. The pipelaying barge, which until then had been a white elephant for the company, found a mission. Its capacity to handle giant pipe made it a natural for work in the Persian Gulf. Its activities throughout the region established a Brown & Root presence in the Middle

East where important contacts were made with other oil companies, such as Gulf Oil and British Petroleum.

Following the Basrah project, the *L.E. Minor* migrated around the Persian Gulf installing other submarine pipeline systems. The *Minor* laid 18.5 miles of 12-, 20-, and 24-inch line from the Arabian American Oil Company's (ARAMCO) Safaniya field to the coast of Saudi Arabia. Next came a major pipeline gathering system and loading line project at Das Island to service production from British Petroleum's Umm Shaif field, located in the offshore concession from the Abu Dhabi. This project was completed in four stages stretching between 1962 and 1966. As the vast field of reservoirs was developed, Brown & Root laid a total of 97 miles of pipe with diameters ranging in sizes from 6 to 26 inches.

The Das Island job presented special pipelaying problems because the pipe could not be buried on the bottom, which was extremely hard and covered with coral. Brown & Root engineers responded by adapting the traditional, round cement pipe coatings to these special conditions. Murphy Thibodeaux, working in the Marine Industries Division under Delbert Johnson, developed a special "trapezoidal weight-coating design" to hold the lines on the bottom and prevent them from flapping on the sea bed. In 1967, Brown & Root returned to Das Island to lay 107 miles of 6-inch to 30-inch diameter lines to connect the island with another major field development, the Zakum field. By 1967, the *Minor* had laid more than 400 miles of pipeline in the Persian Gulf, the Gulf of Oman, and the Mediterranean Sea.[40]

Some of these lines carried onshore production to deepwater tanker loading facilities. In 1961, for example, the *L.E. Minor* traveled from the Persian Gulf to the Libyan port of Marsa el Brega where Brown & Root built a tanker loading terminal and storage facility for Esso Standard of Libya. Esso's Marsa el Brega development consisted of a port area on 7,000 acres of dunes and desert with massive oil storage facilities and a marine terminal to accomodate large tankers. The *L.E. Minor* laid a large diameter oil pipeline 1.52 miles from shore out to a single-buoy mooring device (SBM), a concept developed by Esso engineers for mooring large tankers—in this case, up to 77,000 dwt.[41]

SBMs facilitated economies of speed and scale in the international transport of petroleum. An SBM system consisted of a cylindrical buoy with a mounted turntable to which mooring ropes were connected. During loading, oil flowed from the shore tanks through a submarine pipeline to the swiveling pipes in the body of the buoy, and then along floating hoses to the tanker manifold. By reducing the forces exerted on the buoy, the single-buoy system offered an improvement over conventional buoy methods of mooring a vessel, which typically were subjected to adverse loads from wind and waves that often caused anchoring difficulties. With a single-buoy system, the tanker was always moored with its head to the line of least wind, wave, and current resistance. Single-mooring buoys did have their drawbacks, however. There was only a very short period of correct ship alignment to the buoy during the mating. This created problems with passing the mooring ropes and handling the big hoses. The buoys themselves were not that secure in the water. Storm waves tore up the first SBM installed by Brown & Root at Marsa el Brega. Despite early flaws, the popularity of SBM's grew with the increasing prevalence of mammoth tankers.[42]

Brown & Root spent five years developing Libya's oil export industry. Reeves set up two subsidiary companies to handle the work. Based in Benghazi and Tripoli, these companies operated until 1965. Brown & Root developed production facilities, camps, marine terminals, and pipelines (both onshore and off) for oil companies operating in the Libyan producing regions. In a 1961–1962 joint venture with Arabian Bechtel Corporation, Brown & Root's Libyan team designed an SBM marine terminal and laid a loading line at As Sidrah for a consortium of companies operated by the Oasis Oil Co. Oasis had developed the Dahra field and pumped its crude to As Sidrah via an 87-mile pipeline. During 1963–1965, Brown & Root engineered another major SBM loading system for Mobil Oil of Libya at Ras Lanuf. As Ber Pieper said, "We had worked on all of those terminals up until the time we left, either by ourselves or in joint venture with Bechtel."[43]

As the decade of the 1960s unfolded, the size and geographic scope of Brown & Root's terminal and pipeline projects continued to expand. In 1965, the *L.E. Minor* set a record for laying the largest diameter pipeline in the deepest water. The Iranian Oil Operating Co., a consortium of 15 American, British, Dutch, and French oil companies, had expanded the terminal facilities at Kharg Island, located in the Persian Gulf 25 miles from the Iranian coast. The $50 million second-stage expansion of the terminal, whose first-stage construction began in 1960, included 11 new crude oil storage tanks with a capacity of 7–8 million barrels and a 6,000-foot-long jetty with 10 tanker berths (six for 100,000 DWT tankers and four capable of handling 35,000 to 65,000 DWT). To connect the terminal with Gawanah on the mainland, the *L.E. Minor* laid two 30-inch lines, each 28 miles long, in water up to 160 feet deep. Despite a three-knot current and unfavorable weather, Brown & Root installed the line in just three months. When completed in late 1966, the Kharg Island terminal was one of the largest single marine terminals ever built for loading crude oil. It averaged nearly 2 million barrels per day of throughput.[44]

Brown & Root set new records for large-diameter, deep water pipelines. In 1967, the *L.E. Minor* installed two miles of 40-inch pipe in depths of 130 feet in the Gulf of Oman on a job for the Petroleum Development Co. In 1968, the *Minor* laid and buried what was then the world's largest submarine pipeline, 48 inches in diameter, in 110–120 feet of water. The line ran from the port of Mina al Ahmadi, Kuwait, 10 miles offshore to a tanker-loading facility on an island for the Kuwait Oil Company, which was jointly owned by British Petroleum and the Gulf Oil Company.

J. Ray McDermott and Brown & Root competed for the project. "McDermott engineers said that it could not be done, that it was not physically possible to install a 48-inch pipe, it will collapse," recalled Adi Desai. Although Brown & Root was still developing the theoretical concepts to calculate the stresses and buckling behavior, "we looked at it in different ways, extrapolating, and all that, and we decided, yes, we can lay it." According to Desai, Hugh Gordon gave the engineers the "third degree," asking them if they were sure they could do it. Desai replied, "Well, I'm not sure, but we think we can do it." Brown & Root elected to take the risk.

Using a shorter stinger—to minimize problems with stinger handling and pipe bending—the line was laid without a hitch. Brown & Root imported more than 45,000 tons of pipe, equipment, and other materials from Europe, including heavy aggregate with a high iron content from Sweden to give the line added weight for negative buoyancy. Two crews of 100 working around the clock on the *Minor* laid an average of 61 joints, or 2,440 feet per day.[45] The Kuwait Oil Company line was a pioneering project. Before that, the industry considered a 48-inch line to be fantasy. Now it was a reality. The following year, the Brown & Root lay barge *BAR-265* traveled from the North Sea to Nigeria's Bight of Benin to lay another 48-inch line for the Shell-B.P. Petroleum Development Company.[46] Brown & Root later returned to Kharg Island to install 48-inch lines there as well.

Strapping barrels to 48-inch pipe on the *L.E. Minor* as the barge laid the line for the Kuwait Oil Company

Petroleum companies started ordering larger diameter pipelines to load the increasingly mammoth tankers faster and more efficiently. The Kuwait Oil facility was designed to load oil onto a new class of deep-draft "bigger-than-super" tankers, or Very Large Crude Carriers (VLCCs), with capacities up to 300,000 dwt. The Gulf Oil tankers would then carry their load to Whiddy Island in Bantry Bay, Ireland, where Gulf had constructed its giant Erie Terminal, designed to off-load 2.25 million barrels in a 24-hour period. Oil would then be loaded onto smaller tankers for distribution to Gulf refineries throughout Western Europe.[47]

In the winter of 1968, incidentally, Brown & Root came to Gulf Oil's rescue in finishing the Erie Terminal. The project had fallen behind schedule, due to

The 312,000 dwt *Universe Ireland* berthed at the Kuwait Oil Co.'s newly built sea island terminal. The tanker was on its maiden voyage in September 1968 loading crude to take to Bantry Bay, Ireland

harsh weather and difficulties in driving watertight piles for the terminal jetty into the hard slate and sandstone strata. Many at Gulf worried that the terminal would not be finished for the arrival of the first VLCCs from Kuwait in October 1968. Brown & Root brought its *Atlas* crane barge over from the North Sea to aid the installation of the terminal jetty. The *Atlas* was equipped with a big-hole, skid-mounted, packaged rig that was designed for soil conditions in Alaskan waters. This equipment drilled out the piles and drove them into the hard sea floor. In helping Gulf to complete its terminal on schedule, Brown & Root played an important role in hooking up both ends of a petroleum transport system vital to the economy of Western Europe.[48]

Project Management

By the mid-1960s, Brown & Root Marine had taken on highly diverse engineering and construction assignments around the world and in sundry kinds of natural environments. The pioneering offshore thrust of Brown & Root from the Gulf of Mexico had led to the development of innovative techniques and extensive experience in driving piles, installing drilling platforms and overwater gas compression stations, laying submarine pipelines, and constructing piers, docks, loading facilities, and marine terminals.

These capabilities enabled Brown & Root marine to begin offering total project management contracting, or turnkey responsibility. This meant serving as a single supplier for capital expenditures—designing the project, procuring equipment, workers, and supplies, and managing overall construction. Until then, the company typically contracted for either construction or engineering services. Some projects may have involved both, but not overall management. As John Mackin said, "Project management contracting was not a concept that was well known at Brown & Root . . . It was unheard of."

In 1965, the increasingly unified and fortified Brown & Root Marine first demonstrated its new project management capabilities. Amoco's middle eastern subsidiary, Pan American UAR, had discovered a big producer in the El Morgan field offshore of Egypt in the Gulf of Suez, about 75 miles south of the canal. Amoco and its partners, Gulf Oil and Egypt's General Petroleum Corporation, brought in Brown & Root to design and construct an $18 million offshore/onshore oil exploration and development complex. "We had four international contractors and maybe six to eight Egyptian contractors on the project to coordinate and direct," remembered A.R. Jackson, Brown & Root's project manager on the project.

El Morgan also was the first project for Brown & Root that combined developing an offshore field with offshore tanker loading facilities. John Mackin, head of the Marine Industries Group at the time, noted "El Morgan created a province for us in the overseas offshore field." Working on a fast and tight schedule in 1966 with war approaching between Egypt and Israel, construction crews installed self-contained drilling and production platforms in 100 feet of water connected by pipeline to shore. Onshore facilities included a major tank farm (including the first ever 1 million barrel tank) sitting up on a hill for gravity loading, a small boat harbor, a town, and an airport. Brown & Root linked the tanks via another submarine pipeline to two multiple-point SBMs, one for 100,000 dwt tankers and the other for 200,000 dwt tankers. Brought on stream according to schedule, the El Morgan field initially produced 200,000 barrels of oil per day. In Mackin's words, "it turned out to be Amoco's cash cow for year after year after year," even into very recent times.[49]

Brown & Root subsequently completed other significant offshore/onshore project management jobs. In 1965–66, Brown & Root managed the design and construction of the Lavan Island Complex for the Lavan Petroleum Company's (a consortium of the National Iranian Oil Company, Atlantic Sun, Union, and Murphy) Sassan field located 90 miles off the coast of Iran. The complex included production platforms, a pipeline, and a supertanker terminal to load 250,000 dwt tankers, "which at the time," Mackin said, "was very close to being as big as they could make."[50] In 1967–1968, Brown & Root managed a 12-month "grass roots" program to fabricate and install an offshore production and gathering system and an SBM supertanker terminal for Gulf Oil in the Cabinda Gulf off the coast of Angola. In doing so, Brown & Root aided the development of new and "safer" sources of oil in West Africa as problems touched off by the Six-Day War and the growing solidarity of OPEC challenged the control exerted by the major oil companies over world oil supplies.

Brown & Root also continued to help OPEC nations in the Middle East exploit their offshore potential. During the 1970s, the Persian Gulf experienced an upsurge in offshore activity. The first round of oil price increases quadrupled the revenues of oil-rich states almost overnight. Some of this new-found wealth was channeled into developing large offshore fields. Brown & Root established a fabrication yard in Manama, Bahrain (in joint venture with George Wimpey & Co.) to serve this lucrative market. Although most projects were not technically challenging from the standpoint of marine construction, Brown & Root built major offshore installations for Saudi Arabia, Iran, and Abu Dhabi that added large quantities of crude oil to world supplies.

The size and quality of Middle Eastern fields, both onshore and offshore, assured that this region would emerge as the center of world oil production. Brown & Root and the offshore construction industry hastened the day when this oil entered world markets. In 1969–1970, Brown & Root helped link up Middle Eastern crude with Asian markets by managing the design and construction of Gulf Oil's giant Asian Trans-Shipment Terminal at Okinawa. The company mobilized equipment from all over the world for this terminal, which was built to offload supertankers and reload oil in smaller tankers destined for refineries in Korea and Japan.[52] In the late 1970s, Brown & Root also helped streamline the entry of Middle Eastern and other foreign oil into the United States by designing and building the pumping platform complex for the Louisiana Offshore Oil Port (LOOP), off Freeport, Louisiana. The $800 million LOOP terminal was the first application in the United States of single-point mooring technology and the first U.S. facility capable of handling VLCCs.[53]

By the 1970s, the offshore industry had become truly global. The demand for offshore construction around the world rose considerably in the late 1960s and exploded after the OPEC embargo. One important new area was the Far East, where concessions offshore Indonesia and Malaysia were being explored and developed. Brown & Root gained a foothold into this market by setting up two new fabrication yards: one on Labuan Island off the western coast of Sabah in the South China Sea, and the other on the west end of the Indonesian island of Java on the Sunda Straits. These yards added to Brown & Root's growing capacity to offer a full range of engineering and fabrication services to oil companies operating in the world's most active offshore regions. Brown & Root found extraordinary opportunities in transferring Gulf of Mexico technologies and experience to new offshore provinces around the world. Through overseas work, Brown & Root established new contacts and developed greater organizational capabilities for large projects and technological challenges. Contracts with Europe-based Middle Eastern clients, such as Gulf Oil and British Petroleum, opened the door for Brown & Root's involvement in the North Sea and Alaska. A far-flung network of joint ventures, foreign offices, subsidiary companies, contracting alliances, fabrication yards, and barges enabled Brown & Root to tackle projects just about anywhere and under the most difficult conditions. The company's success at solving engineering and construction problems posed by offshore work in the shallow, calmer waters of the Gulf of Mexico, Lake Maracaibo, and the Persian Gulf paved the way for its future success in meeting the more extreme challenges posed by cold weather, ice, earthquakes, rugged waters, and ever greater depths.

Notes

1. "A $2.3-Million A–E Fee Can Look Small," *Engineering News-Record* (November 15, 1962): 80.
2. This chapter draws from interviews with W.R. Rochelle, H.P. Smith, Basil Maxwell, David Smith, R.O. Wilson, John Mackin, W.B. Pieper, Tim Pease, B.E. Stallworth, A.R. Desai, Delbert Johnson, and A.R. Jackson.
3. T.R. Goedicke, "Areas of Interest in Marine Oil Prospecting," *World Oil* (November 1956): 117.
4. Daniel Yergin, *The Prize: The Epic Quest for Oil, Money & Power* (New York: Touchstone, 1991), 434.
5. G.A. McCammon, "Concrete Pile Foundations for Over-water Drillsites," *Drilling* (October 1958): 69–71; Raymond-Brown & Root C.A., *Maracaibo Offshore Services for the Oil Industry,* brochure, nd.
6. "Marvelous Maracaibo Remains World's Most Prolific Overwater Producer," *Offshore* (August 25, 1959): 34–35.
7. R.B. Leathers, "World's Largest Submarine Gas Line Boosts Lake Maracaibo Conservation," *World Petroleum* (July 1959): 44.

8. Brown & Root, "Compressor Stations—Gas Conservation," Marine Division pamphlet in company possession.
9. Brown & Root, Inc., *Gas Compression Facilities Experience* (Brown & Root, April 1983); Brown & Root, "Compressor Stations—Gas Conservation."
10. Brown & Root, Inc., *Gas Compression Facilities Experience;* Brown & Root, Inc., *Services and Experience in Gas Compression* (Brown & Root, 1970).
11. "There Is Where the Oil Is," *Brownbilt* (Fall 1972): 12; Interview, Bill Stallworth, October 4, 1995. Stallworth claims Brown & Root laid its first pipeline in Maracaibo in 1956. *Brownbilt* says 1954.
12. "Brown & Root Barge Lays the Super Inch," *Offshore* (September 1958): 46; Leathers, "World's Largest Submarine Gas Line," 45–46.
13. "40-Inch Gas Line," *Offshore* (May 1958): 32; Leathers, "World's Largest Submarine Gas Line," 45.
14. R.O. Wilson, Project Book, List of Derrick and Rig Moves.
15. "Marvelous Maracaibo Remains World's Most Prolific Overwater Oil Producer," *Offshore* (August 25, 1959): 34.
16. "Venezuela Venture," *Offshore Drilling* (September 1957): 17–19; "Exploration and Drilling in Venezuela," *World Petroleum* (July 1957): 81–84; and "Lake Maracaibo," *Drilling* (August 22, 1958): 37–38.
17. "Loffland Operates First Mobile Drilling Platform on Lake Maracaibo," *Offshore* (April 1961): 16–17.
18. "Aluminum Drilling Platforms," *Offshore* (May 1958): 23; "An Independent's View of Maracaibo—Superior's Story," *World Petroleum* (July 1958): 58; "Steel Price Hike Adds to Offshore Contractors Burden," *Offshore* (September 1958): 27–28.
19. "Aluminum Work Boats," *Offshore* (May 1958): 17; "Aluminum Pipelines," *Offshore* (May 1958): 19–22.
20. *Engineering News-Record* (August 20, 1956), quoted in Raymond-Brown & Root, *Maracaibo Offshore Services for the Oil Industry;* Gerald A. O'Connor, "Long Concrete Cylinder Piles for Deep-Water Platform," *Civil Engineering* (April 1963): 50–51.
21. "Concrete Pile Supports," *Offshore* (July 1958): 37.
22. "Mene Grande Lets Contract for 50 Offshore Platforms," *Offshore* (July 1958): 72.
23. "Concrete Pile Supports," 37–40.
24. Ed McGhee, "New Maracaibo Gas-Compression Platform Marks Design Milestone," *Oil & Gas Journal* (September 13, 1965): 90–92.
25. "Unigas I," *Brownbilt* (Second quarter, 1966): 19.
26. "Unigas I," 19.
27. McGhee, "New Maracaibo Gas Compression," 91–92.
28. "Unigas I," *Brownbilt,* 19–20.
29. Brown & Root, *Gas Compression Facilities Experience;* "Venezuela," *Brownbilt* (1967–1968): 19.

30. "The Voyage of Menegas I," *Brownbilt* (Summer 1969): 4–6; Alvaro Franco, "New Maracaibo Plant Called 'Most Advanced'," *Offshore* (December 1970): 64–67.
31. "Free-world Offshore Production by Nations," *Offshore* (June 20, 1971): 43.
32. H.W. Reeves, Proposal, "Marine Oil Terminal for Rio de Janeiro, Brazil" (1958), Brown & Root project books.
33. "Offshore Operations in Brazil," *Offshore* (July 1960): 13–18.
34. Brown & Root, "Field Investigation Report Concerning Guanabara Terminal," submitted to Petrobrás, April 22, 1959.
35. R.O. Wilson, Project Book.
36. Preston P. Nibley and Alan C. Nelson, "Economics of Oil Transportation Middle East to Western Europe," *World Petroleum* (November 1961): 56.
37. "New Concessions Spark Offshore Oil Search in Persian Gulf," *World Petroleum* (February 1958): 31–33, 64–66; "Persian Gulf," *Offshore* (August 22, 1958): 45–46; "Parade to the Persian Gulf," *Offshore* (August 25, 1959): 43–45; "Offshore Production to Reach One Million Barrels Daily in Persian Gulf This Year," *World Petroleum* (July 1964): 29–31
38. Michael J. Wells, "Offshore Finds Expand Middle East Oil Potential," *World Petroleum* (July 1966): 20.
39. "En Route to Persian Gulf," *Offshore* (April 1961): 19.
40. "Middle East," *Brownbilt* (1967–68): 12.
41. "Libya Enters the Oil Age," *World Petroleum* (December 1961): 50.
42. Brian C. Hague, "Single Buoy Moorings Can Help Handle the New Mammoth Tankers," *World Petroleum* (November 1969): 26–28, and Part II, (December 1969): 56–62.
43. "Libya's Export Facilities Expanded," *World Petroleum* (February 1963): 34; "Middle East," *Brownbilt* (1967–1968): 12.
44. Russell A. McNutt and Joseph A. Ferenz, "Kharg Island Terminal One of World's Largest," *World Petroleum* (July 1967): 48–57; "Middle East," *Brownbilt* (1967–68).
45. "Middle East," *Brownbilt* (1967–68); "World's Largest Submarine Pipeline," *Brownbilt* (Summer 1968): 18–20.
46. "Second 48-Inch Pipeline Laid Offshore," *Brownbilt* (Fall 1969).
47. Michael J. Wells and A.W. Lindsay, "Bantry Bay—A Difficult Project Nearly Completed," *World Petroleum* (September 1968): 35–41.
48. Wells and Linday, "Bantry Bay—A Difficult Project Nearly Completed," 35–41.
49. Michael J. Wells, "Offshore Finds Expand Middle East Oil Potential," *World Petroleum* (July 1966): 21; Mackin Interview; Brown & Root, Marine Service Brochure, (1970).
50. "Lavan Island Complex on Stream," *Brownbilt* (Winter, 1968–69): 20; Mackin Interview.

51. "Gulf Prepares to Produce Oil off Coast of Cabinda," *World Oil* (June 1968): 111–114; "Cabinda Gulf . . . On Stream and Continuing," *Brownbilt* (Summer 1969): 22–24.
52. "Asian Trans-Shipment Terminal," Brown & Root Marine Brochure (n.d.).
53. "America's First Superport Preparing 1981 Debut," *Brownbilt* (Fall 1980): 17–21.

CHAPTER 7

Mind Stretcher of the Century: Project Mohole

Project Mohole was a U.S. government-sponsored effort in the late 1950s and early 1960s to drill down to the earth's lower crust and upper mantle to learn more about the interior composition and geologic history of the planet. Mohole produced technological and scientific innovations vital to the offshore industry and to the nation as a whole. During the 1960s, many segments of the offshore industry worked with Brown & Root on the project to develop path-breaking drilling technology capable of exploring a realm familiar only to science fiction, the earth's crust. In the process, Brown & Root Marine gained valuable project management experience. As manager of Project Mohole's Phase II, Brown & Root showed off its skills in offshore and onshore engineering and design, and demonstrated how commercial technology can be applied, modified, refined, and invented for scientific study.[1]

Project Mohole captivated the imaginations of both Herman and George Brown. W. H. Tonking, Brown & Root's deputy project manager on Mohole, said that Herman would "come by every Monday morning to get a rundown on the Mohole. . . . He was really interested in this thing. It was his deal." Herman, however, did not live to see either the successes or failures of Mohole. He died of an aneurysm in November 1962. George Brown carried on his brother's passion for the endeavor. As Rick Rochelle pointed out, Project Mohole epitomized the old-style determination, instilled in the company by these two men, to tackle the most difficult technical challenges. It was a matter of Herman and George Brown saying, "Look guys, just forget anything that anybody ever told you and sit down here and say, if you had to do this, how in the hell would you do it? You tell me."

Mohole contemplated something almost unimaginable for the time—drilling in 15,000 to 18,000 feet of water, and then through another 25,000 feet of the earth's crust to the mantle. The maximum water depth drilled in those days, it must be remembered, was only about 200 feet. "So, 18,000 feet," said Adi Desai, "was just completely out of this world." Tonking called Project Mohole "a hell of an engineering problem with a scientific payout. It was no good just to drill a hole. We had to get all of the science out of that hole, too, so this made it more complicated and also increased your costs, your staff and everything." Although politics, scientific bickering, and the Vietnam War brought an end to it in 1966, Project Mohole led to engineering and design innovations that benefited both the study of geophysics and the commercial offshore drilling industry.[2] "I think it was the mind stretcher of the century in marine work," said Rochelle.

The Genesis of Project Mohole

For ages, earth scientists had speculated about the interior composition of their planet. In the early twentieth century, Yugoslavian seismologist Andrij Mohorovčić observed that seismic waves generated by the 1909 Croatian earthquake travelled through the earth at two speeds. One set of waves travelled at a speed the seismologist could calculate based on the familiar composition of the earth's crust. The second set of waves travelled faster than the first, leading Mohorovčić to postulate that the faster waves travelled through a layer of rock more dense and elastic than that found in the earth's crust. Mohorovčić theorized that this rock lay underneath the crust. Later observations corroborated these findings.[3]

In honor of the discoverer, the boundary zone between the crust and the lower layer, or mantle, was named the "Mohorovčić Seismic Discontinuity," Moho for short. Earth scientists theorized that the earth, about 8,000 miles in diameter, consisted of a crust, a mantle, and an inner and outer core. The crust averages about 125,000 feet (25 miles) in thickness; it is thickest beneath mountains and thinnest, about 20–35,000 feet (four to seven miles), beneath the oceans. Geologists estimated the mantle at 1,800 miles thick. The mantle extends to about half-way to the earth's center and constitutes about 80 percent of the earth's volume. The earth's core is divided into an outer core (1,300 miles thick) and an inner core (800 miles thick).

"Mohole" was coined to describe a hole drilled beyond the Moho. A 1956 *Science* article suggested that greater geophysical data could be acquired through drilling such a hole. The following year, the Advisory Panel for Earth Sciences of the National Science Foundation (NSF) displayed interest in expanding earth science research. Members of the advisory panel, in turn, consulted with members of the American Miscellaneous Society (AMSOC), a group of earth scientists who were interested in areas of research that traditional science placed outside of general consideration.[4] AMSOC endorsed the plan to drill deep into the earth. It established a committee in conjunction with the NSF and served as the chief scientific advisor to the NSF on Project Mohole from 1958 through January 1964.[5]

Any attempt to drill down to the mantle had a greater chance of succeeding through a thinner part of the crust under the ocean. No onshore drilling venture even approached the mantle. The deepest well was in Texas, a dry hole drilled in 1958 to a depth of 25,340 feet. This well penetrated deep into the earth's crust, but the hole did not come close to reaching the mantle. Early plans for Project Mohole called for drilling in 12,000 feet of water and then drilling through an additional 23,000 feet of the earth's crust to the mantle. Even though ocean drilling would be more feasible than drilling onshore, nobody had ever drilled in waters close to that depth or tried to put 35,000 feet of drill pipe together. "Should Project Mohole succeed," wrote *National Geographic* in 1961, "it will stand as one of the 20th century's most significant scientific achievements."[6]

The primary objective of Project Mohole was removing samples of the upper mantle.[7] These samples, it was hoped, might yield information on the origin and age of the earth and its oceans, the composition of ocean basins and continents, temperature trends within the earth, causes of earthquakes and volcanoes, the nature of the earth's magnetic field, and plate tectonics.[8] A major scientific tie-in of the project was the scheduled drilling of the Mohole simultaneous with the first moon landing. One theory of the planet's evolution, noted Tonking, suggests that the moon had spun off the earth into orbit. Comparing samples from the moon and from the mantle, some believed, "could have advanced the theories of the origin of the earth and maybe even as far as the origin of life."

Project Mohole presented fantastic opportunities, not only for scientific inquiry, but also for the engineering and design of offshore drilling equipment. It promised to deliver "technological byproducts" that would allow marine drilling to venture deeper and farther offshore than ever before. Over the course of the project, engineers sought to develop a practical semisubmersible drilling rig equipped with a dynamic positioning system, drilling systems that could withstand the vibrations and stresses that would be placed on long strings of drilling pipe, a wireline retractable diamond bit, a subsea riser and buoy system, a deep water hole re-entry system, modifications to conventional pipe handling systems, and larger logging units with improved cable. Project Mohole also addressed specialized drilling issues such as developing a fluid-powered turbocorer, improving drilling bits and methods of handling long strings of pipe, experimenting with special alloy materials for the pipe, and devising sophisticated instrumentation.[9] "Because of the water depth, because of the length of the hole that we were going to drill, and the environment that we were going to do it in," Desai said, "just about every subsystem of the project required a totally new concept or development."

Project Mohole was planned as a \$50 million, five-year program and organized into three phases. The first phase called for a test drilling program and a search for possible drilling sites. The second phase involved constructing the drilling platform and the drilling itself. The third phase would carry out studies on the samples of crust and mantle. As a major federal government project, of course, political controversies and bureaucratic obstacles hindered the project all along the way, from the establishment of scientific priorities to the awarding of subcontracts. Scientists had their own pet projects to push, and some felt that the study of seabed sediments should receive higher priority than drilling to the mantle. Politicians lobbied to have contracts awarded to firms in their districts, and Congress repeatedly held up disbursing funds. By the end, Project Mohole had earned a reputation as a classic example of the "the politics of science."

Phase I

Phase I of Project Mohole—in which Brown & Root did not participate—lasted from 1958 to 1961. The NSF provided funds to AMSOC to carry out Phase I and evaluate the feasibility of drilling from an unanchored vessel in deep water. Authorization for the project gave the NSF responsibility for funding and oversight, while AMSOC assumed an advisory and managerial role.[10]

In late 1960, AMSOC contracted with Global Marine Exploration Company to begin test drilling. Global Marine converted its drillship, the *CUSS I* (an acronym of the original owners: Continental, Union, Superior, and Shell Oil companies), for deep sea drilling. The conversion added a positioning compensation system, which made the vessel more stable by preventing the long drill string beneath the ship from buckling when the ship pitched in rough seas, and an automatic horizontal pipe racking system, which expanded the vessel's pipe storage capacity.

Trial drilling began in April 1961 off the coast of La Jolla, California, in 3,139 feet of water. Following this test, AMSOC relocated the vessel to a deepwater site 40 miles east of Guadalupe Island, about 220 miles south of San Diego off the west coast of Mexico. There the *CUSS I* drilled five holes to a maximum depth of 600 feet in basaltic rock beneath 11,700 feet of water. The only connection between the drilling vessel and the drilled hole consisted of 11,700 feet of 4 1/2 inch drill pipe. No conductor pipe or other lateral supports were used. The *CUSS I* normally maintained its position by means of an anchoring system. But for these tests, the new dynamic positioning system, consisting of four steerable 2,000-hp outboard motors controlled by radar- and sonar-sensing equipment ranging off fixed buoys, held the ship in place.[11]

The test drilling of Phase I produced encouraging results. The cores contained substantial fossil evidence from the Middle Miocene age (15–20 million years ago), and scientists observed that ocean life in the Miocene period was much richer than the ocean life of modern times. Furthermore, two of the holes drilled at the Guadalupe site penetrated, for the first time, the "second layer" of the earth's crust, which was composed of basalt lying beneath about 500 feet of subsea sedimentation.[12]

With Phase I test drilling completed and the exhaustive, and often contentious, site selection of Phase I still in process, the NSF decided to move forward with Phase II. This is when Brown & Root entered the picture.

Phase II

In the summer of 1961, the NSF took prime responsibility for the completion of Project Mohole. Overburdened by its "heavy operational" role with AMSOC in the project, the National Academy of Sciences had persuaded the NSF to hire a prime contractor, most likely an oil firm or offshore drilling company, to manage Phase II. From this point forward, AMSOC was relegated to a purely advisory role in Mohole. AMSOC team members' subsequent demands for more input in the project and control over the scientific agenda led to continual controversies over the next years.

Many companies were interested in managing Phase II of Mohole and participating in such a heralded technological undertaking. Representatives of eighty-seven firms attended the NSF's first informational briefing to potential contractors. The NSF announced a call for proposals to be submitted by September 11. At the last minute, George Brown decided that Brown & Root should submit a proposal, even though the company had not attended the briefing and was not on a list of proposers. Within two weeks, the proposal was ready to go. The proposal emphasized Brown & Root's location and long association with offshore oil operations. Brown & Root's pitch, according to Dick Wilson, who with Tim Pease drafted the proposal, was that "we were sitting in Houston with Rice, Texas A&M, and University of Texas nearby, so we had access to a large amount of help."

The NSF received twelve proposals, including Brown & Root's, and ranked Socony-Mobil as the top bidder. Brown & Root placed fifth. Only the top five bidders received word that their proposals merited additional consideration. Brown & Root was still in the running, but barely. The NSF then scheduled meetings with representatives of each of the five firms to go over the proposals.

Brown & Root's enviable record in the design and construction of large-scale installations on- and offshore impressed the NSF and eventually proved decisive. At Brown & Root's presentation to the NSF on January 16, 1962, in Washington, George Brown made a strong case for Brown & Root's qualifications to mobilize the expertise, manage the subcontracting and site selection, and develop drilling equipment and techniques to advance the project. Based upon an evaluation stressing a wide range of criteria—management and policy, technical and scientific experience and capability, support facilities available or readily obtainable, financial capability, and comprehension and soundness of approach—the NSF chose Brown & Root as prime contractor for Phase II of Project Mohole, which was announced officially on June 20, 1962. (An NSF spokesman later said that another reason Brown & Root was awarded the contract was to "keep any of the oil companies from getting a competitive advantage [over the others]."[13]) The contract stated that

Brown & Root would "furnish or procure all services, material and facilities necessary for the drilling, sampling and logging of a hole through the crust of the earth, for scientific purposes, at a site to be selected in collaboration with the National Science Foundation." The original estimated cost for Phase II was $46,728,000, paying Brown & Root's cost, plus a fixed fee of $1.8 million.

Critics objected to the NSF's decision. Senators Thomas H. Kuchel (R-CA) and Gordon Allott (R-CO) charged that Brown & Root did not have the necessary scientific experience and capabilities. However, George Bush, then president of Zapata Offshore Company and future president of the United States, said, "I think Brown & Root is perfectly able."[14]

Brown & Root made up for what experience it did lack by assembling an impressive team of experts and engineers for the project. The company organized a separate Mohole division and placed 50 people on the staff. "We needed a bunch of expertise in a bunch of areas," said Tonking. "We went out and hired the best engineers and supporting staff that we could find." Adi Desai remembered that the influx of experts and consultants with their creative ideas made the project very exciting. "We used to have lots of outside people come in with ideas about how to do it this way and do it that way. . . . It was really fun going over some of these wild schemes." By the end of the project, 250 people worked for Brown & Root's Mohole division, not including outside consultants and subcontractors. The Mohole team was told that "if you need somebody on the other side of the earth, go see them, and if you learn something from them and they tell you to see somebody else, go see them, but put it all together."

Putting it all together was a difficult and "mind-stretching" task. As Tonking pointed out, numerous planning and operational problems had to be defined and then studied and solved. Brown & Root's Mohole team faced difficult decisions and tasks at every turn, from selecting a drilling location, to determining what kind of drilling vessel to use, to resolving technological and political problems. Brown & Root assumed overall management of Phase II and subcontracted out specialized parts of the project to the most qualified and capable firms. "We had a lot of these jobs where we had outside people running analyses, like at the major universities," said Rochelle. "They'd have these damned computers that would fill half a building, analyzing things like the interaction of the drill string and the casing under multi-dimensional actions of the sea and the drilling vessel." Making the technological leaps called for by Project Mohole required a cooperative effort among many of the nation's most important technology firms, including General Electric, Honeywell, and Schlumberger, among others, who participated in the planning, testing, design, and fabrication of equipment. "It was not a question of extending the frontier," as Adi Desai put it. "It was a question of just having a new frontier totally."

Selecting a favorable drilling site was one important task that dragged out for several years. "We spent considerable time and effort on doing geophysical work in both the Atlantic and Pacific," said Tonking. The list of potential sites was eventually narrowed down to two: one near Antigua at the Puerto Rican Outer Ridge and the Barracuda Fault, which ranged in water depth from 12,000 to 18,000 feet; and the other in 14,500 feet of water in the Pacific Ocean along the Hawaiian arch, northeast of the island of Maui. Ultimately, geophysicists selected the Maui site, which benefitted from a shorter depth to mantle and closer proximity to land and support facilities. Antigua is also in the hurricane path, which made it less attractive. The NSF announced the selection of the drilling site in January 1965, and expected drilling to begin in late 1967.[15]

At the outset, the Mohole team faced a complicated decision regarding the type of drilling vessel for the project. The AMSOC team promoted the idea of using a converted T-2 tanker as the drilling ship, but the project staff of Phase II considered other designs. Maintaining the endurance of the drill pipe and motion characteristics of the platform in unprecedented water and hole depths required a vessel whose heave, roll, and pitch could be minimized to a degree not yet achieved with conventional drillships. Naval architects and marine engineers evaluated all kinds of vessels, but the decision came down to a choice between a ship-shape drilling vessel and an experimental semisubmersible type rig. "Much of that first year's activity," recalled Delbert Johnson, "was spent evaluating the relative advantages of either the ship-shape versus the semisubmersible type structure."

Brown & Root recommended a semisubmersible design, which the NSF accepted. A semisubmersible platform, it was determined, would move less in response to waves and be easier to position over the drill hole. According to Tonking, "No way could we do it with the ship's hull out in an ocean of unprotected waters, and for a long period of time." Mohole engineers then designed a column-stabilized semisubmersible platform with twin lower hulls, 35 feet in diameter and 390 feet long. They modeled their design on a four-column platform built by the Bluewater Drilling Company. "We used that to start our design and then we modified the hell out of it and finally ended up with six columns and with two real submarine lower hulls." The lower hulls supported an upper hull on which the drilling platform rested. The platform design consisted of three deck levels and a double bottom. During drilling, only the six columns would experience wave action, since the lower hulls were fully submerged. The height of the platform from the baseline to the upper deck was 135 feet, and the drilling derrick extended 240 feet above the deck for a total height of 375 feet. At a light draft of 28 feet, the designers calculated that the platform displaced 13,500 tons, and at a draft of 60 feet, it displaced 21,500 tons. The original proposal called for a maximum crew population of 120, later increased to 160. The vessel included a laboratory and storage for 2,000 tons of fuel, 1,200 tons of drilling water, and enough provisions to sustain 60 days of maintenance and operations.[16]

In February 1963, Brown & Root unveiled the semisubmersible, six column-stabilized platform design at the project headquarters in Houston before the Naval Architecture Panel.[17] The platform would cost an estimated $40 million to build and $9 million to operate. This platform presented a novel solution to many of the problems related to motion in deep sea drilling, and one that provided a prototype for future semisubmersibles. Brown & Root's Hugh Gordon later noted that the "original concept of the semisubmersible came out of our old Mohole project before that was killed, that's where it started." Another Brown & Root engineer, Clyde Nolan, said that the Mohole Platform design "was the forerunner of the design that was used for the semisubmersibles by The Offshore Company later."

Still to solve was the challenge of devising a method to keep the semisubmersible stable for a long time in a relatively constant position above the drill pipe and hole. Tonking explained that "We had to come up with some kind of a system to dynamically position the ship. We calculated it would take us about three years to drill the hole because it had to be cored all the way from the ocean bottom to the mantle. Therefore, it had to have the ability to keep its station through some pretty rough weather." Designing a system to keep the vessel in a virtually stationary position for three years in the middle of the ocean was no small assignment. Indeed, it was no more than a theoretical concept at the time. "It never had been proven that you could actually stay on location by just having a bunch of engines keeping you on location without any wires," said Desai.

Brown & Root awarded a contract to develop an automatic dynamic positioning system for the vessel to the Honeywell Corporation, which selected a positioning system that incorporated three baseline sonar systems. Sonar transceivers suspended from the vessel would receive position signals from bottom-mounted transponders. Positioning computers would then solve the geometric problem, determine the platform's location, and then calculate the thrust magnitude and direction needed to keep the unit in position. The computer would send the appropriate thrust commands to electrically powered underwater propulsion units.[18]

One set of propulsion units powered the dynamic positioning system; a second set drove the vessel across the ocean. Engineers estimated that the platform needed to remain within a 500-foot radius circle in 18,000 feet of water for the entire three-year drilling program. To accomplish this remarkable feat, six steerable 7,500-hp right-angle drive propellers, or thrusters, were located below each of the columns. The main propulsion system contained two 6,000 hp engines located at the stern of each hull. These could power the platform at a cruising speed of 8–10 knots. Improved versions of the Mohole system for dynamic positioning and propulsion remained in use long after the project ended.[19]

As the conceptual design of the rig progressed, Brown & Root brought in other subcontractors during 1963 to make detail designs. General Electric analyzed the vibration features of the design, evaluated the effects of waves with respect to vibration frequencies, and made model test recommendations based on these investigations. They developed mathematical expressions to determine the motions of the platform and the forces causing these motions

Model of semisubmersible drilling platform planned for Project Mohole

in all directions. Following General Electric's tests, the U.S. Navy's David Taylor Model Basin, and other model basins at the Stevens Institute and the University of Michigan, conducted model testing. Results from these tests indicated the kinds of adjustments needed to be made in the platform design.[20] In early 1964, Brown & Root contracted with Gibbs & Cox, Inc. of New York, a leading naval architectural firm, to prepare specific plans and specifications for the final platform design.

In early 1965, Brown & Root invited bids from eleven shipyards to build the platform. A west coast firm, National Steel and Shipbuilding, received the $30 million contract for the fabrication of the semisubmersible at its San Diego shipyard.

Although the technical work looked promising, political and bureaucratic hassles began to frustrate progress. "We were on a dog and pony show," said Tonking. Congressional representatives made constant requests on behalf of constituents vying for some of the hundreds of subcontracts let on the project. Relations between Brown & Root and AMSOC had been adversarial from the beginning. Some AMSOC engineers had favored the Antigua drilling site over the one selected near Hawaii, or they still preferred an ocean sediment drilling program over Mohole drilling. They also had pushed for the drillship option over the semisubmersible. AMSOC's marginalization from decision making created tensions. Although the committee was disbanded in 1964, Brown & Root continued to have problems with Washington officials and had to resolve quarrels over the project's scientific goals. Brown & Root's project team chafed at the oversight, auditing, and reporting requirements imposed by Washington. Adi Desai remembered monthly progress meetings lasting three or four days during the latter stages of the project. "It took us two weeks to prepare for the monthly meeting and the presentations," he said. "You spent a week up there and you'd come back, spend another week writing minutes of the meeting and it ends up that you do nothing else but prepare for the meeting, go to the meeting, write about the meeting, and you're back through the cycle. It was that absurd."

Meanwhile, Brown & Root faced a succession of difficult technological challenges. One such challenge emerged during the initial test drilling program. During Phase I test drilling, the crew was unable to place the drill string back in the original hole after removing it. Hole re-entry became one of the most vexing problems confronting the project staff. To solve it, they designed a special funnel-shaped landing base for the drill hole opening. From the hollow core of the drilling pipe extending just above the bit, a sonar transducer would search for the precise location of the landing base. On the platform at a control panel, the location of the bit in relation to the base would be projected onto a monitor. Above the bit and built into the drill pipe casing, a high-powered water jet activated and controlled from three miles above on the platform would then move the bit toward the funnel-shaped landing base. The control operator would adjust the position of the jet so that it could slowly direct the bit toward the funnel. A drilling crew successfully tested this procedure to reinsert a bit into a drilling hole. Subsequently, some offshore oil operators investigated this method for hole re-entry in commercial applications.[21]

Hole re-entry was, so to speak, just one end of the problem. An NSF report summed up the challenges facing the drill pipe design team: "No one has ever suspended six miles of steel pipe—casing or drill string—from an unanchored vessel before, and engineering problems involved in doing this are tremendous."[22] The stresses placed on the drill string by the ocean were likely to shake it to pieces. It was not clear how 35,000 feet of pipe weighing as much as one million pounds would react to the forces placed upon it, by its own weight and wave motion, from the top of the derrick tower through the platform's rotary table and down through the ocean. In addition, the riser had to be strong enough to withstand internal pressure from drilling fluids, the cold deep-ocean temperatures, frictional effects of the rotating drill pipe, as well as ocean currents and movements of the platform. "If I had nightmares about anything on Project Mohole," said John Irons, a member of Mohole's drilling staff, "it was how we were going to keep that riser intact while we drilled."

To address these problems, engineers added buoyancy to the riser. They developed passive buoys of syntactic foam made of glass micro balloons set in an epoxy matrix. The buoys, spaced at intervals along the long pipe, supported 90 percent of the weight in water. Variable displacement buoys offset the great weight of the drilling fluid circulating inside the riser. These buoys also kept the riser straight by adding tension to it.[23] Clyde Nolan remembers that "the tension on the riser itself was a big factor to help you, not so much to pull the vessel back into position but to keep the riser in the position that you wanted it, instead of letting it sag too much in one area or the other." In the event of an emergency abandonment, a back-up buoy could be inflated in a matter of minutes by an underwater gas generator producing a positive buoyancy greater than 400,000 pounds.

New kinds of drill pipe and pipe racking systems also had to be found to handle the heavy weight and miles of pipe that would be needed. The weight of the drill pipe was tapered. Tonking said, "We had the heaviest pipe at the top, of course, and the lightest pipe down at the bottom of the hole." Researchers studied the merits of drill strings made of lightweight metal alloys, such as titanium, and high-tensile yield steels. Rick Rochelle led the team that designed a semiautomatic system to rack the pipe horizontally, so that the tremendous weight of the pipe on the platform did not adversely affect its stationing. Rochelle remembered, "I was brought in to design their pipe rack structure. We redesigned that pipe racker so many times, I felt like it had worn out the steel." The one million pounds of pipe had to be racked differently than on traditional commercial platforms. When pipe sections returned to the platform they had to be placed so that the pieces could be fit back together. "In case we did get off of station for a while, we had to have a way to handle that drill pipe expediently and swiftly to get it back in the horizontal position," said Irons. "We could not have survived with it in the derrick moving the vessel around." Because the pieces were different sizes and weights, the pipe had to be carefully identified. An early version of the modern bar code and scanner technique made this possible.

The drilling itself raised concerns about how the operation would cut and recover a continuous core from the ocean floor to the mantle. Drilling time greatly increases as the bit penetrates deeper into harder rock. For example, drilling from 5,000 to 6,000 feet typically could take from a half day to more than ten days. Between 20,000 and 21,000 feet, typical drill time increases to 38–133 days. At the recording-setting 25,000-foot Texas hole in 1958, the drill bit took one year to reach 18,000 feet. Drilling progress then slowed down considerably. The remaining 7,340 feet of hole took more than a year to drill.

The Mohole would certainly increase drilling time by another order of magnitude as the drill bit reached super hard, very deep rock. As Irons noted, "We were sure that on a project of that size and that depth, we would probably encounter formations that you normally don't get into on the everyday, run-of-the-mill drilling." The Mohole drilling staff decided to address this problem with an innovative drilling system called the "turbocorer." The turbocorer was not a new invention. A Chicago dentist invented the turbodrill concept in the nineteenth century, and Russian geophysicists had patented a version of the turbodrill. Attempts to develop a turbocorer device, however, had confronted two problems: excessive bearing wear, and an inability to monitor the tool's performance at the bottom of the hole. Mohole's engineers sought to develop solutions to these problems.

A two-year joint effort between the Mohole staff and the manufacturer, Dresser Industries, resulted in a new turbocorer, which consisted of a 155-stage downhole turbine utilizing high pressure drilling fluids to turn the bit.[24] Tonking recounted "We realized that we could not drill the hole by rotating a table on the surface of the vessel, turning a pipe five miles below you. The torque in the pipe would be something tremendous. So, we went to the old Russian patents of a turbo drill and from that, developed a turbocorer, because we had to core and we were going to have to core all the way into the mantle."

In 1964, the Mohole drilling team tested the turbocorer and related drilling technologies in a dense area of basaltic rock at Uvalde County in southwest Texas. The particular formation was in the shape of a mushroom that had welled up to within 485 feet of the surface. The strength of the volcanic rock resembled that of mild steel and was approximately 58,000 psi. In an earlier attempt to drill through this basalt structure, three conventional rock bits penetrated a combined total of only 18 feet before the project was abandoned.

The Uvalde test set out to "evaluate newly designed equipment and to observe newly developed techniques in operation under actual field conditions."[25] The turbocorer performed extremely well. Expected to operate for 100 hours without maintenance, the drilling system exceeded this mark by another 103 hours, penetrating rock at the rate of 5-1/2 feet per hour. For the procedure, a 16-inch conductor pipe was set and grouted at 14 feet, and 11-3/4 inch casing was set with 200 sacks of cement at 511 feet. A standard rotary drilling rig operated with 4-1/2 inch drill pipe. The borehole for the conductor pipe was 22 feet deep, beyond which the crew drilled a 9-7/8 inch hole to 458 feet. From 458 feet to 3,479 feet, the staff recovered ninety-eight cores for a total of 1,983.8 feet of coring. Drilling progressed smoothly to 2,492 feet, but the hole exhibited instability and disconfiguration after that. Total depth reached was 3,479 feet. The drill designers used a specially designed diamond bit from Christensen Diamond Products Company that measured 9-7/8 inches in diameter with a 2-inch opening for the coring.

Turbocorer used at the Uvalde test

Nearly 5,000 1/8 carat diamonds studded the bit. Five separately manufactured drill bits performed very well without burning or losing diamonds during the testing period.[26]

The heart of the turbocorer was the rotation system. The bit, reamer, and stabilizer rotated at 600 rpm by the action of pressurized drilling fluid directed through turbine stations housed in the body of the turbocorer. The main body of the corer rotated independently to keep it from sticking to the sides of the hole. The drilling fluids contained minimal solids for reducing friction and bearing wear. It was encouraging news that the turbocorer bearings ran 120 hours beyond their normal life expectancy. Drilling fluid coming from the hole passed through a vibrating screen shale shaker to remove particles larger than 600 microns. In a following stage, sixteen 4-inch hydrocyclones removed all solids larger than twenty microns.[27]

In addition to the drill pipe corer, a sidewall coring tool removed sections of the side of the hole at specific intervals. Tonking described the sidewall coring tool: "If we drilled and had missed core, didn't get any core recovery, or got poor core recovery, we had designed a system that we could go back in the hole and we could cut a triangular-shaped piece of core out of the side wall of the hole, so we would have a continuous core record."

Specialized logging equipment for recording data was essential. Bill Gillingham, Schlumberger's president, called Tonking and said, "'We will be there and we'll give you a good price,' and they did finally. So, this was two winches that would spool 40,000 feet of 5/8 inch double armored cable, electrical cable. These were immense things. They were being built at the time we were canceled." Each logging unit measured 18' × 23' × 9'. The new cable designed for the 35,000 foot project supported much of its own weight and retained a 10,000 pounds pullout capability.[28]

The multiconductor logging cable was designed to lower the drilling core barrel and the instrument package to its proper position where the core barrel sits above the bit. The electronic instrument package measured bit speed, hole inclination, pressure differential, bearing wear, core barrel latch position, and the amount of core in the barrel. The instrument package electronically transmitted this information to a control station on the surface. The overall length of the core barrel and instrument package ranged from 38 to 58 feet depending on the size of the core barrel used.[29]

Replacing the bit through a five-mile drill string was another potential problem confronting the engineers. In typical deep wells, raising the old bit, replacing it, and returning it to the hole bottom could take eighteen to twenty-four hours. For wells deeper than 16,000 feet, studies disclosed that the well bit drilled for only about 50 percent of the time the rig was in operation. The solution was to develop a wireline retractable diamond core bit that fit inside the drill pipe; the drill pipe remained in the hole while the wire line retracted only the core bit.

These developments in drilling equipment were just a few that Project Mohole was working on when it was cancelled in mid-1966. The coring tests at Uvalde were rigorous and the results recorded with precision. Clyde Nolan said, "We had rolls of mud charts rolled down the aisles, down the stairways of the hotel, and we were going to take little points of it, picking times and dates, and we put together, in about two weeks, a daily log of penetration rates for various types of drill bits." But then work came to a sudden halt.

The Termination of Project Mohole

Scientific debates, budgetary dilemmas, and the Vietnam War ultimately brought down Project Mohole. Congress shifted funding priorities during the escalation of the war, canceling projects that might be "expendable." Mohole's original projected cost was substantially less than the 1966 estimate of $127.1 million, making it an obvious target for the congressional scapel.[30] Project Mohole also lost one of its most valued supporters when Albert Thomas, chairman of the House Independent Offices Appropriations Subcommittee, died in early 1966. Representative Joe L. Evins (D-TN), who succeeded Thomas, attacked Mohole as a scientific boondoggle. The NSF continued to support Project Mohole as a major scientific project, but Representative Evins's first major decision as chairman was to kill it.

Efforts were made to revive the project. W. Henry Lambright, in his *Presidential Management of Science and Technology,* wrote that after Evins killed Mohole, "a great deal of lobbying ensued on behalf of the stricken program. Many senators were approached, and [President] Johnson assisted in the effort.[31] The Senate restored funding for fiscal year 1967, but the House of Representatives rejected the $19.7 million appropriation needed to keep work going. Attempts to reconcile in conference committee failed. Meanwhile, media reports charged that the Brown family who had been long-time supporters of Lyndon Johnson, had donated money to Johnson's "President's Club" in return for backing on Project Mohole.[32] George Brown insisted that the money had been pledged "long before this thing came up" and called it "ridiculous" to link the contributions to Project Mohole.[33] The President told reporters, "You can expect to have periodic political charges of this kind until November. They usually come from the party that has been rather strongly rejected by the people."[34] Due at least in part to the controversy surrounding the Brown family contributions, presidential and congressional support for Mohole dropped off, spelling its demise.[35] Project Mohole soon became known as Project "no hole".

The NSF ordered the termination of all activities in October 1966. As Brown & Root president Herbert Frensley said, the company was "shocked and disappointed that one of the nation's most exciting, most potentially productive programs should be so abruptly cancelled."[36] Tonking looked back with regret at having to "stop, cease, and desist." The espirit de corps had been so exuberant, everyone aware that they were contributing to something that had never been tried or even imagined. "What a horrible thing when they cancelled this project for that staff. Oh, it was rough."

The Project Mohole staff could not be completely dismayed. They had made tremendous strides in developing marine drilling technology. Of the dismissed engineers, geologists, and scientists, Tonking said, "There wasn't one of them that didn't have a job within a couple weeks. They were of that quality." The quality of the staff was matched by the quality of the work they did. "As most knowledgeable scientists know the innovations produced by Project Mohole already have justified the research and development costs," noted Herbert Frensley at the time.[37]

Project Mohole enhanced Brown & Root's capabilities in offshore engineering and design, as well as the entire industry's.[38] As of March 1966, Project Mohole "had produced 103 inventions, 88 of which were generated by the staff and 15 by subcontractor personnel." Nearly all of these directly involved applications for offshore techniques and were put to use in commercial offshore drilling applications.[39] "A tremendous amount of technical expertise and equipment expertise was developed on this project," said Rochelle. "It's just mind boggling when you look back on it, how much was developed in so short a time."

By the late 1960s and early 1970s, the offshore oil industry had descended upon the North Sea, where Mohole innovations proved invaluable. As Tonking pointed out "The North Sea would [not] have been developed near as early as it was, because that was rough water, and when you got into the deeper water where you weren't on the bottom—did you notice—they've used nothing but semisubmersibles out there." John Irons testified to the lasting contributions of Project Mohole: "I often look at the trade journals and I see some of the bit designs in there and I think, that goes back to old Darrell Sims on Mohole. He was working on that thing, and here, it's showing up 30 years later out in the industry."

Notes

1. This chapter draws from interviews with W. H. Tonking, W.R. Rochelle, A.R. Desai, R.O. Wilson, Hugh Gordon, C.E. Nolan, and John Irons.
2. See W.H. Tonking, "Project Mohole—Exploring the Earth's Crust,"Royal Society of Arts & Commerce (May 25, 1966): 980–997.
3. National Science Foundation, Project Mohole (NSF, n.d.), 2.
4. See NAS, *The Academy in the Fifities—Beginnings of the Space Age,* 558–563.
5. *Project Mohole,* 6.
6. Quoted in Brown & Root Engineering brochure (1990).
7. William R. Nelson, *The Politics of Science* (New York: Oxford University Press, 1968), 167.
8. See NAS Archives Division of the NRC Earth Sciences AMSOC Committee Mohole Project, 1957–1964 and *The Academy in the Fifties—Beginnings of the Space Age,* 558–565.
9. Mel Hobbs, "Mohole Phase 2 on schedule; cost now up to $100 million," *World Oil* (October 1965).
10. See Willard Bascom, *A Hole in the Bottom of the Sea: The Story of the Mohole Project* (New York: Doubleday & Co., Inc., 1961) for an in-depth review of the early phase of the project before Brown & Root became involved.
11. "'Mohole' Project Aid to Oil Industry," *Offshore* (May 1961): 29–30.
12. "'Mohole' Project to Aid Oil Industry," 29–30.
13. "A $2.3 Million A-E Fee Can Look Small," *Engineering News-Record* (November 15, 1962): 86.
14. Brown & Root, *Project Mohole, History,* 196.
15. Tonking, "Project Mohole," 982–984.
16. Tonking, "Project Mohole," 987.
17. Brown & Root, *Project Mohole, History.*
18. "Deepwater—Positioning Problems Solved," *The Oil and Gas Journal* (September 21, 1964). "Three years may be required. . . . To Drill Through Earth's Crust," *Offshore* (October 1963): 23–24; Tonking, "Project Mohole," 989–990.
19. Tonking, "Project Mohole," 989–990.
20. National Science Foundation, *Project Mohole,* 7.
21. Tonking, "Project Mohole," 990–992.
22. NSF, *Project Mohole,* 10.

23. Tonking, "Project Mohole," 992–994.
24. J.E. Kastrop, "Project Mohole: Tests Tool for Deepest Drilling," *Petroleum Engineer* (March 1965).
25. Brown & Root, *Project Mohole, History,* 375–376.
26. Tonking, "Project Mohole," 994–996.
27. J.E. Kastrop, "Project Mohole."
28. "Schlumberger Custom Designs Two Logging Units for Project Mohole," *Offshore* (December 1965): 101.
29. "Mohole Tool Exceeds its Design Goal," *Oil & Gas Journal* (January 11, 1965); "Mohole Turbocorer Passes Rigid Performance Test," *The Oil and Gas Journal* (November 23, 1964).
30. "Project Mohole 'mistakes' slammed," *Oil & Gas Journal* (June 1966): 56.
31. W. Henry Lambright, *Presidential Management of Science and Technology* (Austin: University of Texas Press, 1985), 160.
32. "Congress Conferees Clash on Project Mohole Funds," *New York Times* (August 18, 1966). Also see "House Rejects Fund for Project Mohole," *New York Times* (August 19, 1966); also see "Big Spending Bill Passed by Senate," *New York Times* (August 11, 1966).
33. "Mohole Contractor is Linked to Gifts Made to Democrats," *New York Times* (August 20, 1966).
34. "Johnson Denies Any Favoritism in the Award of U.S. Contracts," *New York Times* (August 28, 1966).
35. "Congress kills project Mohole," *Offshore* (October 1966): 30.
36. "Project Mohole," *Brownbilt* (Spring 1967).
37. "Project Mohole," *Brownbilt* (Spring 1967).
38. "Mohole Lives," *Newsweek* (January 25, 1965): 89.
39. Don E. Lambert, "Will the U.S. lose race to 'inner space'?" *World Oil* (July 1966).

CHAPTER 8

Inner Space Pioneer: Taylor Diving & Salvage

Underwater diving was an integral but often under-appreciated aspect of the offshore oil industry. The success of offshore operations depended on the assistance of divers. The expansion of the offshore industry, in turn, spurred the growth of diving as a commercial endeavor. By the early 1960s, the Taylor Diving & Salvage Company claimed leadership in the advancement of commercial diving. Taylor Diving established a close bond with Brown & Root beginning in 1960. Eight years later, Brown & Root's parent, Halliburton, purchased Taylor. During the 1960s and 1970s, Brown & Root and Taylor worked hand-in-glove to pioneer the movement of the offshore oil industry into ever deeper waters.[1]

Divers progressed from trouble-shooters in shallow water to highly skilled, underwater technicians engaged in complicated and dangerous tasks at significant depths. Early offshore diving involved brief dives to inspect, salvage or repair structures. As the industry moved into deeper waters, new technologies allowed divers to facilitate deeper offshore operations. Underwater pipelining, in particular, required divers regularly. Taylor Diving gained much of its experience in support of Brown & Root's pipelining work in the Gulf of Mexico. Taylor's innovations in helium-oxygen and saturation diving at its Belle Chasse, Louisiana, research facility resulted in depth records for divers in the 1960s and 1970s. Taylor's development of hyperbaric welding and underwater pipe-alignment techniques were essential to Brown & Root's pipelaying operations.

The impressive technological advances made by Taylor Diving were also made possible by the daring and talent of its divers, who had to become experts not only in the physics and physiology of diving, but in the art of welding and construction. They had to be in top physical condition and well-prepared for the dangers and demands of long hours under pressure. Taylor Diving founded one of the first programs to train commercial divers for the skills they would have to master in the offshore oil business. "Through no coincidence," according to Ken Wallace, former president of the company, "most of the American diving companies which came into being after 1960 were organized by divers who learned the trade while working for Taylor Diving."[2]

Early Diving

Most of the divers who created Taylor Diving learned their trade at the U.S. Navy Experimental Diving Unit (NEDU), operating out of the Navy Yard in Washington, D.C. Organized in the mid-1920s, the NEDU experimented with dives in varied marine environments using different compressed air mixtures for breathing underwater. "With due respect for what this sector of private industry has done," wrote Ken Wallace in 1979, "it would be less than candid [not] to acknowledge the U.S. Navy's prime role in developing modern diving techniques."[3]

The NEDU made two important contributions to early commercial diving. One was the development of decompression tables in the 1930s. Building on the work of British physiologist John Haldane, the NEDU experimented with methods of returning divers to the surface following a staged ascent to avoid decompression sickness or the "bends"—sharp pain in the joints caused by the rapid and harmful release of nitrogen gas bubbles into the blood and tissues where the gas had dissolved under the pressure of greater depths. If a diver were to make a quick emergency ascent, too rapid decompression could cause a fatal gas embolism, a condition where expanding gas bubbles passing through the thin-walled vessels of lung tissue into the blood stream cut off flow to the brain and nerves.[4] In the 1930s, the NEDU formulated a schedule for controlling the ascent, or decompression, from various depths to allow for a slow release of the absorbed gas. The NEDU decompression schedule became a worldwide standard for the industry.[5]

NEDU's other major contribution was the use of helium and oxygen mixtures as breathing gas. When divers submerge beyond about 100 feet of seawater, nitrogen (the largest component of atmospheric air) dissolves more readily in the fatty tissues of the brain and spinal cord, producing an anesthetic effect called nitrogen narcosis. The NEDU found that helium, being one-seventh the density of nitrogen, was less soluble in the tissues of the nervous system, thus alleviating narcosis. Helium also offered less breathing resistance under high pressure. In 1949, after years of experimentation, the Navy started using helium-oxygen gas in "bounce" dives made off the coast of Panama to depths reaching 400 feet to aid in salvaging sunken submarines.[6]

One of the men who made some of these first open sea dives was Mark Banjavich, the founder and driving force behind Taylor Diving. In 1957, Banjavich received his discharge from the Navy after spending ten years in its Submarine Rescue Services. Banjavich learned that the only steady work for divers at the time was in the offshore oil industry along the Texas-Louisiana Gulf Coast. He teamed up with another Navy diver, Edward Lee "Hempy" Taylor, and a French diver named Jean Valz, who had fought with the French underground during World War II. The trio acquired a damaged 85-foot schooner, *The Jesting,* which they repaired and sailed from New London, Connecticut, around Florida and into the mouth of the canal leading into New Orleans' Lake Pontchartrain. They tied the schooner up near a Coast Guard station and opened for business. Doubling as an office and living quarters, *The Jesting* became the first home that Banjavich really ever had outside of the U.S. Navy.

Banjavich displayed an entrepreneurial vision and an understanding of the oil patch culture. When the time arrived to name the new company, Banjavich told his partners that "those Texans are not going to hire a Banjavich Diving Company and they're not going to hire a diving company with the name of Valz. It only makes sense that we give this business an Anglo-Saxon name. So, Hempy, it's got to be your name." As former Taylor Diving engineer Anthony Gaudiano told it, Hempy replied, "Well, O.K., you can use my name but I don't want to be president!" Thus in 1957 Mark Banjavich became president of the Taylor Diving & Salvage Company.

Banjavich and his crew were master divers equipped with skills gained from the Navy Experimental Diving Unit. In 1957, the company attracted the offshore industry's attention by salvaging two sunken drilling rigs. One of those was the jack-up rig *Mr. K,* which had capsized in 30 feet of water during a storm near Grand Isle, Louisiana. Several attempts had failed to right the rig before Taylor was hired. It was a very dangerous job. The rig had been crushed into the mud bottom. Hurricane Audrey passed through soon after the accident and dumped heavy amounts of silt around the wreck. In a series of dives, Banjavich managed to free *Mr. K* from the mud by burning off the forward kingpost and jetting a tunnel under the drawworks and diesels. At any moment, the machinery could have come smashing down on him. As Hempy Taylor reported at the time, "careful work and a constant check on the suspended machinery kept Mark from injury or worse. There is a saying among divers that there are old divers and there are bold divers, but there are no old, bold divers."[7]

Offshore contractors hired divers to perform other tasks in addition to salvage. In the wake of the damaging hurricanes and storms of 1957, operators employed divers to inspect subsea structures for scouring action around the base or cracks in joints and piling. Occasionally, divers made underwater burns using an Oxy-Arch torch, like Banjavich did on the *Mr. K* salvage. They also used torches for underwater welding, but "that was all strictly very temporary salvage stuff." Pipeline installation and repair became the main activity for diving. Divers were indispensable for connecting pipelines to platform risers and jetting trenches for pipe burial, either by hooking up the pipe-burying units or by using hand-jetting equipment in shallow waters. After pipelines had been laid, divers regularly checked and repaired leaks and traced out the course of the pipe to determine if wave and current action had shifted it.[8]

Brown & Root soon crossed paths with Taylor Diving. By the late 1950s, Brown & Root was laying hundreds of miles of pipeline in the Gulf of Mexico along the stretch from Morgan City to Breton Sound. "It was essential for Brown & Root and Taylor to come together," Mark Banjavich said, "and Mr. Minor was the catalyst, without question." L.E. Minor had brought some clients into Lafitte's Blacksmith Shop, a New Orleans bar frequented by the oil field crowd. "If you were in the oil patch," said Anthony Gaudiano, "that's where you went to drink." Jean Valz was moonlighting as a piano player at Lafitte's, and the two men struck up a conversation. That encounter led to a meeting between Minor and Banjavich followed by a steady stream of work for Taylor Diving with Brown & Root on pipelining jobs in the Gulf of Mexico. "I believe Ed Minor was a guy who had the knack of recognizing winners, or people who could do things, who could get things done," said Gaudiano. "I believe he saw that in Mark." By the winter of 1959, Taylor was subcontracting nearly all of the diving for Brown & Root's Marine Operators. Banjavich remembers that Minor asked him, "'Just work for us exclusively,' and I said, 'Certainly,' because he was that type of guy. You could trust his word."

Taylor Diving's work for Brown & Root in the shallow Gulf of Mexico was routine. Then in 1961 Taylor divers accompanied Brown & Root pipeliners to Argentina's Tierra del Fuego for a very difficult job. Taylor Diving handled the underwater aspects of laying a 20-inch pipeline to a tanker anchorage in 50 feet of water. The assignment "was carried out under some of the worst tidal and weather conditions existing anywhere in the world."[9] Banjavich said, "We went down there to do a 30-day job, and it took us 6-1/2 months due to very, very extremely bad weather." The divers certainly earned their pay in helping to lay the "pipeline at the bottom of the world," the first international diving job undertaken by an American contractor.[10]

Plunging Deeper

When Banjavich and his crew returned, they prepared for a new phase in commercial diving. Oil exploration and production offshore Louisiana had moved beyond the 50-foot depth range. The average water depth of new platforms was 100 feet, and record depth already extended to 200 feet. The implications for the diving industry were clear: 1) divers would have to improve methods of overcoming decompression sickness, and 2) they would have to use helium-oxygen gas for deeper dives.

When he returned from Argentina, Banjavich acquired Hempy Taylor's shares in Taylor Diving and took controlling interest of the company. Numerous people who became influential in the history of Taylor Diving soon arrived there, like Banjavich, from the Navy Experimental Diving Unit. Banjavich kept in contact with veterans of the NEDU like Ken Wallace, George Morrissey,

Bob McCardle and others, all of whom came to work for him in the early 1960s. Because "there was really no equipment available at that time to go to those depths," Taylor developed its own, designing the first "recompression chamber" employed in the commercial diving industry and modifying the gear worn by divers.

The original CS-2 recompression chamber, built by Taylor in 1961, was a small vessel with one lock that could accommodate a single diver. The diver entered the chamber upon emergence from the water after making periodic stops during the ascent. With the diver inside, the chamber was pressurized to the same depth as the diver's last stop and then depressurized according to the Navy's depressurization tables. Anthony Gaudiano pointed out that "Of all the commercial diving companies, only Taylor provided chambers on the job. Everybody else required the divers to take their decompression in the water." In late 1961, Taylor designed a larger chamber, the CD-1, equipped with a second lock. A doctor or technician could enter the outer lock without depressurizing the inner lock where the diver was. The chambers offered safer and more practical methods of decompressing a diver from greater depths.[11] "He could be recompressed quickly for the surface stops, then continue his decompression in relative safety and warmth. He could be fed and get medical attention if he needed it," explained Gaudiano.

The next step was to prepare divers to go "deeper and deeper on helium-oxygen." In 1960 in the Gulf of Mexico, Shell Oil sponsored the first commercial helium-oxygen demonstration dive by marine consultant Norman Ketchman.[12] Two years later, Taylor constructed its first depth simulator, similar to those used by the U.S. Navy since the 1920s. In the depth simulator, the company could physically test divers to depths of 300 feet on dry land, but in a wet environment. George Morrissey, who had served as a Master Deep Sea Diver with Banjavich in the Navy, joined Taylor in 1963 as superintendent in charge of developing commercial helium-oxygen equipment, becoming the diving industry's first full-time diving supervisor. He started training men at the research facility, providing divers and equipment to work in the open sea at 200-plus feet.[13]

The progression from compressed air to helium-oxygen necessitated new equipment. Because pre-mixed helium-oxygen was costly and exhaled in greater volumes with increasing depths, divers were equipped with a rebreather apparatus that recirculated helmet gas using a venturi jet taking suction from a carbon dioxide absorbent cannister. "In this manner," said Dr. Robert Workman, the diving physiologist in charge of the NEDU, "exhaled carbon dioxide is removed and the oxygen level maintained in the helmet by a lesser volume of supply of gas than would be required by ventilation with supply gas alone."[14] In designing Taylor's helium-oxygen gear to make it more lightweight, Morrissey replaced the heavy metal helmet used in conventional dry dress diving with a U.S. Navy modified Mark-6 partial rebreather. It was plastic bubble helmet that resembled those used by the Apollo astronauts. The lightweight apparatus proved to be twice as effective as the heavy gear on long duration dives. In place of the standard carbon-dioxide scrubber cannister, Morrissey substituted a compact, lightweight rectangular cross-flow cannister which "proved to be 200 percent more efficient than previous models."[15]

By the mid-1960s, Taylor Diving achieved a day-to-day working capability in the 100-200 foot depth range. But could they go deeper? To do so, diving companies had to find a way to handle greater decompression requirements and enable divers to have greater effective bottom times. The introduction of saturation diving as a solution to these problems revolutionized the commercial diving industry and in the process established Taylor Diving as prime innovator in this technology.

Saturation Diving

Saturation diving, pioneered by the Navy, proved to be a timely development for deepwater construction. It resulted in the spectacular growth of commercial diving. Following the Navy's celebrated "Sea Lab" tests in 1964, Taylor Diving developed its own saturation systems, built a new research facility, and extended the technology to record depths in the late 1960s and 1970s. Mark Banjavich's colleagues regard him as a "visionary" who appreciated the value of saturation diving. Anthony Gaudiano said that "he could see down the road that this technique could be put together and made commercially viable and a fair amount of revenue could be generated from it."

The distinction between conventional and saturation diving was a matter of depth and duration. In conventional diving, also known as "bounce" diving, a diver is on the bottom for a relatively short period of time (usually less than one hour) and then decompressed, which may require a period of several hours. Most working dives are completed in less than one-half hour. The 400-foot dives that Banjavich performed for the Navy off Panama in 1949, for example, lasted no more than about ten minutes, but required about twelve hours of decompression. Naval scientists discovered that body tissue becomes "saturated" with whatever gas a diver breathes at given depths for longer periods until the tissues cannot absorb any more. Because of the large quantities of gas absorbed and the slow rate that it can be expelled, decompressing a saturated diver requires much more time, a matter of days rather than hours. As a rule of thumb, a diver spent a day of decompression for every 100 feet of saturated depth. Researchers found that the time to reach saturated condition decreased at greater depths. They calculated that saturation decompression was required after a diver spent about one hour at depths of 300 to 400 feet and after about one-half hour at depths of 400 to 600 feet. Thus, a diver who worked at these depths for any significant amount of time would need long periods of decompression.[16]

Although decompression time did increase with the length of the dive, saturation diving offered important advantages. It permitted great economies in deepwater work by vastly increasing the ratio of bottom time to decompression time. Once a diver became saturated at a certain depth, decompression time remained the same regardless of time spent at that depth. As one engineer put it, "The decompression period is as long for two hours at 600 feet as it is for one year at 600 feet."[17] In deepwater construction, particularly pipelining, saturation diving presented opportunities to do weeks of intensive work. Banjavich said, "We never would have gone anyplace if we had not used saturation diving. We would have never progressed out into deep water. You had to get down there and had to be able to stay there, of course."

The U.S. Navy had experimented with principles involving tissue saturation as early as the 1930s. In the early 1960s, the Navy's Sea Lab experiments, directed by Captain George Bond, verified saturation diving principles on both animals and humans. In 1965, Marine Contractors, Inc. and Westinghouse, Inc. applied saturation diving for the first time commercially at Smith Mountain Dam in Virginia.[18]

Taylor Diving's SDC-DDC system

As their Gulf of Mexico dives were getting deeper and their jobs lasting longer, Mark Banjavich and George Morrissey sensed that saturation diving would be the wave of the future in the industry. So they designed a two-person diving bell or submersible diving chamber (SDC) that could be mated to a deck decompression chamber (DDC). This allowed divers to be transferred between the two while under pressure.

In operation, two divers were locked in the SDC, which was then pressurized to bottom pressure at a rate of 75 feet per minute. It was then lowered by winch to within a few feet of the bottom. The divers breathed through their helmets, which were fed a helium-oxygen gas mixture from the deck. The chamber was equipped with a carbon-dioxide scrubber and a heating unit to condition the environment and keep the divers comfortable. Once on location, the divers opened a bottom hatch in the SDC and connected their breathing gear to special umbilical lines that led to the surface. Since the gas pressure inside the SDC equalled sea pressure, the interior of the SDC did not flood. One diver exited the SDC to go out and work up to three to four hours, which was a significant breakthrough in work time. The second diver tended the first diver's umbilical while inside the SDC. The divers changed places and continued working for another three to four hours. "So, we were able to get an eight-hour shift out of two divers," Banjavich explained.

When work was completed, the SDC was winched to the surface and mated with the DDC, where the divers were replaced by two others. By using a six-diver, three-team rotation, the group could put in 24 hours of continuous work for several days before they were all decompressed in the DDC. "The diving bell was just an elevator. It was always under the same bottom pressure." The SDC-DDC system offered significant improvements in saturation diving. SDCs allowed divers to work in rough weather and descend through strong currents without exposure. Working out of a controlled environment made possible greater depths and extended work time at the bottom.[19] The Equitable Equipment Company out of New Orleans built the first partial saturation diving chambers for Taylor. They were used aboard Brown & Root's lay barge, the *M-210*. In the summer of 1967, Taylor made its first total saturation dives to install risers for Shell Oil's Marlin System at 320 feet in the West Delta field of the Gulf of Mexico.[20]

"Everything, really, we had to develop as we went along because there was nothing available," said Banjavich. New problems were encountered as divers reached greater depths under prolonged conditions. Life support systems on the diver and in the SDC had to be redesigned. "It was one thing to just take divers down to depth. It was another thing to keep them there and comfortable so they could work." The technological advances to permit humans to live underwater matched some of those that permitted habitation of outer space. "Just one was going under pressure, and the other was reducing pressure," pointed out Ken Wallace. "You had a lot of similar problems and life support activities, they were quite similar."

Using helium-oxygen mixtures at increased depths presented difficulties in keeping divers warm. Because helium has a six-fold greater thermal conductivity than air, divers would be seriously affected by small variations in temperature. "Once you get beyond about 400 feet, the water is just above freezing," Banjavich said. "If you cannot keep those divers warm because the helium is such a conductor of heat or cold, in a matter of minutes, they're frozen out." Taylor Diving addressed this problem with heated suits. The company took the space suits designed for astronaut John Glenn's pioneering trip into outer space and modified them for inner space. Initially, the suits contained plastic tubing with perforated holes for ventilation. Later on, rubber tubing was attached to wet suits. Taylor simply pumped hot water (106 to 110 degrees Fahrenheit) from the surface into the tubes of the hot water suit. The water was heated and pumped with 250,000 BTU/hour heaters. It flowed through an insulated hose to the SDC, and then to the diver through his diving umbilical. The hot water suits worked fantastically well, allowing divers to spend many hours in cold water.[21]

"By early 1967," wrote Ken Wallace, "rapid advances in saturation diving were being made almost on a monthly basis."[22] With the increase in bottom working time, communications with divers became important. Helium atmospheres, however, often caused unintelligible speech transmissions. As anyone who has ever sucked helium out of a balloon can testify, the gas makes one's speech sound like the quacking of Walt Disney's cartoon character, Donald Duck. Radio and electronic communication systems employed by divers had evolved from homemade devices to laboratory designed underwater telephones. In 1967, several electronics corporations researched the "Donald Duck effect," which led to the commercial development of helium voice "unscramblers" that became widely employed in the industry beginning in 1968.[23]

During this period, Taylor Diving engineers improved atmospheric control capabilities in the saturation diving systems (DDCs). They designed environmental controls that enabled divers to live for extended periods in a saturated state. The main lock in the DDC contained toilets, supply locks, bunks, and individual environmental controls. Each DDC had an external control house from which an operator could control power generation and pneumatics in the main lock. Constant monitoring of the gases inside the main lock was crucial. At greater depths under hyperbaric conditions, the threat of gas contamination increased. "Very small contaminants of gases at depth which you could tolerate up here forever," noted Mark Banjavich, "would kill a person instantly at depth." The saturation systems were equipped with oxygen monitor/controllers to regulate oxygen levels. Under higher than atmospheric pressure, oxygen percentage had to be lowered to prevent damaging levels of oxygen in one's central nervous system. The effect is known as "oxygen toxicity." Air contains 21 percent oxygen, but under pressure equivalent to 200-foot depths for example, the oxygen level in breathing gas could not rise above 4 percent. Temperature and humidity control also was mandatory. The Taylor units used cold water and hot water exchanger coils in the environmental control systems. The environment in the chamber was circulated over the coils to condense out excess water vapor and to reheat the gas to comfortable temperatures.[24] Exhaled carbon-dioxide and other gases also had to be removed or "scrubbed out" of the atmosphere. "We scrubbed out everything, including personal body odors, with filters containing potassium permanganate and soda lime," remembered Banjavich. "We changed the whole atmosphere every three minutes."

Taylor engineers also designed an entrance lock through which medical personnel could enter the saturation system in case of emergencies. On one tragic but, in the end, fortunate occasion, an entrance lock on one of Taylor's systems saved a diver's life. That diver almost died when he carelessly opened the discharge valve on the commode while he was still sitting on it. The pressure of the discharge locked his buttocks onto the toilet and pulled several feet of small intestines out of him. The diver had been severely injured but could not be rushed to surface pressure because he was totally saturated at about 250 feet. Taylor officials found surgeons who were experienced in handling war casualites in Vietnam. They were quickly sent through the entrance lock and into the chamber to perform an operation. Because they were in highly controlled, hyperbaric conditions, no ether or anathesia could be used. The surgeons kept the diver alive for fifteen days while he was being decompressed. Miraculously, he survived to tell about it. The accident reminded everyone of the dangers of working in a highly pressurized, saturated environment at great depths. It also stood out as an exception to Taylor Diving's remarkable safety record. As Banjavich told his people at the time, "If we're going to go into these deep waters, one thing that would kill the program is if you start injuring or killing people . . . You have to go to the nth degree of safety to make sure nothing happens."

In 1967–1968, as Taylor started making total saturation dives into deeper water, the company expanded rapidly, building up its research capabilities and training programs. Saturation diving had demonstrated its utility and cost effectiveness to the offshore oil industry. In 1967, Gulf Oil officials estimated that in repairing two Gulf of Mexico production platforms in 210 feet of water, saturation divers required 33 percent fewer hours than surface divers would have, at a 36 percent cost savings.[25]

In 1968, Mark Banjavich sold 80 percent ownership of Taylor Diving to Halliburton Company, Brown & Root's parent. The sale gave Taylor greater financial resources and organizational support to enlarge its personnel, equipment, and research. Several individuals from the NEDU joined Taylor, including Ken Wallace, a diving supervisor who would become Taylor's president, and Dr. Robert D. Workman, a Navy physiologist whom Mark Banjavich called the "grandfather of saturation diving." Workman had enjoyed a distinguished career with the U.S. Navy Medical Corps. As head of the NEDU, he had developed the Navy's decompression tables. Banjavich, who was a close friend of Workman's, recruited the doctor first as a diving consultant and then convinced him to join Taylor full-time when Workman retired from the Navy in 1970. "Everyone was after him," said Banjavich. "He is the reason we could do these very, very deep dives . . . the great safety record we had with the deep saturation diving was attributed directly to Dr. Workman and his expertise. He just controlled all the dives, all the experimental dives that we did."

Innovation in deepwater diving shifted from the NEDU to private industry—and to Taylor Diving. In 1968, Taylor began construction on its new world headquarters and underwater research facility at Belle Chasse, Louisiana, next to Brown & Root's pipe yard. Anthony Gaudiano said, "It was obvious from the feedback we got from Brown & Root, the pipe diameters [would] get larger and the water [would] get deeper. And this was going to happen very quickly." Taylor's depth simulator, the first not owned or controlled by the government, maxed out at 350 feet. They needed a facility that could push the limits of deepwater exploration. Completed in 1969, the research center housed a three-vessel hyperbaric complex, described by the company as "the largest existing facility of its kind in the world."[26] It became the second depth simulator not owned by the U.S. government. The main vessel consisted of a "lower wet pot" and "upper igloo" capable of simulating wet and dry environments similar to those encountered undersea. Although initial research programs were planned to simulate dives under depth pressures of around 1,000 feet, the vessels were designed to withstand pressures equal to 2,200 feet.[27] The Belle Chasse hydro-space research center became the site for state-of-the-art research into underwater diving that pioneered many stunning achievements in offshore construction in the 1970s.

Taylor Diving's main research complex with control console, igloo, annex living compartment, and entry local. In foreground are environmental controls.

Underwater Welding

In the late 1960s, Taylor Diving developed an underwater welding habitat and submersible pipe alignment rig. This equipment empowered divers to make hyperbaric welds (welding in dry atmosphere under pressure) for pipeline repairs and tie-ins. Hyperbaric welding proved to be a pathbreaking innovation in offshore pipelining that found notable success in the North Sea. It was a product of hard work and resourcefulness. "You have to understand that when all of this was going on, and people were working like eleven or twelve hours a day, we didn't have meetings where we sat down and made presentations," explained Gaudiano. "We didn't do the usual planning and had no critical path charts; none of that stuff. You just did it. You got it done . . . People did some very impressive things, innovative things. And I would like to think that the habitat welding was one of those innovative things."

Foreseeing the advantages of being able to weld pipe connections underwater, Taylor engineers began in 1967 to make rough sketches and designs of an eight-foot square structure. It had a quonset hut-shaped roof and lead weights that were used for ballast. Doors in the end could be fitted and sealed onto a pipe. Underwater connections and repairs until then had been mechanical. Divers would use hydraulic tools to bolt flanged connections or mechanical couplings. Welded connections, however, were stronger and more secure than mechanical ones. Creating a dry atmosphere for welding which could be placed around a pipe required sophisticated engineering and control systems. Construction had just begun in mid-1968 when Taylor received a phone call from Brown & Root for a job to repair fuel lines in the Saint Lawrence Seaway near Montreal. Brown & Root had blasted a trench across the seaway and pulled a pair of lines through the trench to the Montreal side. While waiting for inspection of the lines, a ship passed over them and dropped an anchor into the trench, damaging several of the pipes. "So here was an excellent job for the habitat," Gaudiano said.

Taylor Diving and Equitable Equipment, which had been retained to build the structure, hurried to finish the underwater habitat. Engineers designed the gas system so that the welding atmosphere would consist of mostly nitrogen. The U-shaped habitat would be clamped onto pipe by hydraulically operated lower doors which would be brought up against the bottom of the pipe, making a watertight seal. The water would be displaced by the nitrogen. "We . . . use[d] the nitrogen to dewater the welding habitat itself, all the water with nitrogen, just use that as a force, like holding a glass upside down, and just totally got the water out," said Banjavich. Divers would enter the habitat through a trunk. Once inside, they would pull their gear off and put on an oral-nasal mask that provided the proper breathing-gas mixture.

Like most of the equipment developed by Taylor Diving and Brown & Root for the emerging offshore oil industry, the habitat was the product of swift deployment of innovative concepts to meet the practical challenges of marine construction. The unit contained gas sensing systems, hydraulic systems, and

Taylor Diving's first underwater welding habitat

radiographic equipment. "So much of that," recalled Banjavich, "we designed at 2:00 in the morning. I cannot stress that too much. Guys would stay up . . . Things wouldn't work. Motors wouldn't run. The seals would lock up because you had so much pressure on those things." One of the biggest problems in designing the habitat was making sure that the atmosphere was dry and warm. Taylor engineers used nearly twenty tons of air conditioning on the unit to dehumidify the inside, along with 25 kw of inversion heaters to reheat the air to keep the divers warm. A closed-circuit television system allowed monitoring from the outside. "These guys would walk around with shorts on, doing their jobs, just in a shirt sleeve environment, and very successfully." Loaded down with all the necessary control systems, Brown & Root's first underwater habitat measured 11-1/2 by 13 feet and weighed 16,654 pounds.[28]

Taylor Diving rushed the new system up to Montreal. "There was a great haste to get this system up there," said Gaudiano. "I remember a painter was painting the bottom of the habitat as the crane was lifting it up in the air to set it on the truck." Once in place, the divers made the welds using a gas tungsten arc-weld process and radiographed them in the dry while underwater using Iridium 192 isotope in a gamma ray projector. The diver-welders were trained and certified as diver-assistant radiographers. The procedures developed by Taylor are now accepted internationally. The welding process was slow, but the radiographs proved the high quality of the welds, almost as good as surface welds. Hyperbaric welding even offered some advantages over normal surface welding: the absence of oxygen in the atmosphere prevented oxidation; the pressure reduced porosity; and the controlled environment prevented loss of the shield gas from the welder's torch.[29] "Weld quality was a big factor in selling hyperbaric welding from the onset," insisted Gaudiano. "Had the welds been poor or some other problem developed, the process would have gotten a bad name."

The pipeline repair in the Saint Lawrence Seaway proved Taylor's underwater welding habitat, but refinements still had to be made. Employing Brown & Root's "great reservoir of welding talent," Taylor Diving experimented with procedures and consumables. Metal inert gas welding was unacceptable because the arc voltage would change under pressure, and the pressure would cause the arc to constrict. The welders settled on a method that employed metallic stick electrodes with a special coating chemistry to stabilize the arc in hyperbaric atmospheres. "We tried something like twenty or thirty electrodes before we found one that was suitable to weld under pressure," recalled Gaudiano. They ended up finding a low hydrogen electrode, called the Atom Arc Electrode, which Taylor's diver-welders relied on for years.

Taylor Diving took its habitat into the Gulf of Mexico, building two more and performing three pipeline repair jobs in shallow waters under partially saturated diving modes. But the real growth in habitat welding occurred in connecting platform risers to pipelines. In those days, barge superintendents received a bonus for their speed in laying pipe. They would, in effect, make money on the laying of the pipe, but then they would 'lose' it joining the pipe to the riser. The time required to connect the riser rose for platforms in water deeper than 200 to 300 feet and as pipeline diameters increased. Joining the pipelines also became more difficult using conventional methods of coupling or bolted flanges. Banjavich proposed that Taylor Diving complete the tie-ins by making welded connections in the habitat. "You set the risers and lay the pipe," he told Brown & Root, and "we'll go along behind you and tie them in. But we'll need a barge especially set up just to do that." Brown & Root officials agreed to convert and lease to Taylor one of its barges, the *M-280* superintended by Leon Garrett, exclusively for deep sea diving operations.[30]

In August 1969, the company made its first underwater tie-in at 213 feet in Main Pass Block 298 of the Gulf of Mexico. In the middle of the job, with the habitat and divers on the bottom, Hurricane Camille swept through and delayed the operation 13 days. Divers took about 21 days to complete the weld, an inauspicious start for the new endeavor. Over the next year and a half, however, "we were making the same welds in deeper water in three days," Banjavich said. During that interval, Taylor Diving completed fifteen habitat welding jobs in the Gulf, most of them involving riser installation and tie-ins in 300-plus feet of water.[31] The combination of saturation diving with habitat welding provided an extremely valuable service to offshore construction as well as a rich revenue stream for Taylor Diving. In 1968, the company grossed approximately $3 million, an impressive sum for a diving outfit in those days.

As the size of pipelines grew from 12- and 18-inch pipe to 24- and 30-inch pipe, Taylor officials expected increasing difficulties in making pipeline connections and tie-ins. Said Gaudiano, "As pipe diameters, wall thicknesses, and steel grade increased, there would be a point where we would not be able to deform it within the habitat using chain binders and by changing its elevation and moving it around." As bid specifications came in for larger-diameter pipe, Banjavich and his colleagues started working on ideas for some kind of device to maneuver opposing pipeline ends before welding. Ken Wallace claimed that Banjavich had lain awake one night at his home in Florida mulling over the concept of an alignment unit when an inspiration finally struck him. "He got in his car about 3:00 in the morning and he drove in from Destin, Florida, to New Orleans and he called George Morrissey and myself into the office that morning. He got us out of bed. He said he knew how to handle the pipe, how to line up that pipe for the welding."

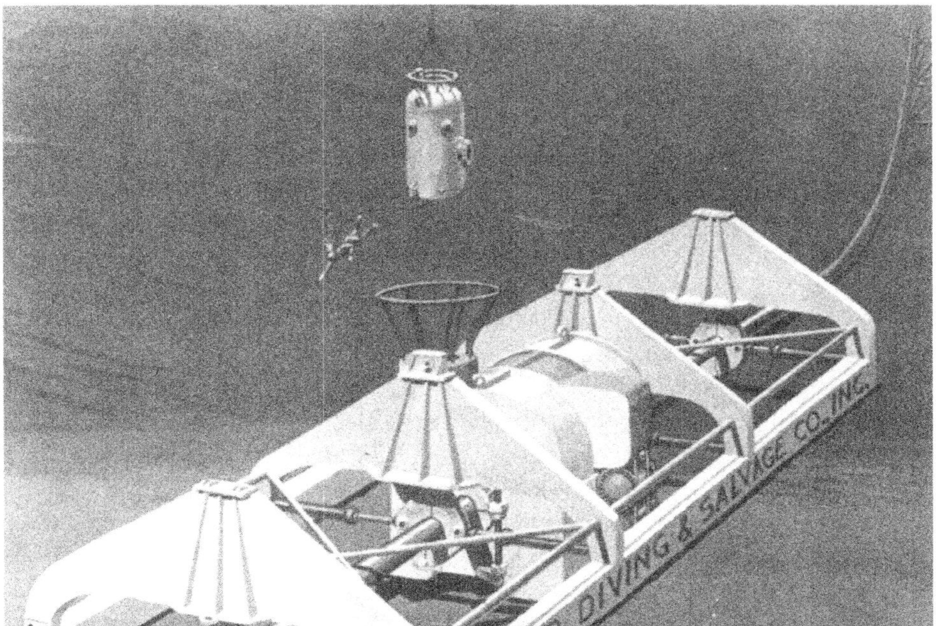

Diagram of Taylor Diving's submarine pipe alignment rig on the sea floor

Banjavich envisioned a frame that would sit parallel over the pipeline ends with the habitat in the center. Previous concepts had called for something that would be perpendicular to the pipe. He wanted a unit that could pick the ends of the pipe off the bottom and then put them into an overbend so that they would be looking at each other. This required a device that ran parallel to the pipe axis. Banjavich then took his idea to Anthony Gaudiano at Taylor's engineering department and said "he wanted something that was hell for stout, you know, that could bend this pipe. Thank God no customer was there because bending the pipe was not what they wanted to hear!"

After several designs and modifications, Taylor Diving produced its patented submarine pipe alignment rig (SPAR). The SPAR was a rectangular unit that consisted of four transverse arches. From the center of the arches, hydraulic clamps would grasp the pipes and pull them up inside a welding habitat that was suspended between the two middle arches. The clamps, operated by remote controls from the habitat, would then push the ends down so that they were facing each other in alignment.[32] Capable of handling 24-inch pipe, SPAR #1 was deployed in 1969 along with Taylor's first underwater welding habitat (UWH-1) to join risers and pipelines in the Gulf of Mexico. "It was superb," said Ken Wallace. "Just nothing like it in the world, anywhere."

A New Era of Deepwater Diving

The marriage of saturation diving and hyperbaric welding by Taylor Diving opened a new era for diving in the offshore industry and fueled the company's expansion. "With the combination of the two fused together, it was just fantastic growth," noted Wallace. Benefitting from the organizational and financial backing of Halliburton and Brown & Root, Taylor Diving hired and trained a virtual army of skilled diver-welders to deploy around the world. During the 1970s, Taylor set a series of diving and deepwater welding records. As Brown & Root's partner, Taylor played a crucial part in developing the great offshore fields of the North Sea.

The opening of Taylor Diving's new Belle Chasse research and training facility in 1969 truly launched the new era of deepwater diving. In November 1970, five of Taylor's divers set a new "wet" deep diving record at Belle Chasse by descending to a simulated depth of 1,100 feet, besting the NEDU's previous record of 1,025 feet. At the time, the French held the record for the deepest simulated dive, at 1,709 feet, but it had been a "dry" dive that did not involve underwater immersion. In Taylor's test, five divers spent eighteen days in confinement, moving between the "igloo" and the "wet pot" and performing exercises at descending depths down to the 1,000 foot level. A complicated system of compressors and heat exchangers in the 7,000 gallon wet pot of the main vessel took the water pressure and temperature to simulated levels. Dr. Robert Workman maintained close medical surveillance of the divers throughout the exercises. The successful record-setting dive,

George Morrissey announced afterward, "helps open the way for oil and gas exploration and production in water depths never before conquered."[33]

Taylor Diving was extending deepwater diving just in time for the oil boom in the North Sea, whose harsh environment and deep water offered the most difficult challenge to offshore construction yet encountered. In 1967, Taylor divers had made their first helium-oxygen dives in the North Sea on a job for British Petroleum to inspect and repair damage to a piling. The current and waves were so heavy, and the water so cold, that divers were sent down *inside* the leg of the jacket as a way to protect them from the elements—a job not suited to the claustrophobic. Taylor ran a steam line inside the leg to heat the chilling water and keep the divers from freezing. The divers intrepidly descended 200 feet to the repair location, but the bottom effect produced by the restless waves of the North Sea still impeded work. Repairs had to be postponed until after the stormy winter season.[34]

After the discoveries of the Ekofisk and Forties fields in the North Sea in 1969 and 1970, oil companies hatched ambitious development plans that called for more total saturation diving assignments. They also contemplated using the new hyperbaric welding techniques for their projected pipelines. Ekofisk and the Forties were located in very deep and tempestuous parts of the central North Sea—in 230 feet and 420 feet of water, respectively. The Frigg gas field was discovered farther north in waters approaching 500 feet depths. Developing these fields required new, large-diameter pipelines in depths ranging from 250 to 500 feet. But could welded connections and repairs be made on pipelines that deep? In 1973, two partners in the Frigg field development, British Petroleum and France's Total Oil, decided to come to New Orleans to find out.

These companies sponsored an experiment by Taylor Diving to make an underwater weld at a depth of 540 feet, well beyond the established record of 340 feet. Taylor made a preliminary test in the hyperbaric research chamber at simulated pressure and atmosphere. Positive results encouraged the research team to move out into the open sea. The Brown & Root derrick barge *H.A. Lindsay* carried two 32-inch pipeline sections, Taylor's new 1,200-foot saturation system, and its SPAR into a deep section of the Gulf of Mexico. Taylor divers pulled off the alignment and welding exercise splendidly at the 540-foot mark. The historic deepwater weld pushed down the depth barrier to marine pipelining another significant increment. Total officials were quite satisfied with the project and announced that they would employ Taylor's welding methods on riser tie-ins to the company's Frigg field gas pipeline.[35]

These developments impressed upon Taylor Diving officials the need to acquire more skilled divers. In 1974, the number of commercial divers worldwide doubled to more than 400 and was expected to double again by the end of 1975. "Where will these new divers come from?" asked *Offshore* in 1974.[36] Diving was still a hazardous occupation that required years of training, and the added need for training in heavy construction and welding made commercial diving a highly skilled trade. Taylor Diving decided to expand its in-house diver training programs. By 1974, 250 trained divers were taught underwater welding, and welders were taught diving. Subjects covered in the diving programs included saturation systems and theory, air and mixed-gas diving equipment, barge operations and layout, compressors, rigging, hydraulics, setting risers and clamps, underwater burning, and the use of explosives. Prospective divers also had to meet stringent physical qualifications and undergo intensive safety training. Training time for saturation divers typically ranged from three to five years, but under intensive and carefully accelerated training schedules, Taylor reduced the time to eighteen to twenty-four months.[37]

In training individuals for hyperbaric welding, Taylor enjoyed the most success with welder-divers. "Underwater welding is a precise science that isn't quickly assimilated by every candidate," said Bob Dykes, training director for Taylor Diving, "so it helps to start with men who are already expert welders."[38] Because pipelining was one of Brown & Root's special areas of expertise in the offshore industry, the company employed many excellent welders who were practically artists with the welding rod. Banjavich brought some of the more experienced welders into the company and "taught them the basics of diving, but always kept experienced divers with them to keep them out of trouble." When Taylor started training welders for saturation diving, Banjavich confessed "I picked the best welders Brown & Root had" and offered the welders "huge sums of money, so they actually all wanted to come on." Construction managers at Brown & Root, understandably, resisted giving up their best workers to Taylor, but they relented in the spirit of mutual cooperation and common purpose. "The beauty . . . between Brown & Root and Taylor," said Banjavich, "was how well we worked together as a family."

The familial bond grew stronger as the two companies tackled the North Sea together. In 1975, Taylor completed its first hyperbaric welding job in the North Sea, connecting a 32-inch pipeline that Brown & Root had laid for Total Oil from the coast of Scotland out to the Frigg field.[39] Taylor Diving received a continuous stream of hyperbaric welding contracts in the North Sea. Taylor hired European contractors to build new SPAR units capable of handling 32- and 36-inch pipes. Ultimately, Taylor had six SPAR units working in the North Sea. They were expensive pieces of equipment, each costing more than $1 million. Gaudiano recalled spending $3.6 million of Taylor's money in one short afternoon. But the investments paid off handsomely.

In 1969, Phillips Petroleum selected Brown & Root to design and install early production facilities for the Ekofisk field, located in the 230-feet depths of the central North Sea. Taylor Diving assisted the installation by Brown & Root's *Hugh W. Gordon* laybarge and hook up of the subsea wellheads. Taylor's most extensive work was in the third phase of Ekofisk's development, after the laying of the pipeline. Based on the demonstrations at its hyperbaric complex at Belle Chasse, Taylor Diving sold Phillips on the superiority of welded pipeline connections over clamped ones. Ken Wallace said that Ekofisk was "where we were able to convince the oil companies, gas companies, that you could weld better than you could clamp and have a more secure operation on the bottom for the rest of the history of the field that they were developing." Phillips hired Taylor to replace all the clamps on their early Ekofisk pipelines and make subsequent ones with hyperbaric welds. Between 1975 and 1978, Taylor divers spent many, many hours in saturated dives making hyperbaric welds and riser tie-ins.[40] Two Brown & Root barges were used almost exclusively for saturation diving on the Ekofisk project. Divers would stay at bottom pressure for 30 days at a time. One diving chamber stayed under pressure for an entire year, never surfacing. Crews rotated in and out through an external lock. "Most of the time," said Ken Wallace, the divers "were either diving or sleeping. Eight hours of working, that usually wiped two guys out and the next sixteen hours, they slept." Upon finally emerging from decompression, one or two of the divers were known to submerge themselves in hard partying at the Gulf Motel in Stavanger, Norway.

The decade of the 1970s was the heyday of Taylor Diving. The company continued to grow and innovate, even after Mark Banjavich sold the rest of his stake in Taylor to Halliburton in 1972 and retired. Taylor's gross revenues rose from $3 million in 1968, to $9 million in 1972, to $30 million in 1973, and peaked out at almost $150 million in 1978. By this point, it was easily the largest diving company in the world, employing 1,300 people, including 500–600 divers.

During the late 1970s and early 1980s, Taylor Diving hauled its saturation systems and habitats to other projects in the North Sea and around the world. Outside of the North Sea, Taylor performed the most hyperbaric welding jobs for PEMEX in Mexico's Bay of Campeche, another major offshore showcase for Brown & Root. "We kept one barge working pretty steady down there," Ken Wallace said, "just doing nothing but hyperbaric welding."

Taylor's growth tapered off by the late 1970s, but the company continued to push the boundaries of deepwater diving. It adopted new, high-tech Remote Control Vehicles for monitoring deep underwater activities by closed circuit television camera.[41] It developed the capability to make long-duration saturation diving down to 1,000 foot depths, and in a 1978 demonstration for Norsk Hydro, Taylor established another world record by welding two sections of 36-inch diameter pipe in 1,036 feet of water offshore from western Scotland.[42] The achievement was tainted by the accidental death of a diver at the original demonstration site in a Norwegian trench, an incident that produced considerable political fallout in Norway. That tragedy, and one other diver death in Mexico's Bay of Campeche, however, shadowed the otherwise outstanding safety record of Taylor Diving.

Like all companies specializing in offshore work, Taylor Diving suffered in the oil price collapse of the 1980s. Major competitors emerged as commercial diving became a big business, employing between 8,000–10,000 people on a world-wide basis by 1978. Gone were the days when Taylor Diving could dominate the industry, not because of Taylor's atrophy, but because of the wider application of Taylor's pioneering achievements. The company's success was due not only to the unique individuals who built up the Taylor organization, but to the close relationship they maintained with Brown & Root. As Mark Banjavich said, "If you go into something that large, I'm talking about between Brown & Root and Taylor, there's not just one person. It's everybody. There has to be huge teamwork to make something like that fly"—or, better yet, swim.

Notes

1. This chapter draws from interviews with Mark Banjavich, A.V. Gaudiano, and Ken Wallace.
2. Ken W. Wallace, "Diving Backs Offshore Growth," *Offshore* (August 1979): 79.
3. Wallace, "Diving Backs Offshore Growth," 80.
4. Susan Thobe, "Divers Play a Big Role in Oil Industry," *Offshore* (August 1975): 53.
5. Robert D. Workman, "Deep Water Diving Calls for Medical and Physiological Solutions to Problems," *Offshore* (August 1973): 43–44.
6. Workman, "Deep Water Diving," 43–44.
7. Edward Taylor, "Professional Diver: New Service Vital to Industry," *Offshore Drilling* (January 1958): 20. Also, "A Diver's Account: Salvaging the "Mr. K," *Offshore Drilling* (October 1957): *32–33.*
8. "For Divers Reasons," *Offshore Drilling* (June 1957): 21.
9. Wallace, "Diving Backs Offshore Growth," 79.
10. Wallace, "Diving Backs Offshore Growth," 79.
11. Wallace, "Diving Backs Offshore Growth," 79.
12. Wallace, "Diving Backs Offshore Growth," 79.
13. Wallace, "Diving Backs Offshore Growth," 79.
14. Workman, "Deep Water Diving Calls for Medical and Physiological Solutions to Problems," 47.
15. George R. Morrissey, "The Taylor Deep Diving System," *Offshore* (August 1966): 90.
16. Workman, "Deep Water Diving Calls for Medical and Physiological Solutions to Problems," 43–44; D. Michael Hughes, "Many Factors Affect Deepwater Dives," *Offshore* (August 1973): 57–58.
17. Hughes, "Many Factors Affect Deepwater Dives," 58.
18. John V. Harter, "Early Dives Pave Way to Progress," *Offshore* (August 1978): 34.
19. Morrissey, "Taylor Deep Diving System," 89.
20. Wallace, "Diving Backs Offshore Growth," 79.
21. O. E. Stoner, "Divers Working at 200 Ft—Six Days on—Six Days Off," Paper Presented at the spring meeting of the Southern District, API Division of Production, March 1967, 198.
22. Wallace, "Diving Backs Offshore Growth," 80.

23. Drew Michael, "Electronics Are Important to Diving," *Offshore* (1971): 36.
24. Workman, "Deep Water Diving," 47–48.
25. Stoner, "Divers Working at 200 Ft," 199–200.
26. "Diving Firm Opens Research Center," *Offshore* (September 1970): 91.
27. "Diving Firm Opens Research Center," 92.
28. "Taylor Diving & Salvage Company—An Inner-Space Pioneer," *Brownbilt* (Winter, 1968–1969): 4.
29. "Taylor Diving & Salvage Company—An Inner-Space Pioneer," 6.
30. "Taylor Diving & Salvage Company . . . Deeper and Deeper," *Brownbilt* (Fall, 1969): 23.
31. Taylor Diving, "Habitat Welding Job Log," Taylor Diving Operating Capability, March 27, 1986.
32. Susan Thobe, "Divers Play a Big Role in Oil Industry," 51–52.
33. "Simulated 'Wet' Dive Goes to 1,100 ft," *Offshore* (January 1971): 86–87; "New Wet Diving Record," *Brownbilt* (Winter 1970–1971): 16.
34. Wallace, "Diving Backs Offshore Growth," 80.
35. "Taylor Divers Complete Historic Deepwater Weld," *Brownbilt* (Fall 1973): 24–26; "Weld Made at 570 ft Depth," *Offshore* (November 1973): 55–58.
36. "Schools Solve a Diving Problem," *Offshore* (1974): 58.
37. "Schools Solve a Diving Problem," 58; Wallace, "Diving Backs Offshore Growth," 80.
38. Quoted in "Schools Solve a Diving Problem," 58.
39. "B&R, Taylor at Work in the Deep," *Brownbilt* (Summer 1975): 20–22.
40. Taylor Diving, "Habitat Welding Log Job."
41. "The Magical World of the RCV," *Brownbilt* (Winter 1978–1979): 12–14.
42. "Taylor Diving: Welding at 1,036 Feet," *Brownbilt* (Summer 1978): 22–24.

CHAPTER 9

Offshore California and Alaska

In the 1960s, Brown & Root Marine ventured into new geographical and technological frontiers, from Venezuela to the Persian Gulf, and from Project Mohole to deep sea diving. During the same period, promising oil discoveries in California and Alaska brought Brown & Root into other environments that presented particularly difficult engineering and construction challenges: creating offshore systems capable of withstanding both earthquakes and ice forces. The path up the West Coast ultimately led Brown & Root all the way to the arctic circle at Prudhoe Bay, where it joined much of the oil industry in finding ways to build the facilities needed to produce oil on Alaska's North Slope.[1]

The ties between California and Alaska were more than geographical, although both west coast regions required logistical adjustments for a company based in Houston. To compete in these regions, Brown & Root had to strike a balance between manufacturing in new fabrication yards on the western side of the Panama Canal and the creative use of prefabricated "packages" of materials shipped from its Greens Bayou yard. The threat of earthquakes in California and Alaska provided the common engineering thread binding the two regions, since seismic forces had to be considered in both areas. California provided a useful stepping stone to Alaska by giving engineers experience with seismic issues in a warm climate before they moved up to Alaska, where earthquakes, ice forces, and arctic conditions had to be considered. Furthermore, by the late 1960s, traditional operations came under intense scrutiny from environmentalists and government regulators. New environmental regulations altered industry practices around the world, but nowhere was their impact more significant than offshore California and on the North Slope of Alaska, where Brown & Root and the oil industry as a whole had to adapt to new regulatory requirements accompanied by long delays.

Offshore California: Steep, Deep, and Shaky

Activity offshore California had to wait for the resolution of the tidelands issue in 1953. Then, after a series of bids on state-owned leases, federal lands offshore California were also opened to oil exploration. Offshore companies found distinct differences in California and the Gulf of Mexico. The most obvious was the steepness of the ocean floor. Unlike the Gulf of Mexico, the Pacific sloped off sharply from the shore. The 600-foot contour off the coast of southern California ranged from only two to fifteen miles offshore. Even sites relatively near the shore could be in deep water. In the Gulf, platforms "out-of-sight-of-land" had marked a memorable milestone, but off California, mammoth structures built to operate at depths that would have placed them far out into the Gulf were clearly visible from the beaches. Along with the beauty of the California coastline, this helped explain the emergence of stronger environmental restrictions in California than in the Gulf. Moreover, whereas hurricanes represented the extreme engineering challenge for platforms in the Gulf, earthquakes were the key design consideration offshore California.

These conditions shaped distinctive approaches to offshore development in California well before the surge in offshore platforms began in the late 1950s. In the 1890s, the discovery of the Summerland field in Santa Barbara county spawned a boom that included the development of an extensive pier system to tap the portion of the field that extended out under the ocean. Over the next half century, piers reached out into the Pacific at sites from the Santa Barbara area down the coast past the Los Angeles area.[2] This approach to the development of what was truly "tidelands" oil became synonymous with California's offshore production before the 1950s.[3] In the early 1930s the first "island of steel" was used offshore California to develop a portion of the Rincon field at a point only 2,700 feet from shore, but in 62 feet of water.[4] Still, piers remained the primary means of reaching portions of onshore fields that extended out under the ocean. The 1938 state law that created the State Lands Commission and gave it jurisdiction over all of the state's tidelands encouraged the continued reliance on piers by severely restricting exploration for oil in areas not adjacent to existing fields.[5]

Although Brown & Root had brief experience with piers in the Gulf of Mexico in the 1930s, it did not take part in such developments in California. Nor did it have a role in the evolution of a second, distinctly Californian approach to offshore drilling, the building of self-contained artificial islands near shore. Several such drilling islands appeared in the 1950s, notably an acre of land created by the California-based Richfield Oil Corporation 3,000 feet from shore in about 46 feet of water near Punta Gorda. A causeway connected this island to the shore, and—as required for a time in the early

1950s by a court decision that was later overturned—all aspects of the structure had to be built from natural materials such as sand and rock.[6] By the late 1960s, the drilling island concept had matured into a complex of four ten-acre islands built offshore from downtown Long Beach by the THUMS group—Texaco, Humble, Union, Mobil, and Shell. Hundreds of wells slanted in all directions from these islands, which used facades to camouflage rigs and equipment and give the impression of real estate developments, not offshore drilling and production facilities.[7] Although drilling islands made of sand and stone and decorated with palm trees were perhaps more aesthetically pleasing to some Californians than were platforms, such islands proved effective only in shallow waters near shore. As the offshore industry moved farther out from the coast, more traditional platforms were required.

The pace of development picked up after 1958, when the State Land Commission of California held auctions on five promising parcels between Point Conception and Santa Barbara. This auction broke existing records for offshore bids, with winning bids totaling more than $55 million for almost 4,000 acres of offshore leases. The winning companies then expanded exploratory drilling in the region, using a fleet of barges and self-propelled drilling boats. Included was the *CUSS I,* a drilling ship with a 98-foot derrick that later set numerous drilling records as a part of the preliminary work on Project Mohole.[8]

In an article aptly entitled "Drilling Off California Will Be Tough,"[9] the *Oil & Gas Journal* summarized the differences between this region and the Gulf of Mexico: "California offshore locations encounter higher every day waves, greater water depths, fewer adequate harbors and onshore facilities." Driven by westerly winds, waves could reach heights of up to 26-feet for periods of 12 to 36 hours. Such extreme wave action called into question the viability of the launching techniques developed in the calmer waters of the Gulf of Mexico. Initial calculations suggested a basic design criteria of decks at least 35 feet above the high tide point and structures capable of withstanding winds up to 75 miles per hour. The lack of well-developed harbors or experienced offshore fabricators in the region meant that offshore construction companies faced difficult start-up problems for West Coast operations. After cataloguing all of these differences, the *Oil & Gas Journal* nonetheless concluded its article on a characteristically optimistic note: "Oil companies show no fear of the problems peculiar to drilling off California. And before the year is out, the companies will be unveiling some techniques to lick these problems."[10]

The most serious of these problems was earthquakes. The Santa Barbara Channel, for example, had sustained numerous small to medium-sized quakes in the twentieth century, with two in the channel—in 1925 and 1941—topping 6 on the Richter Scale. Several others measured greater than 7 in nearby areas onshore and offshore. As the oil companies prepared to move deeper offshore California in the 1950s, they had limited data on the impact of these previous quakes on onshore oil facilities. The shock waves from moderate earthquakes had seriously damaged oil storage facilities and pipelines, and the tsunamis or tidal waves produced by quakes had destroyed loading docks and associated storage. In designing offshore structures, engineers could draw on the work of such organizations as the Seismology Committee of the Structural Engineers Association of California (SEAOC), which had helped to establish the earthquake loadings needed so that "structures could be analyzed elastically and designed based on maximum allowable stresses." The benefits of translating onshore codes to offshore structures was questionable, however, since offshore structures have different dynamic responses because of the "flow of water around the members when the structure is displaced and the flexible pile groups which connect the structure to the ocean floor."[11]

Brown & Root became involved in this work when the company's increasingly analytical engineering department took on a small contract for Shell Oil on the technical feasibility of putting a platform in 250 feet of water offshore California. With little data on seismic loadings offshore, the group of young engineers used their recent university training in the new structural analysis to supplement the rule of thumb approach to design. As in the analysis of all the various forces—wind, water, ice, seismic activity—operating on offshore structures, the earliest earthquake design calculations of necessity progressed with almost no useful experience gained from the actual impact of earthquakes on offshore structures.

The offshore industry "relied on a number of university professors at Cal Tech and the University of California to be the source of guidance in this area," but even "as late as the early 1970s . . . the knowledge of earthquake design was not all that good." In the mid-1960s, however, two factors encouraged the rapid development of seismic analysis of offshore structures: the advent of more powerful computers with more advanced structural analysis capabilities, and the construction of platforms offshore California. Practical experience in computer-assisted design of offshore structures in earthquake-prone regions produced an accepted state of the art and practice of seismic design that could be outlined and subsequently modified by new knowledge in the API's bulletins on recommended practices.[12]

Two key concerns dominated the work of seismic designers: provisions for "strength level earthquake" and those for "ductility level earthquake." Strength level provisions sought to assure that the platform was designed to "maintain all nominal stresses safely within buckling or yield for the maximum level of earthquake activity expected during the life of the structure."[13] Ductility level provisions sought to assure that should an earthquake occur which was beyond the designed earthquake, there would be enough structural resilience to absorb the energy of the more severe earth tremor and prevent the loss of life. Such ductility calculations were difficult to establish, so they required conservative designs that could be as much as twice as demanding as the strength level guidelines. Design and construction of platforms offshore California went forward using the best available seismic design considerations, and subsequent minor earthquakes—such as a 5.1–5.7 quake in Santa Barbara Channel in August of 1978—had "only minor effects on these platforms."[14]

In the first round of construction after California leased large blocks of lands in 1958, Brown & Root asserted its traditional leadership in this new region by constructing a major platform for Texaco in 94 feet of water near Santa Barbara. At the time, Platform Helen was the largest offshore platform in the world, with a lower deck measuring 124 feet square and an upper deck 55 feet above mean water level measuring 130 by 138 feet. Built at a cost of $2.25 million, Helen had space for a maximum of forty wells.

Construction of a massive platform in the Pacific Ocean required adjustments for a Houston-based company. The large size of the jacket raised problems for shipment through the Panama Canal, so Brown & Root decided to build the decks at its Greens Bayou yard while subcontracting the jacket at a fabrication yard in Long Beach, California. The giant decks required two barges to make the 4,500 mile, 40-day journey through the Panama Canal. Once the jacket had been pinned to the bottom by steel piles driven about eighty feet into the ocean floor through the jacket's hollow legs, the two decks were set in place. By September of 1960, the platform was completed.[15]

Brown & Root remained active offshore California throughout the 1960s, fabricating many of the jackets in California and most of the decks in Houston. The company established a strong presence in this area. Los Angeles-based Union Oil was an important client. One project, Union's "Little Eva" platform in the Huntington Beach field in 1964 and 1965, was billed as the "smallest platform in the west," and its history shows how fast offshore work moved before the coming of stricter environmental regulation. Union bid on the lease for this platform in July 1963. Three days later the first wildcat on the lease was spudded. A platform was ordered from Brown & Root in October of the same year, and the fabricated platform was loaded on a barge in Houston in January 1964. A month later it arrived in Long Beach; by April it was installed, and in May drilling began. Only 27 months after the lease date, 30 wells had been drilled and completed from the $3.65 million platform.[16]

Such rapid construction of an offshore structure slowed dramatically in the years after 1969, when a major oil spill in the Santa Barbara Channel helped launch a national political movement for stricter environmental controls. On January 28, 1969, a blow-out during developmental drilling produced an offshore oil spill estimated at 21,000 gallons per day. Twelve days passed before the spill could be controlled, and 50,000 to 70,000 barrels of oil flowed into the channel. The national outcry helped convince the U.S. Congress to pass a wave of strict environmental regulations and the spill focused strong public sentiment in California against the further development of offshore oil deposits. Things were never again quite the same offshore California. Even when leasing and development opened up again after the furor over the Santa Barbara oil spill subsided, the permit process had become so complicated, convoluted, and time consuming that the twenty-seven month turnaround on Little Eva in the mid-1960s seemed like some sort of mythological event from a far distant past. The delays and controversies surrounding offshore development in California after the Santa Barbara oil spill convinced many companies to look elsewhere for opportunities, and Brown & Root joined others in the industry in moving up the West Coast to Alaska.[17]

Cook Inlet and the Monopod

The first giant step into Alaskan waters came in the mid-1960s, when oil companies invested heavily in the platforms and pipelines needed to recover oil from under the icy waters of the Cook Inlet. Subsequent to Cook Inlet came the extreme challenge of producing from Alaska's North Slope and the construction of the Trans-Alaska Pipeline System (TAPS) to transport this oil to market. Brown & Root took important roles in both Cook Inlet and the North Slope. Although TAPS-related construction and engineering were not, strictly speaking, offshore projects, the company's Marine Division used both technology and client relationships initially developed in other regions to

Cook Inlet and TAPS

become an important part of the spectacular projects on the North Slope. In both the Cook Inlet and the North Slope, Brown & Root's engineering expertise overcame the challenge of a cold environment: ice forces and tides in the inlet; permafrost and low productivity on the Slope.

Brown & Root had previous experience in arctic construction—including work on installations for the DEW-line system which spread from the Bering Strait to the Bay of Fundy—but nothing had prepared the company for the combination of ice forces, tides, and earthquakes in the Cook Inlet, where, according to *Offshore,* "operating conditions are worse than in any spot on the globe."

Development began in the Cook Inlet in 1957 with the discovery of the 200-million-barrel Swanson River field on the Kenai Peninsula about 80 miles southwest of Anchorage, but the pace of oil development quickened after Alaskan statehood in 1959. Alaskan law encouraged unit agreements as an effective means of developing oil deposits in a demanding physical environment, and exploration and development work in Cook Inlet advanced under a number of unit agreements. In the decade after the discovery of the Swanson River field, approximately 120 exploratory wells were drilled in Alaska, with most located in Cook Inlet. After several years of discouraging returns, a series of strikes in 1962–1967 made the inlet one of the hottest plays in the United States. By the late 1960s, six major fields had been found in Cook inlet.

The most obvious new challenges posed by Cook Inlet were ice forces up to 150 kips per foot of horizontal projection, a huge tidal change projection with 30–35 foot tidal variations, and earthquakes as severe as California. Special steels and welding techniques would be required to handle the minus forty degree temperatures. Anything built to stand in Cook Inlet would have to handle much greater lateral loads than previous offshore structures, as well as the continual encounters with blocks of ice in the spring thaws.

In the rush to produce oil from the new fields, traditional small platforms were temporarily installed in 1962, 1963, and 1964. One such "Lake Maracaibo-type tender drilling operation" survived for two winters in the inlet before having its diagonal bracing members sheared off by ice forces. These temporary structures gave engineers platforms from which to test the force and thickness of ice forces in preparation for the construction of permanent platforms. Their studies suggested that "3-foot 6-inch ice sheets with an average crushing strength of about 200 pounds per square inch could exert a lateral force of 100 kips per lineal foot of diameter." Such sheets moved at a pace "about as fast as a man can walk."[18] The bleak conclusion of these initial studies was that "the lateral wind, current and ice force acting on the typical four-legged self-contained type of platform in Cook Inlet was from three to four times as great as for the wave and wind force on a self-contained platform in the Gulf of Mexico."[19]

To stand up to these extreme forces, Brown & Root and others in Cook Inlet developed a new generation of four-legged, self-contained platforms adapted to withstand the ice forces and the tidal changes in Cook Inlet. At a cost "several times as much as the normal offshore platform," four strong legs could be built of low-temperature steel and "designed to withstand a total sideways thrust of 5,000 to 6,000 tons—twice the thrust of the Saturn V rocket used in the Apollo moon project."[20] These platforms had relatively large legs from fourteen to twenty-eight feet in diameter using low temperature steel and no diagonal supports in areas subject to ice forces. Designed to contain all well conductors and risers, these legs bore the brunt of the ice for the entire structure. Enclosed decks allowed workers to avoid extreme temperatures.

Brown & Root constructed several large-legged platforms, including one for a joint venture of Union and Marathon and another for Mobil Oil. The Union-Marathon structure was ordered in December of 1965 for use in the McArthur River field. Designed to drill up to forty-eight wells in depths up to 125 feet, this structure was the first in Cook Inlet with sufficient space for two drilling rigs, a feature that facilitated rapid development during the mild weather of the summer months. The jacket was built for Brown & Root on the west coast by American Pipe & Construction in Portland, Oregon. The large leg design allowed the jacket to be floated on two legs for towing from Oregon to Alaska, where a most difficult setting was required to rotate the jacket into place amid the powerful currents of Cook Inlet. The decks for the Cook Inlet projects were fabricated at Greens Bayou and transported through the Panama Canal for a 5,000 mile voyage to Alaska. This pattern of construction allowed Brown & Root to build the jackets and decks for Alaska in established yards in warm climates where work could go on year around.

While conducting its preliminary studies for the Union-Marathon platform, Brown & Root explored every available design to find a cost-effective engineering solution for the severe environment of Cook Inlet. Using computer programs to study loads, the company examined tender-type structures that could be removed each winter, caisson structures of various designs, four-legged towers similar to that being installed by others in the inlet, a unique A-frame structure, a "monopod" design with one large diameter leg attached to pontoons that rested on the floor of the inlet, and a grounded barge approach. The study concluded that for Union's proposed site in the Trading Bay field, the single caisson monopod design held clear economic advantages over the other approaches.[21]

Thus was born one of the most interesting platform designs in this era of Brown & Root Marine's history. The monopod was fully developed from preliminary sketches to a finished operating complex in thirteen months from August 1965 through October of 1966, and it was installed at Trading Bay in the summer of 1966. Designed for use in about 60 feet of water and an additional 30 feet of tidal variation, the monopod sought to minimize the use of steel while providing maximum resistance to ice forces. The 5,000 ton structure stood 130 feet tall and featured a single giant leg, 28.5 feet in diameter. To spread the tremendous loading from ice forces at the bottom of the leg, engineers used eight large diagonals reaching out from the leg to two pontoons 24 feet in diameter and 174 feet long. An important component of the design was the reinforcement with two-inch-thick low-temperature steel

Monopod under construction

on zones of the upper leg subject to recurring ice forces. The ice moving back and forth in the inlet "cut like sandpaper," and the single leg built of especially strong steel offered both a relatively low cost and a relatively sturdy structure.

The unique jacket and its 105-foot square deck was built at American Pipe and Construction in Vancouver because it was too large to be shipped from Houston through the Panama Canal. Brown & Root used two test models to study forces that would affect the structure during its long tow out and launch. The first model predicted the drag on the jacket during its upright tow from Vancouver to Cook Inlet. Towing such a large structure for long distances and launching the monopod in the tidal surges of Cook Inlet would be risky. Initial plans called for sinking the monopod straight to the bottom, using a pneumatic system to monitor and adjust valves and keep the jacket in balance. Model tests, however, revealed that the monopod's natural buoyancy might prevent it from remaining stable as it sank, so engineers developed an alternative approach that set the structure by placing the tip of one pontoon on the bottom before completing the flooding needed to settle the structure. After much planning using the data obtained from model tests, the tow out and launching went smoothly in the summer of 1966.[22]

Because of the interest generated by this one-of-a-kind structure, the *Oil & Gas Journal* examined its performance several years after its installation. The *Journal* found that the structure had been conservatively built, with its design loading far exceeding the real forces measured on the platform. The single-leg approach had numerous benefits: the location of all wells in the center of the platform, the ease of loading and unloading, the ease of towing in an upright position, and the low initial cost relative to other designs because approximately 70 percent less steel was required to build a single large leg than four smaller legs. The article concluded that the monopod had performed well and that the design made sense in very harsh environments with large horizonal loads in water depths up to 100 feet.[23]

Retrospective analysis of the design within Brown & Root was somewhat more critical. One of the Brown & Root engineers who worked on the project later recalled, "At the time, it was the best thing we had and seemed like the logical way to do it, but, obviously, it way oversimplifies what you have when you have these big cylinders joined together." Another of the platform designers concluded that very conservative construction of the structure had compensated for design problems. In theory, the monopod promised to be a very stable platform in a most unstable environment, but A.B. Crossman recalled that in practice it "was not a real comfortable platform to be on, because the ice would raft up in big, huge chunks and . . . would . . . torque it, and then it would break loose and give it a jerk." This made the monopod "not so user friendly. Evidently, it wouldn't pass the 'toilet test.' You know, if you sat on the toilet there, you'd tend to get all wet because it would give you a good splash." The final word on the monopod design rested on a practical measure of all engineering: "It has managed to survive, so I guess it wasn't too bad."

By the most practical of all measures, however, the monopod was not a rousing success, since the design did not become a standard approach by Brown & Root or other major offshore construction companies. Had today's sophisticated analytical techniques been available in the sixties, the "non-user friendly" aspects of the monopod's behavior could have been predicted and corrected. Not having these methods available, Brown & Root focused on the construction of more traditional four-legged structures modified to withstand the extreme horizonal loading of Cook Inlet.

That new equipment would be needed in this harsh region was most evident in the installation of a four-legged platform for Mobil Oil for the Granite Point field in the inlet. The company agreed to a tight schedule for the delivery of a major platform for Mobil, and it used a relatively new design for floating the jacket from the fabrication yard in Astoria, Washington, down the Columbia River, and up the Canadian coast to Alaska for installation. Jacket design to withstand ice forces had produced structures unsuitable for launching from barges, and in response, Brown & Root and other companies active in Cook Inlet had developed a new generation of self-floating jackets. Built on their

Four-legged platform in Cook Inlet

sides and then launched like ships, these jackets were designed to float on two of their four large-diameter legs. In order to reduce the area of the structure exposed to the ice, Brown & Root had designed the Mobil structure with smaller diameter legs than required for a safe tow. To correct for this, a special flotation device consisting of 110-foot length of pipe about twelve feet in diameter was attached to the permanent structure with latches designed to be opened in the early stages of installation so that the flotation tank could be removed before the final setting of the jacket. Once this temporary tank had been separated from the permanent structure, plans called for a controlled flooding of the jacket to settle it down over an existing wellhead under more than 100 feet of water pushed to and fro by currents.

The plan unraveled quickly. After the jacket arrived at the Inlet, hooks designed to hold the flotation tank to the structure pulled loose under the extreme pressures of the tidal currents, releasing the tank prematurely and causing the "whole bottom end" of the jacket to plunge downward. The structure tried to "turn crossways in the current, breaking the cables where we had it moored to the derrick barge." Suddenly the chase was on, as the currents pushed the structure steadily away from its proposed resting point. Confronted with this runaway platform, the crews could only hang on and wait for reinforcements—stronger equipment to control the partially flooded structure in a tug-of-war with the tidal currents of Cook Inlet. For seven days Brown & Root crews used up to seven tugs and a derrick barge to hang onto the jacket and guide it away from collisions with existing structures in the inlet. To prevent additional flooding, workers rigged an air compressor to maintain air pressure in the legs. John Irons, who had the unenviable assignment of "riding" the jacket, later said: "That fell to my unhappy lot. I'd get on that thing out there 8–10 hours with a long tow rope. . . . I rode the structure to keep the compressor going and to keep checking the pressures in those legs. . . . It was a dangerous and a lonely job, because you could have fallen off in that cold, swift water up there." After a week-long struggle fraught with danger and uncertainty, the crews finally brought the runaway structure to heel and installed the platform. Although Brown & Root had agreed to install the platform on top of an existing wellhead, all involved in this ordeal were content to find a permanent resting place for the platform in the right neighborhood and get on with the job of drilling a new well.

Recalling this episode might be amusing now, but no one thought it was funny at the time. Brown & Root was belittled as "Clown & Boot" and depicted in cartoons as Texas cowboys trying to lasso a runaway jacket. This was a stinging blow to the pride of a self-professed "can do" company, but it reminded everyone of the difficulties of working in a new environment: "People tended to underestimate the dangers of these new areas because everything was extrapolated from such a nice calm place such as the Gulf of Mexico that is so easy to work in." No one involved in the desperate effort to control the runaway jacket in Cook Inlet would ever again underestimate the challenges offshore Alaska.

New, stronger equipment was needed to answer these challenges. While undertaking a series of pipeline projects in the region, Brown & Root faced conditions it had never before experienced. Eight-knot currents, thirty-two-foot tides, fast-rising storms with winds up to seventy miles per hour that produced swells as high as eight feet, and a very short working season made pipelining in the Cook Inlet an adventure. Moreover, the area lacked places to house and supply crews, and the distances from fabrication yards were long.

Many of the unique aspects of Cook Inlet pipelining could be seen in Brown & Root's construction of thirteen miles of dual, 8-inch gas pipeline for Phillips Petroleum in the northern sector of the Cook Inlet in the summer of 1966. These lines connected a platform still under construction in the inlet to an existing 16-inch onshore line at Moose Creek, about forty-five miles south of Anchorage. Careful planning was used to "beat the clock" and complete this project in the brief window of opportunity provided by the relatively favorable conditions from early July to late September, the primary construction season in this section of Alaska. Supplies and equipment were

Stinger in Cook Inlet

brought to the construction site the previous winter and left there for the construction scheduled to begin the next summer. Included in the equipment was a 350,000 pound stinger plough made of special high strength steel to guide the pipe to the bottom of the inlet while also using 60 psi high-pressure water jets to dig trenches. This stinger was simply laid on the bottom of Saldovia Bay during the winter and then raised to the surface when construction began. As workers arrived at the distant site by boat in June 1966, they found waiting for them a 192-bunk quarter's boat that would be their home for the duration of the project. The crew spent a spartan summer of twelve hour work days laying the pipe as rapidly as conditions allowed.

The severe currents carrying glacial silt back and forth dictated the use of two 8-inch lines instead of one larger line. Parallel lines provided an extra measure of security, since one would continue to operate even if the other had to be shut down for repairs. The smaller diameter also gave the pipelines a smaller profile to lessen the force of swift currents during installation.

Brown & Root began its work at the platform, where it connected the two lines through a J-tube to a $20 million platform being constructed with a capacity to drill thirty-two gas wells. As Brown & Root worked from the platform toward the shore, Locher Construction worked outward from the shore to 1,000 yards offshore, where the two parts of the line ultimately met. Locher had been acquired by Halliburton in 1968, and it worked with Brown & Root on major projects in Alaska.[24] Work on the Phillips pipelines proved most difficult, since glacial forces had packed the earth so hard that even the largest available trenching equipment could not dig the five-foot deep trenches needed to bury the lines in the tidewater section near shore. The solution was to bring in explosives experts and blast the trench to the required depth.

As Locher pushed outward under very trying conditions, Brown & Root's lay barge steadily moved toward the shore. Conditions in Cook Inlet forced numerous adjustments in traditional pipelaying procedures. The use of the giant stinger plough was the most obvious of these adjustments, but others occurred underwater and out of view. Huge boulders deposited by glaciers over the bottom of the inlet plagued pipelaying crews. A sonar scanning instrument aboard a survey vessel could scan forward from 200 feet to 3,000 feet, searching for boulders and other obstructions along the pipeline's projected route.[25] Some of these were removed by cranes, but others, too heavy to lift to the surface, were attached to a rock grappler hung from the beam of a 150-ton crane and then swung underwater to new locations. This procedure had to be completed ahead of the pipeline crews, and it had to be done quickly, since construction costs of $3,000 per hour gave everyone a very strong incentive to avoid delays.

To keep the pipeline on its planned path in surging water, Brown & Root used two unconventional procedures: a laser system and a transponder system. Set on shore and pointed toward the lay barge, the laser provided a quick and easily read bearing. The pencil-thin red light pointed out the path; if the lay barge wandered very far off course, those on deck could no longer see this beam of light. Anchor operators could simply aim down this laser beam to stay on the proscribed course. Added accuracy came from radio signals sent out by two transponders that were triangulated to plot the "as-built" location of the pipeline within a few feet. The two locational systems helped crews winch the lay barge forward in a relatively straight line using a spread mooring system of two-inch thick, 3,000-foot-long anchor cables attaching the barge to the four 30,000 pound anchors and eight 20,000 anchors. This arrangement gave it the stability needed to accomplish its task.[26]

Perhaps the most difficult job required at Cook Inlet was that of the divers, who inspected the pipeline once it was laid. The glacial silt in the inlet clouded the water, obscuring underwater vision. In places, the divers had no alternative except to run their hands over the pipe, carefully feeling for flaws in an exacting, inch-by-inch inspection. No doubt, these divers were not disappointed when the project was completed ahead of schedule in early September, well before the severe winter conditions made further work impossible.

The North Slope and TAPS

Cook Inlet had attracted major oil companies to Alaska, but in the late 1960s, a massive discovery at Prudhoe Bay shifted the focus of oil development to the North Slope. Everything about this new prospect was larger than life. With an estimated recoverable reserve of almost ten billion barrels of oil, Prudhoe Bay was the largest field ever discovered in the United States, but its location in the arctic circle meant that it was far away from established infrastructure and plagued by the most extreme weather as yet encountered by the petroleum industry. Its projected production of up to two million barrels per day promised a measure of relief to the United States from its growing dependence on foreign oil, but efforts to reach this level of production faced strict scrutiny from a new wave of environmental regulations passed in the late 1960s and early 1970s.[27] In this era, Alaska's North Slope became the most significant petroleum development in the United States, and Brown & Root took a front row seat for this big show.

The show was held in a very cold tent. Temperatures dipped to minus 70 degrees Farenheit, and winds reached 90 miles per hour. An arctic desert, the North Slope did not experience heavy snowfall, but high winds whipped up snow and ice. The petroleum industry had never before worked in such conditions, which threatened the physical and psychological well-being of workers and the functioning of equipment. Successful oil production on the North Slope required innovations that protected both workers and machines from one of the harshest and most isolated working environments in the world.[28]

The arctic environment was harsh, but also fragile. Under the tundra's shallow layer of surface soil lay up to 1,800 feet of permafrost—material frozen solid for at least two years. New sources of heat from surface operations could upset the natural equilibrium and cause serious environmental problems as well as practical problems for construction. Thus, arctic construction needed to protect the permafrost, and this concern affected almost every phase of the design and construction of production, gathering, and processing facilities at Prudhoe Bay and of the the 800-mile trans-Alaska pipeline system (TAPS) built to take oil to Valdez, an ice-free port in southern Alaska from which the oil could be shipped to market in ocean-going tankers.

Brown & Root Marine became deeply involved in work at Prudhoe Bay and on TAPS, although this was not, strictly speaking, marine work. While not exactly the same as platform and pipeline design and construction, the creation of the permanent production facilities on the North Slope made use of several of Brown & Root's strengths: creative design, prefabrication, the capacity to complete demanding jobs on a strict timetable, assembly of structures in harsh environments, and project management. The company also benefited from its ties to British Petroleum, one of the major companies active at Prudhoe Bay.

Brown & Root had completed demanding projects for BP at Kharg Island in the Middle East and in the North Sea. Thus when BP and its partner Sohio sought a contractor to engineer and construct half the facilities at Prudhoe Bay, the companies looked to Brown & Root. ARCO had responsibilty for building the other half of the permanent complex needed at Prudhoe Bay, and it, too, negotiated for a time with a joint venture including Brown & Root for this work. Indeed, the call from BP to invite company representatives to London to discuss the Prudhoe Bay project came to Houston during a conference with ARCO on the same topic. Key Brown & Root executives excused themselves from the conference to take the call, then hopped on an all-night flight to London. One of those who made the flight recalled "We landed in London that morning, and about four days later we had the job."

The Prudhoe Bay complex was massive and difficult to construct. As oil companies drilled wells in the region, Brown & Root moved forward aggressively with the design and construction of all of the pipelines and buildings needed to produce and transport oil from these fields to the main pipeline. Included were the three gathering centers in BP's half of the field, the power plants and electric grids for the entire area, a large headquarters/control building, and other outlying buildings. Additionally, 200 miles of pipelines were needed to transport crude oil to the gathering centers and then to ship separated oil to the northern end of TAPS while reinjecting natural gas back into the field. In all of this work, the key design considerations included protecting workers and equipment from the cold, protecting the tundra from possible damage, and minimizing the time spent on construction.

Prefabrication of most of the facilities in Houston and on the west coast allowed Brown & Root to minimize the delays and costs of arctic construction, and the North Slope work for BP took prefabrication to an unprecedented level. Entire buildings bound for Alaska and weighing up to 600 to 1,000 tons were built in Houston and shipped in modular form through the Panama Canal, up the Canadian and Alaskan coasts, around through the Bering Strait and on over the Beaufort Sea to Prudhoe Bay. This journey of almost 10,000 miles took approximately two months, and it had to be planned carefully so that the barges reached the North Slope during the six-week period from late July to mid-September when ice usually subsided enough to allow barges to complete the trip to Prudhoe Bay.

In May of each year from 1970 through 1977, barges left Greens Bayou bound for the North Slope. Other modules left Seattle. Even the best planning could not prevent problems when winter came unusually early. As the company pushed for completion of the processing center at Prudhoe Bay in the summer of 1975, many of the barges bound for Prudhoe Bay had to turn back because of ice. Those that had reached the site earlier were frozen out in the harbor, and a bridge made of ice had to be rigged to unload their cargoes. The vital materials in the barges that turned back had to be transported across Alaska by more expensive overland and air routes, but most of the shipments managed to reach their destination before the winter.[29]

Once barges reached Prudhoe Bay carrying their prefabricated modules, the real fun began. These extremely heavy structures had to be transported as

Prefabricated structures bound for North Slope via Panama Canal

many as twenty-five miles inland to locations on the tundra. The solution was a tractor-like contraption known as a "creepy crawler" or simply a "crawler." Designed on the same lines as the moving platforms used by NASA to transport the Saturn rocket to the launch gantry, these crawlers had four 250-ton hydraulic jacks that lifted modules eight inches off the transportation supports and moved them at a steady rate of .5–.75 miles per hour. In Houston, crawlers loaded the modules on barges sunk to the bottom of a special loading harbor. After deballasting, the barges began their long journeys under tow to Alaska. On the North Slope, crawlers unloaded the barges and moved their cargo over gravel roads to inland locations, where they were mounted on preconstructed piled foundations supported in part by "ice piles" made by drilling a hole in the ground, filling it with a mixture of water and sand, and waiting for this mixture to freeze.

The prefabricated structures embodied many innovations in design. Although the Russians had published a limited body of literature on arctic construction, when it came to design, Brown & Root's engineers had "to do it the hard way," going "back basically to heat transfer and all kinds of things" not generally considered by structural engineers. One key to arctic design was a near-obsession with insulation. The modules for Prudhoe Bay were built around a steel frame using poured-in-place concrete and wood panels enclosed in a metal skin. Doors with air seals worked like "refrigerators in reverse,"

Creepy crawlers at Prudhoe Bay

sealing the warm interior space against the bitterly cold wind that pushed ice and snow against the structures. To minimize the buildup of snow from drifts, the structures used aerodynamically designed exterior walls with curved surfaces and no exposed ninety degree angles. All designs stressed enclosed work spaces, including modular designs that allowed pieces of buildings to be joined together from the inside out. Designers incorporated high-strength steel and rubber to minimize the effects of weather on equipment, and they devised means for treating waste and protecting water and fuel. Even the simplest tasks required creative adaptations in the arctic regions.

While keeping the cold out of the structures, designers also went to great lengths to keep heat generated by the facilities away from the fragile permafrost: They had to examine the effect of flares on surrounding soil and predict the effect of heat projection out of the buildings. Protecting the permafrost required using thick gravel pads underneath structures, constructing structures and pipelines off the ground on piles so that cold air could dissipate heat, and limits on many types of construction to winter months when the soil was hard frozen. The lessons learned in building the permanent production complexes on the North Slope subsequently could be applied to the construction of the northernmost sections of TAPS, which used innovative multiple insulation around buried sections of pipe while elevating about half the pipeline above ground.

The completed BP complex was a mini-city, sturdy and self-contained in the remote arctic. The centerpiece was the operations center, dubbed by many "the BP Hilton." This 95,000-square-foot, three-story building contained the electronic controls and monitoring equipment for the western half of the field and the gathering and processing centers, permanent dormatories for 140 workers, and an indoor recreation center. To help combat "cabin fever" among hard-working people in a remote and desolate environment, the interior sported bright colors, an indoor swimming pool, and a glass-enclosed arboretum "complete with flowers and Prudhoe's only trees." ARCO's similar facilities in the eastern half of the field held permanent housing for about 200. The 5,000 to 6,000 workers who developed the field and helped build the production facilities found housing in temporary structures and in "camps" such as the one maintained by ACI in Deadhorse, south of Prudhoe Bay. Efficient, if not luxurious, accommodations were the order of the day for those working in most severe conditions.[30]

The facilities at Prudhoe Bay ultimately fed up to two million barrels of oil per day into TAPS, a 48-inch pipeline with twelve pumping stations. Brown & Root participated in two ways in the construction of TAPS. With Williams Brothers Alaska, it consulted for project design review and preconstruction planning. The company also was a member of a joint venture named Arctic Constructors (Brown & Root, the Ingram Corporation, Peter Kiewit Sons, Williams Brothers Alaska, and the H.B Zachry Company) that built the northernmost 223 miles of the line from Prudhoe Bay to the Yukon River. In both capacities, Brown & Root worked for ALYESKA, a consortium headed by Sohio, BP, ARCO, Exxon, and Mobil, which owned and operated TAPS.

Brown & Root began its work as a design consultant in 1972, during a four-year delay in pipeline construction due to legal challenges and problems with native land claims. Key issues examined included seismic design, heat transfer along the line, and wildlife crossings. Severe earthquakes had to be considered in pipeline design in Alaska, and TAPS incorporated innovative measures to prevent damage in the event of major quakes along the pipeline route. A distinctive zig-zag design combined with the use of high-stength steel and strong vertical support members gave the pipe extraordinary "survivability" for earthquakes much stronger than any recorded in Alaskan history. To prevent damage to the permafrost from heat transfer from the hot oil flowing through the cold pipeline, designers used multiple layers of insulation and aboveground construction in unprecedented ways. Finally, they designed crossings to avoid obstructing migrating wildlife. The legal and regulatory challenges by environmentalists encouraged strict attention to such issues, and "there was a lot more attention to the right-of-way and to servicing this thing than I think the original folks had intended."

Work finally began on the pipeline in the spring of 1975 and continued at a frantic pace until the completion of the line in the summer of 1977. In 1976 and 1977, as many as 19,000 workers pushed the pipeline rapidly through long stretches of forbidding terrain. This dangerous, demanding task progressed with ten-hour shifts and schedules of 50 days on followed by 14 days off for most workers. Hazardous conditions took their toll even on Brown & Root executives. In January 1976 an airplane crash in Anchorage following an inspection trip to Prudhoe Bay took the lives of four Brown & Root executives, G. A. Dobelman (Senior Group vice president), Warren T. Moore (president of ACI), Wolf Pabst (vice president of ACI), and Vic H. Abadie, Jr. (manager of engineering for the company's San Francisco office).[31] The danger to those who built the pipeline was especially clear to workers from ACI who built forty-three microwave towers—twenty-seven of them on remote mountain sites serviced by helicopters—to house the microwave communication system used to monitor the operations of TAPS. There was no easy work on TAPS, whose completion was a monumental achievement.

Brown & Root Marine crawled onto land for its work on the North Slope and TAPS, but its marine experience proved useful in completing this demanding and exacting work. Project management was the order of the day, as BP focused on its own developmental work and relied on Brown & Root to take charge of the construction of its Prudhoe Bay facilities. Cook Inlet served as a useful introduction to the severe weather encountered on the North Slope. Previous work with prefabricated structures gave Brown & Root a running start in organizing the flow of modules to Prudhoe Bay, and this work, in turn, helped to revitalize the company's Greens Bayou yard. In assuming a significant role in one of the great petroleum "plays" in U.S. history, Brown & Root joined many major oil and oil-related companies in Alaska. Its experience there helped prepare the company to undertake other large and demanding projects offshore Mexico, in the North Sea, and around the world.

Notes

1. This chapter draws from interviews with P.H. Moore, J.B. Weidler, Marshall Cloyd, A.B. Crossman, John Irons, Edward Tallichet, W.B. Pieper, and R.O. Wilson.
2. R.F. Yerkes, H.C. Wagner, and K.A. Yenne, "Geology, Petroleum Development, and Seismicity of the Santa Barbara Channel Region, California: Petroleum Development in the Region of the Santa Barbara Channel," *Geological Survey Professional Paper 679-B* (Washington, D.C.: GPO, 1969): 15.
3. L.P. Stockman, "Eight Percent of California Crude Produced from Tideland,"*World Petroleum* (December 1947): 42–46.
4. Wallace A. Sawdon, "Deep Sea Drilling on an Island of Steel," *The Petroleum Engineer,* Vol. 4, No. 3 (December 1932): 26–27.
5. Ralph G. Frame, "California Offshore Petroleum Development," in *Summary of Operation, California Oil Fields, 46th Annual Report of the State Oil and Gas Supervisor,* Vol. 46, No. 2, 6–7.
6. "Drilling Island Nears Completion," *Offshore* (July 1958): 68–69.
7. "West Coast Oil Future is Bright," *Offshore* (June 20, 1967): 70–71.

8. "Barges Coring Off California," *Oil & Gas Journal* (July 7, 1958): 116.
9. William T. Smith, "Drilling Off California Will Be Tough," *Oil & Gas Journal* (February 17, 1958): 68–70.
10. *Ibid.,* 70.
11. I.Demir Karsan and Albert Koehler, "Design of Offshore Platforms Subject to Seismic Loadings," Proceedings of OE Con II, 1969, 285–314.
12. See, for example, American Petroleum Institute, Recommended Practice for Planning, Designing and Constructing Fixed Offshore Platforms, RP2A, 7th Ed., Dallas, Texas, January 1976.
13. Joseph Kallably, "Considerations for Analysis and Design of Piled Offshore Structures in Severe Earthquake Environment," OTC 2748, May 2–5, 1977.
14. Richard K. Miller and Stephen F. Felszeghy, "Engineering Features of the Santa Barbara Earthquake of August 13, 1978," Earthquake Engineering Research Instittue, UC-Santa Barbara, December 1978.
15. Brown & Root Marine: Engineers, Constructors, undated pamphlet, Brown & Root Archives; Simon Cordova, "Cuarta Offshore Oil Field," *58th Annual Report of the State Oil and Gas Supervisor,* California Summary of Operations, 1972, Vol. 58, No. 1.
16. Ted Armstrong, "Little Eva: A Fast Reputation Out West," *The Oil and Gas Journal* (August 9, 1965): 58–59.
17. The Santa Barbara oil spill has been discussed in numerous books. See, for example, Robert E. Kallman and Eugene D. Wheeler, *Coastal Crude in a Sea of Conflict* (San Luis Obispo, Cal.: Blake Printery, 1984) and Dwight Holing, *Coastal Alert* (Washington, D.C.: Island Press, 1990).
18. *The Oil and Gas Journal* (January 13, 1964).
19. J.B. Daigle, "Cook Inlet Drilling and Production Platforms," *Proceedings of OE Con III, 1968,* 20–47.
20. "Conquering Cook Inlet," *Shell News* (November–December 1966): 16.
21. Brown & Root, "Cook Inlet, Alaska: Drilling Structures for Union Oil Company of California," July 1964, copy in Brown & Root Archives.
22. Marshall P. Cloyd, "Monopod," *Civil Engineering-ASCE* (March 1968): 55–57.
23. "How Union's Alaskan Monopod Has Performed," *The Oil and Gas Journal* (March 2, 1970): 68–69.
24. "Locher Company . . . Growing with Alaska," *Brownbilt* (Fall 1971): 18.
25. "New Sonar in Subsea Work," *Offshore* (September 1966): 68.
26. Brown & Root, "The Challenge of Cook Inlet," copy of film in Brown & Root Archives.
27. For a contemporary view from an oil industry perspective, see James P. Roscow, *800 Miles to Valdez* (Englewood Cliffs, New Jersey: Prentice-Hall, 1977). For a more recent account of the environmental controversies, see Peter Coates, *The Trans-Alaskan Pipeline Controversy* (Anchorage: University of Alaska Press, 1993).
28. F.G. Larminie, "The Arctic as an Industrial Environment," *Journal of Petroleum Technology* (January, 1971): 19–26.

29. "Ice Blocks Cargo Shipments for Trans-Alaska Pipeline," *Engineering News-Record* (September 18, 1975): 42.
30. Nancy Rueth, "Prudhoe Bay: Oil Boom Town," *Mechanical Engineering* (November 1975): 22–27.
31. "Plane Crash Claims 4 Builders, Injures 2 Others," *Brownbilt* (Winter 1975–76).

CHAPTER 10

A Crash Program in Mexico's Bay of Campeche

As Brown & Root confronted the technical challenges of the most extreme kinds of environments, it also gained political and organizational experience in working with multiple governments, contractors, and fabricators. The company's work in Latin America, the Middle East, and Africa, its management of Project Mohole, and its leading position in the North Sea and Alaska demonstrated the marine unit's wide-ranging expertise and capabilities. By the mid-1970s, it seemed that the only question regarding Brown & Root's talents in offshore development was not "Can you do it?" but "How fast can you do it?"[1]

The demonstration effect of Brown & Root's pioneering efforts worldwide, and in the North Sea in particular, earned the company new offshore business in Mexico in the late 1970s. Mexico's long-standing oil autonomy had finally weakened. The country's oil production declined and its reliance on imported crude oil had risen. Mexico's "crisis in self-sufficiency" motivated Petroleos Mexicanos (PEMEX), the state oil company, to explore for new sources of domestic oil reserves.[2] In 1977, President José Lopez Portillo staked his new government's oil development strategy on suspected sources of oil offshore in Mexico's Bay of Campeche. Because of Brown & Root's proven leadership in the North Sea, and the strategic contacts it made with important Mexican individuals, PEMEX chose Brown & Root to manage the development of offshore Bay of Campeche.

The most challenging aspect to the Bay of Campeche development was time. For Brown & Root, Mexico was a straightforward project involving Gulf of Mexico technology. Depth and weather were not major problems, the continental shelf area was similar to the northern Gulf of Mexico, and the region was relatively tranquil. Money also was not an issue—PEMEX was willing to underwrite almost any cost. But Lopez Portillo and PEMEX did want to see oil production developed as rapidly as possible. The Bay of Campeche turned out to contain major offshore fields. Even if the technology was not as exotic as that used in the North Sea or in the deepwater Gulf of Mexico, the rapid development of those fields required a tremendous amount of effort and coordination in the procurement, fabrication, and installation of the system's structures. It also demanded the creative energy of Brown & Root engineers to try to contain a major well blowout that gushed for nine months in 1979–1980.

Brown & Root's operating arrangements with Mexico were complicated. PEMEX is the state oil company, whose mission is to preserve the country's oil independence. To work in Mexico, Brown & Root had to form joint venture arrangements with Mexican firms, one in engineering and one in construction. Brown & Root officials developed a relationship of trust and cooperation with its Mexican partners and with PEMEX. During 1977–1982, oil production in the Bay of Campeche soared to slightly less than two million barrels per day. Brown & Root engineer Chuck Osborn reminisced that the Bay of Campeche was "a once-in-a-lifetime project [in] that you never see something like that again because it was hurry up and build it and get it installed."

The Mexican Oil Play

Established in 1938, PEMEX symbolized Mexican nationalism. The company underwrote Mexican industrial growth after World War II by supplying oil to the domestic market at subsidized prices.[3] By the late 1960s, however, Mexican oil reserves had dropped, and by 1971, Mexico had became a net importer of crude oil. The shock of soaring world oil prices caused by the OPEC embargo in 1973 plunged the nation into a prolonged economic crisis, which peaked in 1976 with the devaluation of the peso.[4]

The Mexican government responded in 1974 with a three-year, $3 billion, onshore and offshore exploration program.[5] The government hoped to regain national self-sufficiency in energy and expand exports to take advantage of high world oil prices. The discovery of the Reforma fields in the states of Tabasco and Chiapas stimulated a new program of oil exports. By the spring of 1975, Reforma was producing 315,000 barrels per day, almost 50 percent of the country's output. But as the *Oil & Gas Journal* observed in 1975, "Reforma alone . . . won't be able to bear the entire burden. . . . especially when the rest of the country's fields have at best been stabilized through repressurization or have a marked decline trend."[6]

PEMEX officials soon shifted their attention offshore. PEMEX had developed wells during the late 1960s and early 1970s offshore Tuxpan,[7] but in the fall of 1975, the company announced that it had encountered signs of significant oil and gas from cretaceous structures tapped by a wildcat well, Chac No. 1, drilled 43 miles north of Ciudad del Carmen in the Bay of Campeche.[8] Further drilling, seismic studies, and mapping in the area led PEMEX to suspect that the oil came from an offshore extension of the Reforma fields 100 miles away, one that contained "reserves comparable to those of some fields in the Middle East."[9] Taking office in December 1976, the new government of José Lopez Portillo unveiled a six-year program that called for $15.5 billion in capital expenditures for petroleum exploration and development, part of which would be focused on marine areas. PEMEX enlarged its offshore rig fleet by leasing

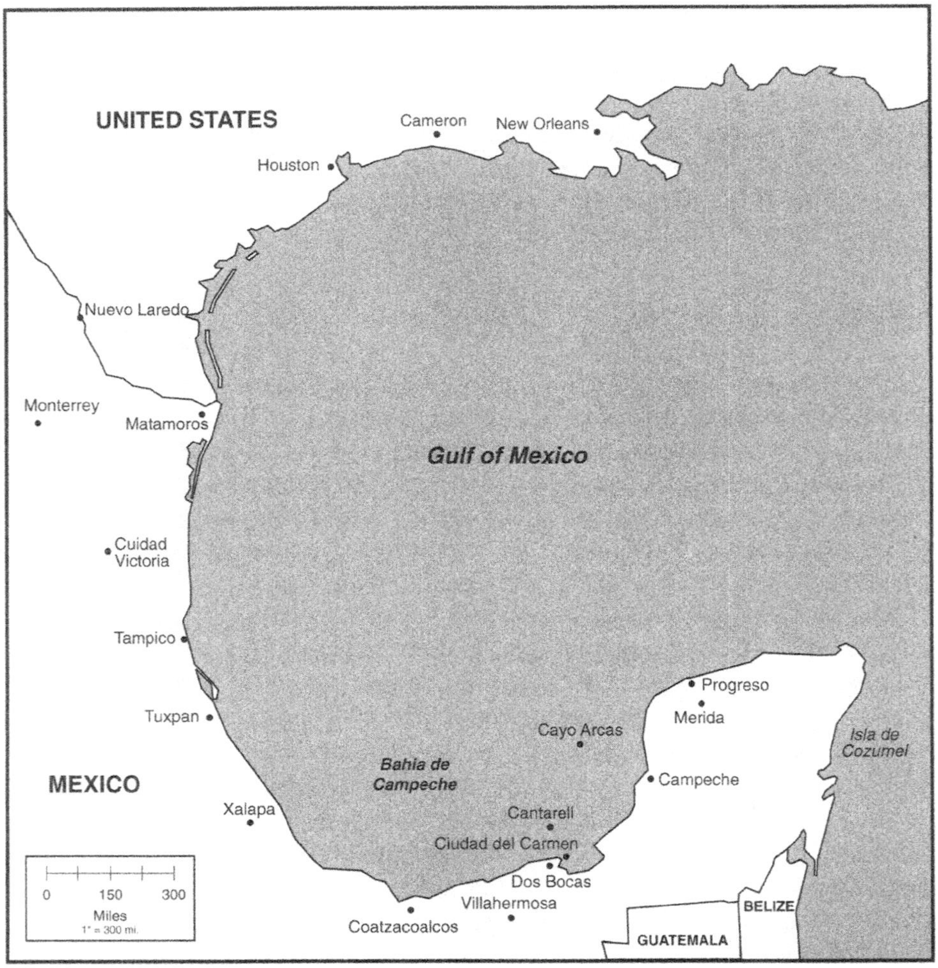

Southern Gulf of Mexico

two jackups and the semisubmersible *Sedco 135*, with which it planned to drill exploratory wells and build up an offshore production capacity of 360,000 barrels per day by 1982.[10]

Brown & Root seized an early opportunity to provide engineering services for developing oil in the Bay of Campeche.[11] PEMEX had not yet made a big discovery in the bay, and so it had not drawn up plans or made any public announcements about it. The objective of expanding production for export was "still very unsettling to most Mexicans."[12] But Jorge Diaz Serrano, the new director general of PEMEX under Lopez Portillo, approached Alfonso Barnetche Gonzalez to prepare for a possible major development in the bay. Engineer Barnetche had retired from PEMEX as subdirector of drilling and production. Barnetche contacted Halliburton, Brown & Root's parent company. Ed Paramore, president of Halliburton, put Barnetche in touch with Dick Wilson, a Brown & Root executive who had returned from the North Sea and now had worldwide offshore responsibilities. Barnetche told Wilson that PEMEX intended to develop offshore fields in the Bay of Campeche and he proposed a partnership with Brown & Root to manage the engineering for the development.

In late January 1977, Barnetche and Wilson took PEMEX representatives to the North Sea to demonstrate Brown & Root's handling of major projects. In February, they escorted a group to the London offices of Brown & Root, U.K. After London, the group moved on to Aberdeen to see the engineering work being done at the end of the Forties project. They went offshore to the Ekofisk complex, which involved enhanced Gulf of Mexico technology with structures in 200 feet of water. The group traveled to Bergen to view the firm's work in the Norwegian sector on the Statfjord B project. At the end of the trip, director-general Diaz Serrano joined the group in visiting the Paris office, where Brown & Root was designing the Frigg field development for Elf and Total.

According to Jamie Dunlap, project director of the Paris office, the Frigg project convinced PEMEX to accept Brown & Root as the project manager for the Bay of Campeche development. Brown & Root developed a "large amount of project management technology" for the Frigg field, the largest gas field that had ever been developed offshore up to that point. In that project, the company completed the first pipeline pull into the base of a large structure. The most complicated aspect of the development, however, was the fact that the field was located on the median line between the U.K. and Norwegian sectors. Brown & Root had to split design work between the sectors, under different regulations and for different certifying agencies. The Paris office housed a staff of 1,300, who coordinated work in seven countries and at eleven fabrication sites. The office developed innovative scheduling techniques and possessed the only Brown & Root computer link to the United States from the continent at the time. Dunlap remembered taking the PEMEX group into the computer room, which contained a Data 100, a unit the size of two desks stacked vertically and supported by 40 or 50 key punch operators. Dunlap asked one operator to "pull me up Houston. She typed in 'Hello Houston,' and in a couple minutes received the reply, 'Hello Paris.'" This computer technology was impressive for the time and was "one of the things that really sold [PEMEX] on the job."

In April 1977, Wilson and Barnetche negotiated a contract with PEMEX modeled on the contract that Brown & Root had with Elf and Total for engineering and project management services. Barnetche and Brown & Root established a fifty-fifty joint venture, which they named Proyectos Marinos (Marine Projects), as the official project manager for the Bay of Campeche development. As early as May 1977, Brown & Root started transferring people to Mexico, many from the Paris office. Design work began in the summer. Proyectos Marinos opened an engineering office in Mexico City that was run by Jamie Dunlap on the Brown & Root side of the joint venture and by Alfonso Barnetche and his brother Alberto on the other side. "The Barnetche brothers and Jamie, and about five or six [other] people all sat in one big room at the very beginning," said Osborn. The Mexico City office eventually expanded to a staff of more than 1,000 people. All of the basic engineering was done in Houston initially, until much of it was moved to Mexico as part of a technical services agreement to train Mexican engineers in offshore design.

In the summer of 1977, Brown & Root engineers started designing production platforms for about 50,000 barrels per day and wells for 3,000–5,000 barrels per day, but no major well had been discovered. "Talk about being ahead of the game," Wilson recalled. "[We had] no indication what size the wells would be or if there would even be any wells." Development in the Bay of Campeche was driven largely by PEMEX's faith that oil existed there and by its desire to produce it as quickly as possible to help meet Lopez Portillo's six-year goal to increase total Mexican oil production from 900,000 barrels per day to 2.25 million barrels per day. PEMEX's hopes for the bay were exceeded. The first major discovery well, Akal, which was drilled in late 1977, came in at 20,000 barrels per day. Proyectos Marinos's preliminary designs were too small. They had to be modified and upgraded for wells that averaged 8,000 barrels per day. "They weren't expecting North Sea-size fields," said Wilson, "but North Sea-size fields are what they got."

In fall 1977, PEMEX announced that Akal, and a subsequent discovery called Nohoch, indicated not a marine extension of the Reforma producing area, but a new offshore structural trend in Paleocene limestone pays. PEMEX named the combined Akal and Nohoch discoveries the Cantarell Field. "We seem to be facing another startling development," marveled Graciano Bello, PEMEX's southern zone manager, "It's like the nurse coming out of the delivery room to tell the anxious man he's the proud father of a baby . . . only to come back a while later to tell him it's twins."[13] The announcement that the Bay of Campeche might be another North Sea jolted the world oil industry. Marine design companies descended on Mexico City seeking a piece of the action, but they were six months too late. Brown & Root already "had engineering locked up."

Brown & Root also enjoyed an advantage in construction. The Barnetches preferred to stick exclusively to engineering, but with the brothers' help,

Brown & Root found another partner, Felix Cantu. Like Barnetche, Cantu had worked for PEMEX and then owned a construction company. In 1969, Cantu had supervised a pipeline that Brown & Root had laid off Tuxpan to the Morsa Platform. Ed Tallichet, who had managed the project for Brown & Root, had introduced Cantu to Wilson. In the summer of 1977, Brown & Root formed a new joint venture with Cantu called Corporacion de Construcciones de Campeche (CCC).

PEMEX eventually subcontracted a substantial part of the marine work to CCC. A portion of the fabrication was performed in Mexico, and CCC set up a yard in Tuxpan, where several structures were built, and an operations base at Cuidad del Carmen. Much of the major fabrication, however, happened at Greens Bayou. Coupled with existing business on the U.S. side of the Gulf of Mexico, the Bay of Campeche development kept Greens Bayou completely busy during 1978–1982. "I had never seen a project like that before or since," said Osborn. "Our fabrication yard here at Greens Bayou was full of jackets and we'd have meetings twice a month with about fifty people to coordinate everything." PEMEX did give some work to other marine construction outfits, such as J. Ray McDermott and Heerema. For the most part, though, the Bay of Campeche was a Brown & Root show. "We had so many people involved in that, both in Mexico and in the yards," remembered Joe Rainey, "that it was sort of mind boggling at times."

Fast-Track Development

Brown & Root's scheduling ability was one of the qualities that had convinced PEMEX to contract with the company for the Bay of Campeche development. Sage Burrows claimed that "a lot of our work came because we did have the ability to produce schedules and stick to them." Brown & Root soon found out, however, that with PEMEX the concept of scheduling flew out the window. "You just worked seventy hours a week as fast as you could on everything you could and put as many people on it as you could," said Burrows. PEMEX "just didn't accept any schedules that you gave them. They wanted you to get out there and work full-blast, full-time, and get it out the quickest that it could be gotten out."

The Cantarell development was a crash program from the beginning. After a string of discoveries in 1977–1978, PEMEX shifted into high gear, announcing plans to drill twenty-four wildcats. At the end of 1977, PEMEX unveiled plans to build a $150 million, deepwater oil-export terminal capable of handling supertankers at Dos Bocas on the southern coast of the Gulf. CCC received the contract to lay a 36-inch pipeline to connect the Cantarell complex with Dos Bocas. Exploratory drilling revealed twenty new fields and eight extensions in the complex.[14] PEMEX developed a two-year plan for Cantarell that entailed picking the five best structures, setting two twelve-slot drilling platforms, and planting twenty-five smaller drilling platforms and seven production units in waters extending to a maximum depth of 230 feet.

Although PEMEX kept a low public profile about its discoveries, it pushed hard privately to make sure that the development would live up to expectations.[15] By the fall of 1978, those expectations were growing. Word spread that Mexico's new offshore reserves were "monstrous." Some analysts even compared the size of Mexico's oil fields to those in the Middle East.[16]

Brown & Root mobilized its engineers and fabrication yards to get the crude flowing and meet PEMEX's urgent demands for early production platforms. But by mid-1978, when Brown & Root started building structures and scheduling equipment, bottlenecks appeared within PEMEX. Proyectos Marinos officials realized that they "weren't going to meet schedule commitments if decisions weren't made then, that week." Barnetche and Wilson explained the problem to engineer Adolfo Lastra, PEMEX subdirector of primary production, who appointed Clemente Beltran to be the PEMEX project director for offshore. Beltran's appointment sparked meetings at Proyectos Marinos, often lasting until two or three o'clock in the morning, where Beltran made firm commitments on schedules, barges, and equipment. The new unified command at PEMEX was a watershed for the project, allowing it to advance.

By early 1979, the Cantarell development had gained momentum. Brown & Root vessels soon arrived from the North Sea. The company eventually employed five major pieces of equipment in the Bay of Campeche: the combination derrick-lay barge *L.B. Meaders,* the lay barge *BAR-280,* and the derrick barges *H.A. Lindsay,* the *BAR-297,* and the *George R. Brown.* In March 1979, the lay barges completed construction of the 102-mile pipeline linking the Akal fields to Dos Bocas. PEMEX wanted state-of-the-art pipelines, so CCC brought in Taylor Diving to perform saturation diving and make hyperbaric welds. Welded underwater pipeline connections incurred higher costs and longer installation times, but they mitigated the possibility of leaks. In early 1979, Brown & Root built and installed four drilling platforms and had scheduled other early production facilities for installation in the Akal, Nohoch, and Abkatun reservoirs between April and June. The CCC staffing report for May 1979 listed 3,075 people employed in the Cantarell development: 80 in pipelining, 2,572 in hook-up, 389 in diving/welding, and 28 in administration, all working on day rates.[17]

Offshore production from Cantarell appeared as the savior of the Mexican economy. "Oil reserves and production is (sic) not only the symbol of Mexico's independence," wrote *Offshore* in May, "but will likely be the only solution to chronic problems in its economy."[18] PEMEX director-general Diaz Serrano stated that growth in Mexican oil reserves depended on what was found offshore. Cantarell continued to yield large wells that improved the prospects for Mexico's oil-export development strategy. PEMEX found the Akal and Nohoch fields to have a much greater spread than originally thought, prompting PEMEX to double its original estimates of the productive capacity of the area. In June 1979, Diaz Serrano said, "The North Americans think we have 60 billion barrels, and they are right."

The growing potential of Cantarell fed pressures from PEMEX to get the oil flowing. Proyectos Marinos scrambled to keep up with the fast-track scheduling. As of June 1979, the joint venture had been awarded the engineering for twenty-five drilling platforms, three riser platforms, five temporary production platforms, four permanent production platforms, various miscellaneous structures, and design work and field services for the Dos Bocas port. Clemente Beltran would fly to Houston once a month and "bang on the table," insisting that design and fabrication move faster. Chuck Osborn remembered one meeting with Beltran and Wilson concerning a deadline that "seemed like an impossible task." Osborn was asked, "Can you get all that done?" As he thought about it, Herb Nelson, sitting next to him, said, "Yes, he can do it." After the meeting, Nelson advised Osborn that "in the future, when Beltran asks you a question like that, say yes, and then worry about how to do it later."

Throughout the spring and summer of 1979, Brown & Root and its Mexican partners mobilized enormous resources to prepare the Akal and Nohoch fields for early production. Declining business in the North Sea had freed up a lot of people and equipment that could be transferred to the Gulf of Mexico. Brown & Root's home base in Houston, just across the Gulf from the Bay of Campeche, enabled the company to coordinate the rapid fabrication and

Three PEMEX platforms under fabrication at Greens Bayou

delivery of structures to meet PEMEX's deadlines. Brownbuilders fabricated the largest structures in Houston and towed them across the Gulf straight into the fields where they set the jackets. By June 1979, PEMEX prepared to connect four wells from Akal and Nohoch to start production at 30,000 barrels per day.[20] Plans were made for a festive inauguration of its new offshore development, but a dramatic event in early June overshadowed those plans. "A blowout of unexpected proportions in Campeche Sound," reported *Offshore*, "tempered PEMEX's pending announcement of what is expected to be one of the most productive reservoirs found in the Western Hemisphere's recent exploration history."[21]

The Ixtoc Blowout

On Sunday morning, June 3, 1979, Dick Wilson and a group of Brown & Root representatives flew to Ciudad del Carmen to arrange for opening the valve for the first production from the Cantarell complex. It was a clear day with miles of visibility. Wilson's group took in a panorama of the Gulf coastline stretching from Texas to the Yucatan peninsula. As they approached Ciudad del Carmen, they sighted a thin plume of smoke shooting skyward from the middle of a group of barges and platforms. The plane descended and circled the area. Confirming their worst suspicions, they saw a semisubmersible drilling rig sitting at a thirty degree angle with a raging fire on board. The Brown & Root group landed and went to their scheduled meeting. The agenda, not surprisingly, had changed from planning the inauguration to "What do we do about this?"

In the early hours of the morning, the *Sedco 135* semisubmersible, leased to PEMEX, had been drilling a delineation well called Ixtoc-1 at 11,863 feet under 150 feet of water. At around three a.m., escaping gas from the hole spread to the hot pump motors on the rig and ignited. The drilling crew apparently had pulled the drillpipe from the well after experiencing a loss of circulation, but failed to connect the safety valve on the drillpipe and pump compensating amounts of mud into the hole.[22] As the fire engulfed the rig, the marine riser disconnected from the blowout preventer stack, and the platform structure gave way, dropping twelve thousand feet of drill pipe and equipment overboard. The fire destroyed most of the mast and drilling deck of the Sedco rig as it moved off location. Two Norwegian work boats came to the rescue, pulling the rig to shallow waters and evacuating sixty-three crew members safely. The $22 million rig, however, had been damaged beyond repair and was scuttled a month later in a 6,000-foot trench off Campeche.[23]

The rig had made a large discovery of light crude, which was now spewed into the Gulf, accompanied by high-pressure gas that had ignited after reaching the water surface. PEMEX was criticized for its drilling procedures, decision-making, and handling of the blowout,[24] but the company's biggest problem over the subsequent months was bringing the blowout under control. PEMEX estimated the uncontrolled flow to be 30,000 barrels per day. Ixtoc, the Mayan name for "fire god," now released his flaming fury in the Bay of Campeche.[25]

Aerial view of Ixtoc blowout

PEMEX deployed a virtual armada of workers, ships, planes, and helicopters to the site to kill the well and clean up the spill. The company introduced floating barriers and chemical dispersants to contain the spreading oil.[26] Despite these efforts, oil slicks flowed northward in the direction of the Gulf Stream.[27] The blowout attracted international media attention and negative publicity for the offshore oil industry. A diplomatic incident was possible as the creeping slicks threatened to contaminate shrimp and oyster beds as well as parts of the U.S. coastline.

PEMEX contacted oil-spill control experts around the world for help. "It seemed like there were professors and people from all over the world coming out there and crawling around," said Jamie Dunlap. PEMEX brought in the famous blowout specialist Red Adair to direct wellhead shutoff efforts. The well was a subsea completion that could be killed only by some kind of underwater operation. When it "became obvious that traditional control methods [could not handle] this seafloor situation, the call went out to Martech International, Inc. of Houston, experts in subsea aspects of the petroleum industry." Martech arrived with its TREC unmanned submersible, a television camera-equipped vehicle that was used to survey the area around the well head and identify a safe path through the debris for divers to appoach the stack and attempt to shut off the well.[28] Tons of drillpipe had settled around the well, creating a "veritable underwater junkyard." One of the early divers on the scene likened it to a "platter of spaghetti."[29]

In late June, the shutoff effort almost succeeded. Operating out of a bell, divers managed to find their way to the BOP stack and hook in choke-and-kill lines leading from a flat-top work barge called *Able Turtle.* The barge pumped seawater and 10-pound mud into the well through the lines to bring the pressure down. The BOP rams closed off the flow, the well showered down, and the fire went out. However, another break occurred in the wellhead. As pressure built, the rams were reopened to release the flow, and the fire was reignited to consume as much oil as possible.[30]

After this failure, crews tried a very unconventional method to control the blowout. PEMEX gathered up nearly 100,000 steel, iron, and lead balls and fed them into the well. They hoped that the balls would be heavy enough to form a bridge that would collapse the sides of the well, but the supercharged vent just spewed out the balls. Divers reported that the expelled balls were stacked like cannon balls all over the sea floor surrounding the well. The idea sounds bizarre in retrospect, but PEMEX was desperate to find some way to snuff out the well.[31] Professor Jerome Milgram, an oil spill expert from M.I.T. who monitored the blowout, said at the time that the plan "was motivated more by emotion or desperation than by reason, because straightforward engineering calculations show that the balls will be blown right out of the well."[32]

This desperation led to tragedy. Taylor Diving lost a diver who had been sent down to tie in a device for feeding the balls into the wellhead. Diving

Diver preparing to descend with surface fire from Ixtoc blowout in the background

around the stack was a difficult and dangerous job. Ken Wallace described the noise underwater near the well "like about 20–30 locomotives going down a railroad track all at the same time. It was such a roar. Communication was very difficult because of the roar that was coming over the communications to the diver." One diver, Alan Anderson, swam underneath some wreckage near the vortex of the blowout. The tremendous force of the flow, intensified because its fluids were at a lower density than the surrounding water, swept him up into it. His lifeline air hose caught on a pipe and left him hanging in the vortex. When Anderson was finally pulled out, his wetsuit and equipment had been completely stripped off. This unfortunate incident revealed the awesome power of the blowout and the hazards of trying to deal with it on the seafloor.

The failure of the efforts to kill the well in June and July forced PEMEX to rest its hopes on two directional relief wells drilled by jack-up rigs. PEMEX began drilling the first of the wells immediately after the blowout. The second well was started in early July. It would take some time, however, before the wells reached Ixtoc. In early August, PEMEX estimated that the blowout would flow until October.[33] To contain the spill in the interim, PEMEX implemented "Operation Sombrero."

Operation Sombrero

Operation Sombrero was an innovative solution from Brown & Root to capture oil from the blowout. Engineers suggested placing a large, inverted funnel over the the wellhead. The sombrero-like funnel would capture the oil-gas-water mixture and send it through a pipe to a separator. Harris Smith, who had been instrumental in designing the gas compression systems for Lake Maracaibo, instigated the scheme by proposing the concept of a gas lift to bring the fluids to the surface for separation. "We had used the principle in a number of platforms to lift cooling water in large quantities just by injecting gas in a large pipe and lifting the water." Smith maintained that they could apply the same principle with the natural gas coming out of the well.

The idea evolved. As the story goes, Smith was standing on the *BAR-289,* looking at the blowout, when he said, "Why don't you just put a big sombrero over that damned thing and smother it out?" Director-general Diaz Serrano, standing nearby, replied, "That's a good idea. Why don't you do it? It's your job." Brown & Root went to work on it right away. Jay Weidler said, "We knew we couldn't set the thing straight down over the wellhead because the force was so tremendous." The first concept for the funnel called for building a cone into a jacket, setting it on the bottom with giant skids, and pulling it over the wellhead, but all the debris on the bottom made this impossible. Then a group of engineers, including Smith, Wilson, Roy Jenkins, and Joe Rainey, flew over the area and noticed that the *Meaders* was anchored next to the edge of the oil slick spraying the fire about 50 yards away. They were inspired to set a jacket next to the blowout with a cone attached to it by a cantilevered boom. The boom could swing the cone over the top of the wellhead like a giant door. The group sketched out the design and presented it to subdirector Adolfo Lastra after they landed. Lastra took one look at the sketch and said, "OK. Do it."

Like the entire crash program for the offshore development in the Bay of Campeche, the greatest challenge to Operation Sombrero was time. Brown & Root had to design, model test, and build the thing very quickly, based on scant information about the volume of oil and gas that the well was producing. As Smith said, Operation Sombrero was "another one of the famous Brown & Root can-do-type things—we did something that was almost impossible to do in a short period of time." In June, while PEMEX was still organizing its directional drilling program, Brown & Root officials promised to have the sombrero ready in three months. PEMEX agreed to the schedule and set the operation in motion.

Jay Weidler assembled a team in Houston to design and test a funnel structure. They decided on a four-pile jacket with hinges for a swinging assembly consisting of a long, triangular truss structure that held the funnel on the end. A pipe, the upper chord of the truss, would run from the sombrero up to the platform for separation, and the oil would flow across a bridge to a second platform equipped with burners. Heinz Rohde said, "We were in the dark a little bit about capacity on this funnel." Weidler's group designed some extra capacity in case estimates turned out to be too low. The group tested the model in plastic tanks in the parking lot garage of the Brown & Root building on Waugh Drive in Houston that simulated the flume from the well. The 40-foot diameter, 300-ton octagonal funnel was fabricated at Greens Bayou and hauled down to the site on schedule in early September 1979.[34] A jacket had been found in Corpus Christi and installed before the sombrero arrived.

One of the most difficult aspects of the project was positioning the jacket so that the funnel would swing directly over the well. The jacket "had to be very precisely set and oriented, otherwise you would miss the well." Brown & Root crews lowered the jacket into the water with a temporary boom that had TV cameras mounted at the tip. Before driving the piles into the jacket, positioning could be done remotely just by watching the TV cameras. "That was when Murphy's Law took over," said Smith. The night before this was to happen, the boom fell back against the jacket and crushed all the coaxial cables leading to the television cameras. After efforts to repair the cables failed, the installation crew resorted to "plan B," which was to use Taylor Diving's underwater RV TV camera to see the action. But in trying to get close enough, the RV was ripped away by the plume. The final option was sonar. Even though its accuracy was questionable under the blowout conditions, "the sonar mechanism was what finally positioned the jacket."

Other setbacks delayed the operation. In early September, bad weather interrupted the scheduled deployment of the funnel assembly. Just when the construction barge *Sarita* had suspended the big cantilevered boom by crane to hook it up to the jacket, a hurricane blew in. Crews lashed the boom to

Sombrero and cantilevered boom ready for installation

the side of the barge and fled to the calmer, leeward side of the Yucatan Peninsula to wait out the storm. Once they returned and attached the boom to the jacket, a mechanical problem with the hinges developed. Joe Rainey remembers telephoning Jay Weidler in the middle of the night asking him to come down and figure out what went wrong. Weidler arrived the next day, took one look at the contraption, and said, "It broke." In late September, after the boom was unhooked and the defective equipment was replaced, a crack was discovered in the funnel. It then had to be returned to Houston for repairs.[35]

Finally, on October 15, the repaired funnel and boom were reattached, and a tug boat with cables pulled the sombrero into place. The funnel was maneuvered to a position about 10 feet over the wellhead, and the system started to produce oil and flare gas. It worked as it had been conceived. However, the funnel was greatly underdesigned. Oil still streamed off into the sea. According to engineers, the flow from the blowout was four to five times as much as the 7,500 barrels per day figure estimated by PEMEX. Jerome Milgram, who had rigged a device to measure flow velocities onto the funnel, estimated the flow to be 40,000 barrels per day at the end of October. Milgram argued that the sombrero collected only about 10 percent of the escaping oil, not 80 percent as PEMEX claimed. Half of what was collected, he added, was rejected by the separation system because of the high emulsion of oil and water created by the turbulence under the funnel.[36] "I have a bottle of [Ixtoc oil] sitting at home," said Heinz Rohde twenty years later, "and the water is still coming out of it." The turbulence from the blowout reverberated through the sombrero platform. Weidler remembered that "The period of the water hammer that we thought we were going to get, instead of being around six seconds, it was coming in at about two seconds, which was about the natural period of the system, so it was kind of exciting."

The result of Operation Sombrero was a disappointment. According to Rohde, "People had expected that this would work like a vacuum cleaner and once it was in operation, there would be no oil, and that wasn't the case." In late December, strong waves battered the sombrero, breaking a hinge on the boom and forcing PEMEX to abandon the structure.[37] Although it did not live up to expectations, the sombrero represented an innovative attempt, under severe time constraints, to harness the violent forces of nature underwater to productive ends. The Ixtoc blowout served up an unusual and imposing challenge, made more difficult by the limited information available to Brown & Root engineers. Jamie Dunlap claimed that had the sombrero "been six times the size it was and covered six times the area, then it would have probably worked." Still, it did help collect some of the oil and it gave PEMEX a psychological boost to say that extraordinary measures were being taken to rein in the runaway well.

Ixtoc blew for nine months, from June 1979 to March 1980. Whether one estimated the flow to be 7,500 barrels per day or 40,000 barrels per day, "you're talking about quite a few gallons of oil," as Weidler said. Some of the oil lapped onto the Texas shoreline during fall 1979, soiling more than 140 miles of beachfront and prompting legal action from many quarters. Lawsuits against PEMEX, Sedco, and other companies totaled nearly $400 million by the time the blowout was killed. PEMEX spent more than $225 million to bring the well under control.[38] The two directional relief wells finally reached Ixtoc and pumped in enough mud and water to kill it. In the final chapter of the story on March 22, PEMEX "took one of its floating rigs, made sure nobody smoked, went right over the well as gas was still coming up, dropped a pipe down into the well and cemented it up."[39]

Two Million Barrels Per Day

The Ixtoc blowout did not deter PEMEX from expanding offshore oil development in the Bay of Campeche. "I can remember being on the *Meaders* right there beside the fire coming out of the water," said Chuck Osborn, "and down below, having meetings with PEMEX about where the next platform was going to go." Mexico's two most prolific oil wells ever drilled had gone into production in October in the Akal field, producing 60,000 and 50,000 barrels per day. "In spite of the tremendous costs suffered by PEMEX to recover the lost crude and cap the Ixtoc well," reported *Offshore,* "PEMEX officials feel the discovery more than made up for the loss, Ixtoc's reserve figured substantially in boosting Mexico's offshore reserves in the past year."[40] By the end of 1979, nine producing fields had been discovered in the Campeche area. These discoveries, announced PEMEX, would enable Mexico to meet its 2.23 million barrels per day oil production goal by 1980, two years earlier than previously planned.[41]

By mid-1980, Brownbuilders had installed four permanent production platforms, twelve drilling platforms (each with an eight-leg, twelve-well design), and most of the flow stations and connecting lateral lines in the Cantarell complex. During 1980, PEMEX spent $237 million on platforms alone for the Campeche fields. Over the next several years, the area became congested with structures, pipelines, and flowlines. As work progressed on the installation of forty platforms linked by a 418-mile pipeline network, the hazards of crowding increased.[42] Brown & Root had set up about five shore stations and placed Argo systems on each barge to survey and lay all the pipelines. "We had quite a crew down there for quite a while," said Peter Cunningham. The lines started crowding each other and many were damaged by anchors. Brown & Root finally set up a radar system on one of the central platforms, which acted like a big control tower that pinpointed any vessel out in the water relative to everything else. Before a vessel could drop anchor, it had to call the control tower for clearance. "They would be watched on this big screen and all the pipes were superimposed on it, and then they could be told whether they could or could not drop anchor or whatever they wanted to do."[43]

During 1980–1981, Brown & Root completed another fast track project for early production from the Abkatun field. This one involved converting a giant 160,000 dwt tanker to a floating storage and production system. "We mobilized all our engineers," recalled Allen Johnson, "moved them down to Harbor Island, Texas, and stayed there about eight or ten weeks. We did all the engineering, bought all the equipment, put it on the tanker, repiped it, and hooked it up." The tanker was moored permanently to a single point mooring buoy. It had to be fitted on one side with large, pneumatic, cell-type fenders to handle 60,000–80,000 dwt tankers that arrived to offload the oil. One of the difficult and innovative aspects of the job was equipping the conversion with hydrogen sulfide, sulfur dioxide and combustible gas detection systems to combat the high-sulphur gas from the Abkatun field.[44]

When oil started flowing in large volumes from the Campeche offshore system, a question arose about where it could all go. The Bay of Campeche yielded 1.3 million barrels per day by the end of 1980, half of Mexico's total production. With growing reserve estimates and production, President Lopez Portillo announced plans to increase crude oil exports to 55 percent of the country's total.[45] Much of the Campeche crude traveled from the Akal and Nohoch fields through two 36-inch pipelines to Dos Bocas. An offshore terminal at Dos Bocas was undergoing development at the time to bolster exports. However, the northern storms that blew in across the Gulf created large waves and strong currents that rendered Dos Bocas and other port locations along Mexico's southern coast practically useless for loading tankers during winter.[46]

PEMEX and Brown & Root developed a solution to Mexico's port problem by installing an export monobuoy with a one-million-barrel storage tanker moored to it near the Cayo Arcas coral islands, located 230 kilometers offshore Ciudad del Carmen northeast of the Bacab field. Brown & Root also laid three 36-inch pipelines to connect the loading facility to the fields in the Cantarell complex. Offshore terminal facilities were eventually completed near Dos Bocas, but Cayo Arcas became the loading place of choice for tankers carrying oil from offshore fields in the bay. In the larger scheme of the Bay of Campeche offshore development, Wilson said, "Cayo Arcas was very, very important."

In the span of four years, Brown & Root and its joint ventures in Mexico—Proyectos Marinos and the Corporacion de Construcciones de Campeche—had developed crude oil production from the Bay of Campeche to two million barrels per day, making the area one of the most prolific in the world. The Cantarell fields emerged as a showcase for Mexico's oil-based economic development strategy under President José Lopez Portillo. Since they were so abundant and promised to be such a valuable source of national revenue, Lopez Portillo and PEMEX were willing to pay a premium for their rapid development. Although the development involved little technological innovation, the time frame and the local working arrangements did present new challenges for Brown & Root. Dealing with the Ixtoc blowout also tested the design ingenuity and improvisational skills of the company's marine engineers.

Brown & Root was rewarded for its efforts. The Bay of Campeche provided much work for Brown & Root equipment and personnel just as the major North Sea projects were winding down. At the end of 1980, during the height of Brown & Root's involvement in the Bay of Campeche, the joint ventures employed more than 3,000 people and operated five major pieces of equipment. Brown & Root laid more than 1,000 kilometers of pipeline (including three major lines to Cayo Arcas and two to Dos Bocas) and installed close to fifty structures. Despite the Ixtoc blowout, Brown & Root assisted in making the Bay of Campeche into a stunning success for PEMEX.

The growth of the Mexican offshore oil industry stalled after 1982 because of decline in the price of oil and severe economic problems. The huge foreign debt contracted under Lopez Portillo to finance its petroleum-based development policies dragged on the Mexican economy as interest rates rose. The new Mexican president, Miguel de la Madrid, scaled back petroleum expansion.[47] Most of the engineering work that Brown & Root performed for PEMEX was finished by around 1982. Construction lasted until 1984, and the joint ventures dissolved in 1986. The Bay of Campeche was one of the last major offshore oil developments in the world before the oil-price collapse of the mid-1980s ushered the offshore oil industry into the doldrums.

Notes

1. This chapter draws from interviews with C.D. Osborn, R.O. Wilson, Jamie Dunlap, T. Smith, III, Sage Burrows, H.K. Rohde, Ken Wallace, H.P. Smith, W.R. Rochelle, J.B. Weidler, Peter Cunningham, and A.C. Johnson.
2. See Isidro Morales, "PEMEX and the Crisis in Self-Sufficiency," in Jonathan C. Brown & Alan Knight (eds), *The Mexican Petroleum Industry in the Twentieth Century* (Austin: University of Texas Press 1992), 233–235.
3. Morales, "PEMEX and the Crisis in Self-Sufficiency," 233.
4. Morales, "PEMEX and the Crisis in Self-Sufficiency," 252.
5. Alvaro Franco, "Mexico Moves to Close its Oil Deficit," *Oil & Gas Journal* (January 28, 1974): 98–99.
6. Alvaro Franco, "Mexico's Crude-Exporting Role May be Short-Lived," *Oil & Gas Journal* (May 26, 1975): 26.
7. "Big Oil Province Shapes up off Mexico," *Oil & Gas Journal* (July 27, 1970): 72–74.
8. "Worldwide Drilling," *Offshore* (September 1975): 5.
9. "Pemex Sees Reforma Extension Offshore," *Oil & Gas Journal* (March 7, 1977): 78–79.
10. Alvaro Franco, "Bay of Campeche May Rival Reforma Area," *Offshore* (January 1978): 43–45.
11. "Pemex Sees Reforma Extension Offshore," 78.
12. Leonard LeBlanc, "The Rising of an Oil Powerhouse," *Offshore* (May 1979): 136.
13. Quoted in Alvaro Franco, "Giant New Trend Balloons SE Mexico's Oil Potential," *Oil & Gas Journal* (September 19, 1977): 82.
14. Franco, "Bay of Campeche May Rival Reforma Area," 45; "Pemex Shifts into High Exploration Gear," *Oil & Gas Journal* (August 20, 1979): 106–108.
15. "Campeche Rumors Fly Amid Pemex' Silence," *Offshore* (September 1978): 67–70; "Mexico Aims for 3 Million B/D by 1985," *Oil & Gas Journal* (June 25, 1979): 52–53.
16. Robert G. Burke, "Bay of Campeche is the Focus of Future Oil," *Offshore* (September 1978): 14.
17. CCC, Manpower Report, May 19, 1979.
18. Leonard LeBlanc, "The Rising of an Oil Powerhouse," *Offshore* (May 1979): 146.
19. "Mexico's Record Still Without Equal," *Offshore* (June 20, 1979): 188.
20. "Cantarell Expands to 23 Drilling Units," *Offshore* (May 1979): 154–157; and "Bay of Campeche Nohoch and Akal Being Rapidly Developed," *Oil & Gas Journal* (March 26, 1979): 163.
21. "Pemex Battles Campeche Blowout," *Offshore* (July 1979): 80.
22. "Pemex Slapped for Handling of 1 Ixtoc," *Oil & Gas Journal* (December 17, 1979): 36–37.
23. "Pemex Battles Campeche Blowout," 80–82.

24. "Pemex Slapped for Handling of 1 Ixtoc," 36–37.
25. "Ixtoc-1 Blowout Draws Massive Control Effort," *Offshore* (August 1979): 43–48.
26. "Pemex Deploys Hundreds to Campeche Blowout," *Oil & Gas Journal* (June 18, 1979): 64.
27. "Ixtoc-1 Blowout Draws Massive Control Effort," 43.
28. "Ixtoc-1 Blowout Draws Massive Control Effort," 45–46.
29. "Divers Working to Control Blowout off Mexico," *Oil & Gas Journal* (June 25, 1979): 46.
30. "Ixtoc-1 Blowout Draws Massive Control Effort," 45–48.
31. "Pemex Claims 1 Ixtoc Oil Flow Cut to 10,000 b/d," *Oil & Gas Journal* (August 26, 1979): 62.
32. "Pemex Slapped for Handling of 1 Ixtoc," *Oil & Gas Journal* (December 17, 1979): 37.
33. "Pemex: Ixtoc Blowout May Flow until Oct. 3," *Oil & Gas Journal* (August 6, 1979): 56.
34. "Sombrero-Like Device Used at Mexico Blowout," *Oil & Gas Journal* (September 3, 1979): 33; "Bad Weather Interrupts Operation Sombrero," *Oil & Gas Journal* (September 24, 1979): 75.
35. "Pemex's Operation Sombrero Status Unclear, *Oil & Gas Journal* (October 8, 1979): 35.
36. "Pemex Slapped for Handling of 1 Ixtoc," 37.
37. "Pemex Cuts Flow from 1 Ixtoc to 1,200 b.d," *Oil & Gas Journal* (December 24, 1979): 40.
38. "Campeche Production Aims for Million B/D," *Offshore* (June 20, 1980): 126.
39. See "Pemex Kills 1 Ixtoc Blowout after 10 Months," *Oil & Gas Journal* (March 31, 1980).
40. "Campeche Production Aims for Million B/D," 126.
41. "Mexico Moves Oil Spotlight to Campeche," *Oil & Gas Journal* (November 19, 1979): 208.
42. "Mexico Adds Punch to Progress," *Offshore* (June 20, 1981): 189.
43. "Radar System Protects Campeche Wells, Pipeline," *Offshore* (January 1985): 19.
44. "SPM Terminal Aids Oil Flow Start Off Mexico," *Oil & Gas Journal* (Journal 19, 1981): 34.
45. "Gulf of Campeche Boosts Mexico's Oil Output," *Oil & Gas Journal* (November 10, 1980): 162.
46. "The Offshore Terminal at Dos Bocas," *Offshore* (February 1981): 8.
47. "Mexico Trims Oil Expansion to Help Balance Budget and Prop Economy," *Oil & Gas Journal* (August 30, 1982): 87–99.

PART 3
The Challenge of the North Sea: Rough Waters, Hostile Conditions

CHAPTER 11

Confronting a Monster: The Early Natural Gas Industry

Experience in demanding conditions around the world prepared Brown & Root to take part in one of the greatest offshore booms in history, the development of the North Sea in the 1970s and early 1980s. Brown & Root was the first offshore construction company in the region in the 1960s, and after the surge in oil prices in the early 1970s, it was well-positioned to tackle the severe, but profitable, challenges of recovering oil and gas from under the cold, rough waters of the North Sea. Harsh, demanding working conditions in these waters forced Brown & Root and the industry as a whole to make fundamental adjustments in most aspects of offshore operations.[1]

Initially, Europe had little of the infrastructure traditionally used by the offshore industry. As the region's offshore industry matured, strong European concerns emerged, presenting competition for American companies such as Brown & Root. In a region jointly controlled by European nations skeptical of American domination of their energy development, Brown & Root faced the additional challenge of managing as best it could the political forces that shaped economic options.

The technical challenges of the North Sea unfolded in three stages. In the 1960s, Brown & Root developed several gas projects in the calmer, shallower waters of the southern North Sea, although the high winds, choppy waters, and hard clay bottom lengthened installation times and suggested greater difficulties to come. The second stage of North Sea development began in the early seventies, as companies moved to produce petroleum from new oil and gas fields, including Ekofisk (Phillips), the Forties (BP), Beryl (Mobil), and Brent (Shell). These fields in the harsher, deeper waters of the central North Sea proved conclusively that the North Sea held massive oil deposits, and a wave of technical innovations enabled their subsequent development. Brown & Root assumed a central role with British Petroleum in the Forties and played a key part in Phillips Petroleum's work at Ekofisk. In so doing, the company expanded its technical capacities in many directions at once, making enormous strides in its ability to design, fabricate, and install platforms and pipelines. In the third stage, many companies rode the high oil prices of the late 1970s into massive projects throughout the North Sea. Although the company no longer could claim a dominant position in offshore construction in the region, Brown & Root completed significant, techically ambitious projects before the price collapse of the early 1980s slowed development dramatically. In the twenty years after Brown & Root Marine built its first platform in the North Sea in 1964, this region played a critical role in the company's evolution.

The North Sea's extreme weather limited most offshore activities to the warmer months, placing a premium on equipment with larger capacity and better response to the sea. Working in these conditions compelled Brown & Root to develop large, technically advanced vessels. Fabrication of the needed platforms obliged the company to establish a new yard at Nigg Bay in Scotland. The design of such structures required the growth of a substantial engineering office in London, and the commitment to project management further expanded the Brown & Root staff. During the boom years of the 1970s, a torrent of money flowed into the North Sea, and a torrent of oil and gas flowed out. The high price of oil and the urgency of developing non-OPEC sources of crude generated extraordinary investments in new technology. The tone of the times matched the long-held culture of Brown & Root: any project seemed possible in the heady days of North Sea expansion.

The Monster to be Tamed

The North Sea covers 185,000 square miles bounded by the United Kingdom, Belgium, the Netherlands, Germany, Denmark, and Norway. Until the early 1960s, its commercial value was based on the local fishing industry: herring, haddock, and cod. Fisherman were well-acquainted with what oil men would soon discover: the North Sea is a foreboding place. Although conditions in the lower North Sea are generally tamer than those in the deeper, colder waters to the north, the entire region experiences weather much more severe than other areas traditionally worked by the offshore industry. Work seasons were brief as offshore projects were planned around the especially harsh winter conditions.

Two dominant pressure systems, a thermic high pressure area near the Azores and a low pressure area near Iceland, usually create unstable and unpredictable weather. On occasion, two weather reporting stations located relatively close together might seem worlds apart, with one noting gales and the other calm winds.[2] The volatility of North Sea weather makes forecasting difficult, adding uncertainty to even the most careful planning of offshore work.

The average daily maximum wind during the winter months is 35 mph. Winds higher than 40 mph are common during about 25 percent of winter days. Winter storms can produce hurricane-like conditions, with gusts of up to 120 mph and 80 mph winds sustained for as long as an hour. These storms can arise quickly out of calm seas and clear skies, producing ferocious, 100-foot waves whose destructive power is increased by tidal currents. Because the seas are almost constantly agitated, the relentless pounding of waves introduces serious metal fatigue problems for offshore structures.

Unreliable information about conditions in the North Sea plagued offshore pioneers in this region. A.B. Crossman of Brown & Root said, "Nobody had ever tried to predict what you design something for in the North Sea. They would say it's a bad place, but they really had no observations of wave heights out in the North Sea." Fishermen and coastal dwellers could not provide useful or reliable data, for "they knew when a storm was coming, and not many people were caught out in the middle of the North Sea." Merchant vessels had sailed the North Sea for hundreds of years and their observations provided a useful starting point, but the roll and pitch of vessels made it difficult to measure wave heights, directions, and periods. "Wave-Rider" observation buoys provided measurements as much as 20 percent more accurate than those previously recorded from ships.

Brown & Root also turned to wave and weather specialists from New Orleans, who had provided wave forecasts for the Gulf of Mexico. Two European weather and wave statisticians—MAREX of the U.K. and KNMI in The Netherlands—provided reliable information on the North Sea from the 1960s onward. Brown & Root made extensive use of their data to understand and anticipate conditions that would have to be overcome to build and operate offshore structures in the region.

Even before the exact height of North Sea waves was known, it was clear that conditions there posed extraordinary problems for offshore work. Using the conventional equipment of the era, the typical work season in the North Sea lasted only from April to September, and pipelaying was possible for only about 60 percent of this period. The winter months were simply too dangerous for offshore work, and, at times, even for carrying out normal supply operations. Offshore companies learned quickly that they had to work faster and build bigger in the North Sea.

Conditions on the ocean floor also hampered offshore operations. In the southern parts of the North Sea, ocean depths are 30 to 150 feet, but in the extreme northern section, they increase to 650 feet. Along the Norwegian coast lies the Skagerrak trench, which plunges to depths of 2,200 feet deep. This trench presented a formidable barrier for pipeliners and encouraged the growth of offshore oil storage facilities in the Norwegian sector of the North Sea. The floor of parts of the North Sea is composed of clay, shale, and sand, and strewn with boulders.[3] Because it is both harder and rougher than subsea surfaces in most other offshore producing areas, the North Sea's floor presented special problems in driving piles and burying pipelines.[4]

Beneath the rough waters and hard ocean floor lay an ancient desert whose sands hold tremendous quantities of crude oil and natural gas. Europe's growing dependence on imported oil and manufactured gas in the post-World War II years created a strong incentive to explore all domestic sources. But the development of offshore deposits in the region awaited two closely related developments: the creation of a technology capable of recovering oil and gas from this tempestuous region, and the coming of a price for oil that made recovery economically feasible.

The Origins of North Sea Oil and Gas

The first significant indication of hydrocarbons in the North Sea region came in August 1959, when a Shell/Esso venture found the massive Groningen gas field, an onshore field near the coast of northeast Holland. Small gas fields in Holland and England had been discovered as far back as the 1930s, but Groningen marked a sharp departure. This twenty-mile long field had a capacity of six billion cubic feet per day (Bcf/d), making it one of the largest natural gas fields in the world. The unexpected discovery of this natural gas near the energy markets of Europe focused the attention of oil and gas companies on surrounding areas. Despite the obvious hardships of exploration in these rough, cold waters the prospects of major new finds encouraged a wave of exploratory drilling in the region. Speculation that the seabed of the North Sea might have structures similar to that of Groningen set off a round of exploration offshore.[5]

Before drilling in the North Sea could proceed, however, territorial rights to the offshore lands had to be established. During the 1950s, the governments of nations bordering the North Sea began to debate what guidelines should be used to determine territorial jurisdiction over the North Sea. Each nation bordering the North Sea possessed exploration rights within its own territorial waters out to three miles from shore, but political and legal uncertainties remained regarding national rights to the remainder of the vast North Sea.

The United Nations addressed these concerns at its Conference on the Law of the Sea in 1958, which adopted the Geneva Convention on the Continental Shelf. This convention stipulated that each country with a North Sea coastline had sovereign rights over its continental shelf up to a determined subsea boundary. The continental shelf was defined as the seabed and its subsoil extending from the termination of a nation's existing territorial waters to a depth of 656 feet (200 meters), or deeper if natural resources could be extracted.[6]

The Continental Shelf Act required twenty-two signatures before it became law. While awaiting ratification in the early 1960s, Shell/Esso conducted four unsuccessful drilling operations in Dutch coastal waters before suspending further operations until the resolution of the Continental Shelf Act. On May 14, 1964, Britain became the twenty-second signatory, and the act

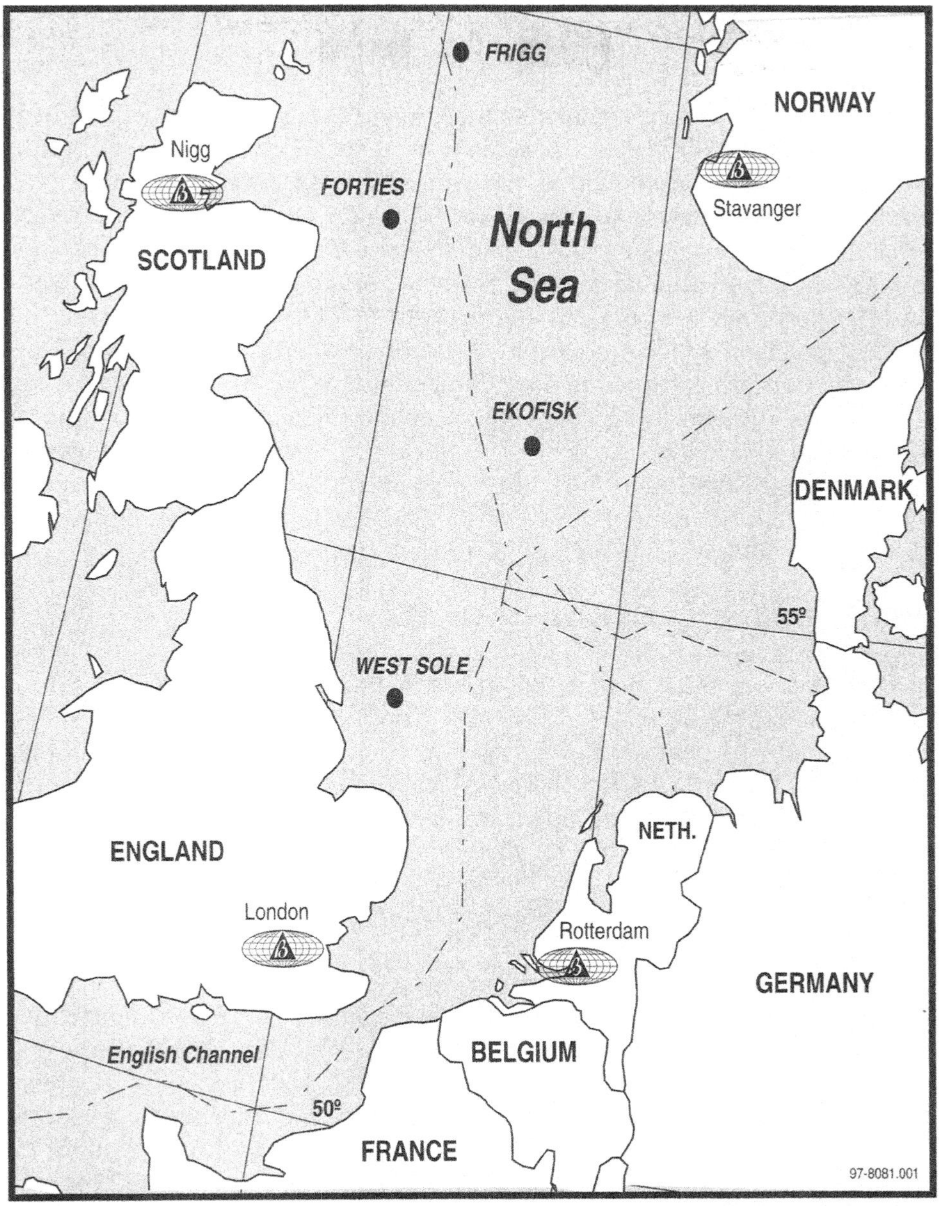

The North Sea

was ratified. Under this agreement, the U.K. gained rights to about 40 percent of the North Sea; Denmark 10 percent; the Netherlands 11 percent; and Norway 27 percent. What was then West Germany acquired rights to a yet smaller percentage, but later negotiated sovereignty over a part of the Dutch and Danish Continental Shelf as well, based on the fact that its coast line is convex inward. With boundaries legally defined, concessions could be granted to drill in these waters.

The rush to encourage exploration began immediately. Denmark and Germany granted the first North Sea concessions. In 1964, Germany granted drilling rights for three years covering the entire German continental shelf to an eleven-company consortium. Denmark followed by awarding an unusual 50-year exclusive concession to Danish shipowner A. P. Möller. Möller later brought in other oil companies as partners. Britain then entered the race to license North Sea territory, issuing licenses to 51 companies involving 348 blocks, each of which measured approximately 100 square miles. These concessions represented approximately one-third of the British North Sea, and set the stage for the initial exploration of large segments of the region. As one British oil man said in 1964, the North Sea was a "massive gamble . . . played with multimillion dollar dice." As energy companies began mapping the subsurface of much of the North Sea, the industry contemplated the prospects of this newly emerging offshore province.[7]

Brown & Root in the Early North Sea

On the eve of the Groningen discovery in 1959, Brown & Root was already an internationally active construction company and a leader in offshore construction and engineering. In 1958 and 1959, Brown & Root and Texas Eastern had participated in a proposed venture to build a gas pipeline under the Mediterranean Sea from Algeria to France to bring large volumes of natural gas to Europe. The discovery of the Groningen field and the threat posed by political instability in Algeria combined to convince Brown & Root to drop out of this venture and concentrate instead on potential developments in and around the North Sea.[8]

After the discovery of gas at Groningen in August 1959, Brown & Root decided to open an office in Europe to be closer to any new developments in the North Sea and to facilitate its work in the Middle East. London was chosen as the most logical site, and the new office in Pall Mall opened in the summer of 1959. To direct the London offices, the Browns chose Sir Phillip Southwell, a man of considerable standing in the oil industry and British politics. Sir Phillip's association with Brown & Root came near the end of a long and distinquished career. After service in World War I, he had earned a degree in petroleum engineering before working for the Anglo-Persian Oil Company, predecessor to British Petroleum, from 1930 to 1945. Then followed a long stint as managing director of Kuwait Oil Company Limited, also affiliated with BP, from 1946 until he came to work for Brown & Root. He had been knighted in 1958, and his abilities, stature, and connections made him a valuable point man for the company's London office.

In 1960, Brown & Root's Hugh W. Gordon joined Sir Phillip in London to organize what became a wholly owned subsidiary, Brown & Root (U.K.)

Ltd. Then in 1963, as an agreement on territorial rights in the North Sea seemed imminent, the company added R.O. "Dick" Wilson, who had offshore experience in the Gulf of Mexico, Venezuela, and Brazil. His initial responsibilities included developing business contacts with potential North Sea clients.

In 1963, before the ratification of the Continental Shelf Act and the issuing of production licenses, Brown & Root formed a joint venture with the Dutch contractor, Pieter S. Heerema. The creation of Brown & Root-Heerema SA was a logical alliance, combining Heerema's equipment with Brown & Root's industrywide contacts. Heerema was Dutch by birth, but he had moved to Venezuela after service in the German army made him unwelcome in his native land after World War II. Heerema entered offshore construction in Venezuela, competing in the 1950s with Raymond-Brown & Root in the Lake Maracaibo concrete pile business. Brown & Root had worked with Heerema as a subcontractor, and Wilson had developed a good working relationship with Pieter Heerema. By the early 1960s, Heerema had built up a personal fortune, and returned to Holland. As Wilson explored business options in the North Sea from the London office, he talked informally to Heerema about the possibility of combining forces with Brown & Root in the region.

Heerema had a vessel ready to begin work in the North Sea and a strong desire to undertake offshore work in the region. Upon returning to Holland, Heerema had built the *Global Adventurer*, a 500-foot, 14,000-ton converted

tanker with a 200-ton, American Hoist R-40 revolving crane. This vessel was unlike anything Brown & Root had ever used or even seen. It was a ship, not a barge, but it was designed and modified to perform the heavy lifting required in offshore operations. Heerema felt that a ship-shaped, self-propelled vessel was better suited for conditions in the North Sea than a conventional derrick barge. As the name *Global Adventurer* implied, Heerema wanted a vessel capable of pursuing opportunities around the world, not just in the North Sea. After its conversion in the winter of 1963, the *Global Adventurer* moved to the top of a short list of one as the most desirable vessels for work in the North Sea. Although Brown & Root officials remained skeptical of this departure from the tried-and-true barge designs of the Gulf of Mexico, they recognized the desirability of getting construction equipment into the North Sea as quickly as possible while teaming with an established European partner.

In December 1963 Brown & Root and Heerema agreed on the terms of their joint venture. Heerema would contribute the equipment, initially the *Global Adventurer*, while Brown & Root would pay Heerema $1.5 million for half interest in the vessel and use its regional contacts to help develop business. Additional equipment would be jointly acquired when needed. Beyond these basics. The contract outlined few details about the management or the business strategy of the joint venture. This marriage of convenience did not include a clause specifying the rights and responsibities of each partner, and a bitter divorce ensued after only two and a half years.

The Brown & Root-Heerema joint venture was contentious from the start, in part because of the lack of clear lines of management authority. Pieter Heerema was an old-fashioned entrepreneur whose approach to business combined Herman Brown's passion for personal control and George Brown's salesmanship. He disliked waiting for corporate deliberations, preferring to strike out on his own. Brown & Root's management in Houston and London initially seemed willing to allow Heerema considerable leeway in guiding the joint venture from his offices in The Hague, but tensions grew between the Dutch entrepreneur and the Texas-based corporation.

The prospects for conflict were evident in the joint venture's first major project, the installation of a platform in the North Sea for the construction of a "pirate" television station. In fall 1963, before the joint venture agreement, Heerema had become one of three prime investors in a plan to build "REM-island," a platform outside of any nation's territorial waters from which unregulated, commercial radio and television broadcasts could be made. The Dutch had previously permitted radio broadcasts from ships outside territorial waters in the North Sea, but the construction of an "extra-territorial" platform for television transmissions raised fundamental questions of international law and national policy toward the relatively new medium of television.

Amid intense governmental debates about this project, Brown & Root-Heerema pushed on, using the *Global Adventurer* to construct a six-legged platform and install the deck and broadcast tower in 54 feet of water just outside Dutch territorial waters off Noordwijk. In July 1964, Radio North Sea went on the air, followed the next month by TV North Sea. These broadcasts were very popular, but against the backdrop of international agreements on national territorial rights in the North Sea, opposition parties pushed the ruling Conservative party to dissolve parliament. The government fell in November 1964. A new government under Labour was formed, and in December 1964 Holland moved to shut down the stations, sending Dutch Federal police and Navy supported by helicopters to occupy the REM-island and seal off the broadcasting equipment. Having followed Heerema into this controversial project, Brown & Root was glad to finish its brief detour into platform construction for offshore broadcasting, which indirectly had hastened the fall of the Dutch government, and return to the oil and gas projects it knew best.[9]

In 1964 and 1965, a wave of exploratory drilling opened the first era of gas development in the North Sea. As the only offshore contractor present in the area, Brown & Root-Heerema assisted in this exploration and then constructed, transported, and installed the platforms and pipelines needed to bring first gas and then oil to market. Germany encouraged exploration in its coastal waters even before the completion of the Geneva Accord on rights to the North Sea, and the joint venture got its first taste of offshore construction in the "German Bight" region. Rights to explore this region were granted to a consortium of ten companies. In early 1964, Amoco Hanseatic, a representative of this consortium, called on Brown & Root-Heerema to help prepare for its first exploratory drilling. The consortium had arranged for the towing of the mobile drilling rig *Mr. Cap* into the North Sea, and it contracted with the joint venture to build and install four-pile well protectors through which *Mr. Cap* could drill. On this project, Brown & Root-Heerema installed the first offshore platform in the North Sea. The *Global Adventurer*, based in Rotterdam, set the platform in 108 feet of water, securing it to the ocean floor with piles (21-1/2" outside diameter × 1" wall thickness) driven 55 feet into the subsurface floor.[10] In mid 1964, Brown & Root built an identical platform in water 98 feet deep using piles driven 114 feet into the subsurface flooring.

Brown & Root-Heerema built and installed eleven fixed platforms and one that was reinstalled in 1964 and 1965. All except two were installed in the German Bight, and the structures ranged from four-pile well protectors to sixteen-pile production platforms in waters ranging from 60 feet to 120 feet.[11] When severe weather stopped work in the North Sea for the winter of 1964–65, the *Global Adventurer* certified its versatility by making a twelve-day journey to Malta and then on to Es Sider off the coast of Libya. There it installed a mooring system for the pipeline laid by the *L.E. Minor* before returning to the German Bight.

This work off the coast of Germany presented unique hazards, since only shipping lanes had been swept to remove mines left over from World War II. Before each platform was installed, the German Navy needed to sweep the site and a path from the shipping lanes to the site. On one occasion, a jacket being installed floated away into a potential mine field, and the captain of the tow barge refused to retrieve it. Only when the anchor handler on the barge volunteered to fetch the platform could installation go forward.[13]

The demonstrated success of the *Global Adventurer* in the German Bight prompted Brown & Root-Heerema to expand its fleet of barges and ships for North Sea service. The first new barge built for the joint venture and dedicated for operation in the North Sea was the *Atlas*, a derrick barge built in Holland and designed along the lines of Brown & Root's largest existing barge, the *Foster Parker*. Launched in the summer of 1965 with dimensions of 350' × 100' × 25', the *Atlas* seemed large enough to handle the rough waters of the North Sea. Original plans called for a 500-ton revolving crane to be manufactured by the American Hoist Company, but the wreck of the train transporting this crane across the United States caused a long delay in its delivery. So Brown & Root-Heerema rigged out the *Atlas* with a temporary 1,100-ton "A-Frame" sheer leg with steam powered winches similar to lifting gear used in Rotterdam harbor. Using such cranes from a floating vessel offshore was much more difficult than using them from fixed structures in a harbor.

The projects completed by the *Global Adventurer* and the *Atlas* in 1964 and 1965 were a part of the first flush of exploration in the North Sea. Early returns from the exploratory wells drilled in the German Bight seemed promising, with much excitement resulting from a dramatic blow-out of a gas well in the summer of 1964. Halliburton, Brown & Root's parent company, made its first appearance in the North Sea when it aided Red Adair's efforts to bring this blow-out under control. However, tests showed high levels of nitrogen coming from the well, and it never produced commercial quantities of gas. Indeed, the much anticipated gas boom in the German Bight never materialized.

A Fixed Platform for Conoco

By late 1965, gas discoveries by British Petroleum and Conoco had drawn attention to the U.K. sector. At the time, mobile drilling rigs were in short supply throughout the world. The first rigs from the Gulf of Mexico had been towed into the North Sea in the summers of 1964 and 1965, while several European yards also began construction of mobile drilling rigs. Leading the way was BP's *Sea Quest*, a state-of-the-art rig fabricated at Harland & Wolfe's yard in Northern Ireland from 1964 to 1966.[13] Still, companies such as BP and Conoco sought creative ways to speed the pace of exploration of the U.K. sector. Brown & Root-Heerema joined Conoco in one such project, testing the viability of a movable exploration platform.

BP and Conoco found different answers to their common need for rigs for exploratory drilling. BP located a ten-legged jack-up construction barge in Le Havre, France, built by Hersent-De Long. The barge was converted by George Wimpey & Co., equipped for offshore drilling, and put to work as the *Sea Gem* drilling exploratory wells. Conoco worked with Brown & Root-Heerema to develop a movable fixed platform. The idea seemed simple: design a platform that could be pinned to the ocean floor with "removable" piles that could be blown loose once exploratory drilling at a site had been completed, thus freeing the platform to be towed to a new location and reinstalled using new piles. In practice, this idea proved unworkable.

As an alternative, Conoco contracted with Brown & Root to design Conoco I, a fixed self-contained exploratory drilling platform for the U.K. sector. Its design adapted Gulf of Mexico approaches to North Sea conditions. "We were in Dick Wilson's house one Sunday," a Brown & Root engineer recalled, "and we ended up sketching something that was heavier than the Gulf of Mexico but using some Gulf of Mexico drawings to go by and then using those drawings to try to get fabricators to build the thing." Technologically, it was a simple platform made to rest on the seabed. If and when hydrocarbons began flowing, the exploration platform would be moved to another site and a permanent production system would be installed.

Brown & Root-Heerema managed the fabrication of the $1 million platform using de Groot Zwijndrecht—a contractor that had built bridges and other construction from its yard near Rotterdam—to fabricate the jackets and decks. Well-known to Pieter Heerema and located near the action in the North Sea, de Groot had built many of the platforms installed by the joint venture in the German Bight, and it was one of several European yards then adapting existing facilities for offshore construction. In fact, Heerema liked the yard so much that in the late 1980s he bought a controlling interest in de Groot.

Conoco I consisted of two adjacent platforms mounted on ten tubular legs (one six-leg structure and one four-leg structure); each leg was 39 inches in diameter and 180 feet long, and the platform deck measured 66 × 144 feet. The combined weight of the jackets and deck was 1,650 tons, making Conoco one of the largest structures as yet installed in the North Sea, though it was not particularly large by Gulf of Mexico standards of the time. The land rig mounted on the platform could drill beyond 20,000 feet, and Conoco planned to drill its first exploratory well from this platform in about 80 feet of water to a depth of 12,000 feet.[14]

Such plans did not go smoothly. Brown & Root's Joe Rainey described the project as "snake bit." The misadventures began during the initial tow-out of one of the jackets in May 1965. The weight of the six-legged section exceeded the lifting capacity of the *Global Adventurer*'s crane, so the structure had to be installed by launching instead of lifting. When they tipped the jacket off the barge into the sea, the crew made a rude discovery. "They had given us the wrong water depth," said Larry Starr. "We put it over the side and it disappeared. There was about ten feet difference in water depth." What was built was built, so Conoco's geologists chose an alternative site 1/2 mile away in about 65 feet of water. Although rough waters hampered efforts to position the jacket, it was put in place by moving the anchors and shifting the position of the crane vessel on its moorings.

The composition of the ocean floor added to the installation problems. In soft-mud sites, a jacket normally sinks under its own weight until the mud mats, attached to the bottom of the legs, mobilize enough soil resistance to balance the weight. In the fine, hard-packed sands of parts of the North Sea, however, jackets often do not sink evenly under their own weight or penetrate to a sufficient depth. The Conoco I legs had to be worked into the sand until the jacket was standing level. Once the platform was secured with 36-inch diameter, 280-foot long piles, exploratory drilling could begin.[15]

Brown & Root-Heerema faced a more difficult test in the spring of 1966, when the joint venture sought to relocate the Conoco platform for the first time. As the A-frame crane on the new *Atlas* lifted the deck, disaster struck. Problems with one of the steam winches controlling the lifting cables caused one end of the load to plunge downward. The living accommodation module, containing the crew's sleeping quarters, a restaurant, and a self-service laundry,

Broken A-frame with Conoco deck. Photo courtesy of Dirk Blanken

slid off into the ocean, leaving the rest of the deck hanging perilously from one of the *Atlas*'s two lifting hoists. Help was needed immediately, but no other large cranes were available worldwide to take the deck from the disabled crane.

With no obvious solution at hand, those in charge of the operation drew upon their collective experience to find a way to save the deck. The nearby Humber River offered one option. The extreme tidal changes in the Humber might be used to remove the deck from the crane cable, if a site could be found to "park" the deck temporarily until the *Atlas* could be repaired or a new winch found. Calls to local authorities inquiring who might give permission for this operation yielded a practical answer: "Don't ask, just do it." Using piles intended for the installation of the platform, workers constructed a temporary jacket in the middle of the Humber River. Then the *Global Adventurer* guided the deck onto its makeshift resting place and tidal changes allowed the *Atlas* to release its hold. Later these same tidal changes pushed the *Global Adventurer* toward the temporary jacket, pinning a tug boat between it and the *Global.* The accident damaged three of the piles and dumped several of the tug's crew into the cold, dark, and powerful waters of the Humber. All-night search parties pulled several crew members from the river, but hope waned for two of the crew still missing in the darkness. Morning brought a happy ending, however, revealing the two missing men still asleep in their bunks in the tug boat. Accustomed to the extreme swells in the North Sea, they had slept through the commotion surrounding the near capsizing of their tug. As the weary search crews laughed, people along the Humber River awakened to find an offshore platform sitting in the middle of the river. In this era before closer regulation of the oil industry, neither public officials nor the news media pursued the matter, and Brown & Root-Heerema repaired the *Atlas*, retrieved the deck, and removed the pilings.

The relationship between Brown & Root and Pieter Heerema did not fare as well as Conoco's deck. Arguments about the responsibility for this disaster and for choosing the A-frame crane on the *Atlas* intensified continual tensions over lines of authority. Heerema's unilateral decision to purchase the ex-Shell tanker *Isocardia*, for conversion into a vessel similar to the *Global Adventurer*, disconcerted his Brown & Root partners, who favored more traditionally designed barges over ship-shaped vessels. These disputes generated heated exchanges in which Heerema threatened to scuttle the *Global Adventurer* in Rotterdam harbor before he would compromise. Instead, he flew to Houston, where he reportedly said to George Brown: "We both went out to rob a bank. You bought a hammer and I bought a cold chisel. After chiseling all night, the doors to the vault open and the gold is ready to be brought out. You now pay me two dollars for the hammer and tell me to go home? No way. I demand substantial compensation for using me and my equipment as a stepping stone in the North Sea." Brown & Root ended its joint venture with Heerema by buying his interest in the *Global Adventurer*, the *Atlas*, and two cargo barges also built for the joint venture by de Groot. Brown & Root also granted

Heerema an annual payment for two years to replace foregone profits. In exchange, Heerema agreed to stay out of the crane ship business for those two years.

As for the Conoco platform, it finally reached its new drilling site, where it found natural gas, though not in commercial quantities. Conoco then leased the platform to another company, Rycade, which hired Brown & Root to move it twice more for exploratory drilling.[16] All in all, the experiment with this "mobile" drilling platform was no great success. Such makeshift approaches to exploratory drilling were expedients pending the completion of rigs under construction to meet the demands of drilling in the North Sea. Brown & Root's work for Conoco left many questions unanswered and one ironic footnote to history: four years after the inadequate height of its jacket forced Conoco to leave its initial drilling site, Conoco returned to the site and discovered the massive Viking gas field about 70 miles off the coast of England in about 85 feet of water. Had Conoco I been able to drill in the site for which it was intended, Conoco would have discovered the first commercial oil or gas deposits in the North Sea.

The Birth of a North Sea Gas Industry

Instead, that honor went to British Petroleum. In November of 1965, BP discovered a large deposit of natural gas at West Sole—55 miles from Conoco's original drilling site—in U.K. waters off the southern coast of England. After several years of high hopes and false starts, the North Sea had finally yielded commercial quantities of hydrocarbons. *The Sunday Times* caught the enthusiasm in a feature article about the rush to produce gas by announcing that "among all the North Sea explorers there is now a tremendous surge of confidence."[17]

Brown & Root dominated offshore construction in the North Sea during this initial gas boom, building the bulk of the platforms required to produce this gas and most of the pipelines needed to carry it to markets onshore. The company boasted an impressive fleet built primarily in Holland for use in the North Sea, including the *Global Adventurer*, the *Atlas* and its sister vessel the *Hercules*, the *H.W. Gordon* (a pipelay, combination barge), the *M-228* (a bury barge built in the United States), the *BAR-279* (another bury barge), and five cargo barges. Indeed, until the early 1970s, Brown & Root enjoyed a near monopoly on the ownership of large modern construction vessels in the North Sea. Moreover, the company brought to the region a solid reputation among the major oil companies active there, long experience in other offshore provinces, and the early presence established in its joint venture with Heerema. Even for a company as experienced as Brown & Root, however, the North sea would bring special challenges.

That the North Sea was demanding and unforgiving was evident in the first flush of excitement brought by the discovery of the West Sole field. BP contracted with Brown & Root to design two platforms and a four-pile well protector to be placed around the conductor at the discovery site. Fabrication of the jacket well protector raced ahead in Rotterdam and was completed in only four weeks. But as Brown & Root towed the jacket out toward the site where the converted construction barge *Sea Gem* was drilling, disaster struck. On December 26, 1965, as the jacket moved through calm waters, one of the legs of the *Sea Gem* suffered a brittle fracture. In the first major disaster for the oil and gas industry in this region, the *Sea Gem* collapsed into the sea, and thirteen persons lost their lives.

Pressing on despite the *Sea Gem* accident, BP awarded the contract for building platforms A and B for the West Sole field to Brown & Root in January 1966. Events moved rapidly as the companies prepared to make efficient use of the coming summer construction season. Knowing of the availability of a Sante Fe drilling rig that had recently finished a job for Mobil Oil in Holland, Brown & Root's engineers in London quickly sketched a layout for a platform capable of using this rig. After receiving BP's approval for the layout, Dick Wilson walked out of BP's London offices at Longbow House and stopped next door at the Moorgate Tube Station. There, he phoned Houston with the details of the layout and placed an order for materials for the pilings and jacket legs, which were to be built to Gulf of Mexico and Conoco I standards. The fabrication work was carried out on a rush basis in Dutch and German shipyards, enabling the first jacket to be completed for launch in early May 1966.[18]

Installation did not go quite as smoothly as fabrication. Brown & Root's A.B. Crossman later recalled the difficulties driving piles into the hard bottom of this section of the North Sea:

> Somebody gave me a copy of the old *Engineering-News Record* "Pile-driving Formula" and sent me out there. . . . We found that it was completely different than what people might have thought because again, our crews and our barge superintendent and so forth were used to driving piles that were in the Gulf of Mexico-type soil and they go down easily. And they got down about six or ten feet or something and hit the boulder clay or whatever it is, the first layer of hard, hard clay, and they stopped.

The *Atlas* and the *Global Adventurer* were both available to install these platforms, and they had 60,000 foot pound Vulcan S-60 steam hammers. Although considered large for the time, these hammers could drive piles no deeper than 50 feet into the sea bottom; 100 feet was the design depth. A larger diesel hammer borrowed from the British Admiralty also proved inadequate. According to Crossman, the short-term solution was a labor-intensive effort to "under-ream and drill them out," which required "quite a lengthy time to put in those piles." Using an air-lift waterjet system to wash out the core material of the pile then became the common method for ensuring adequate pile penetration. The long-run solution was the development of "much, much larger hammers and much better ways to drive piles." The massive Menck steam hammers that became standard operating equipment in the North Sea were initially developed in response to the problems in penetrating the hard clay bottoms at West Sole and other early gas fields.

The hard clay bottom and the rough waters of the North Sea also required improvements in pipeline laying equipment. George Brown said: "Some months we are able to work only three or four days. Any sea can get rough, but the North Sea lays it on. During one period we expected to lay forty miles of pipeline. We laid only nine."[19] Brown & Root had to adapt to such conditions to maintain its traditional leadership in offshore pipelines in this new environment. One writer concluded in 1973 that Brown & Root had conducted "the lion's share of North Sea pipelaying."[20] Hugh Gordon gave more personal testimony: "We tended to specialize more in the pipeline side than the offshore platform side. We did a lot of that, but we pushed the pipelines harder just because we liked it more."

In 1965, Brown & Root had laid a short pipeline from a loading terminal in the North Sea to a Shell refinery in Denmark, but the company's work at West Sole marked the first substantial North Sea pipeline work in the open sea. The 45-mile, 16-inch line from West Sole to the onshore facilities of the British Gas Council at Easington, East Yorkshire, lay under up to 105 feet of water. Bob Avery, British Petroleum's manager of contracts stationed in London, noted that laying pipeline in the North Sea at any time except during the summer months was "financial suicide." He felt that traditional lay barges would be too small to complete the West Sole job, since strong crosscurrents in the North Sea would buffet them about, buckling the pipe as it was laid.[21] The completion of this gas line—which was relatively short and small in diameter in comparison to lines laid by Brown & Root in other parts of the world in the 1960s—required Brown & Root to make major new investments in equipment. In the late 1960s, the company invested $20 million on North Sea vessels.

To lay pipe in the North Sea, Brown & Root had to build stronger, faster lay barges. The first generation of purpose-built lay barges, developed for work in the Gulf of Mexico and the Middle East, lacked the strength and speed demanded by conditions in the North Sea, which resulted in unacceptable downtime due to weather. Traditional pipe-laying processes could be used in the North Sea, but new equipment was needed to meet the demands of laying larger-diameter and thicker-walled pipes in a region where much had to be done in a short work season.[22] The first major investment was in the *Hugh W. Gordon*, a combination derrick and pipelay barge. Named after the firm's vice president for marine work, the barge was built in Rotterdam by De Rotterdamse Droogdook Maatschappij NV. The *Gordon* marked a departure for the offshore construction industry; looking back more than a decade after its launch in 1966, *Offshore*'s twenty-fifth anniversary issue in 1979 labeled the *Gordon* the "first (lay) barge of the second generation."[23]

The *Hugh W. Gordon* was similar in many respects to earlier barges, but its size set it apart. At 400 × 100 × 30 feet. and with a displacement of 8,300 gross tons, it was the world's largest operational barge. The *Gordon* was also

Hugh Gordon and *Atlas* at work in North Sea

the first pipelaying barge in the North Sea, and it laid all of the early gas lines there. A newspaper account in the summer of 1966 captured the sense of its massive size: "On board there is very little feeling of movement, even in the long North Sea swell, and the great expanse of deck makes one think more of a fair-sized factory floor than a seaborne vessel."[24] With comfortable accommodations for about 250 workers, the *Gordon* could stay out in the North Sea and work for long periods when conditions allowed. Then when the work season ended with the onset of winter, the vessels's air-conditioned quarters gave it the flexibility to work in tropical regions.

Numerous adaptations required by the North Sea were evident onboard the *Gordon*. A stronger hull increased the strength and the weight of the vessel. A mooring system sized for rough seas included ten lines, each with 5,000 feet of two-inch cable and 30,000 lb. anchors. A 250-ton Clyde revolving crane gave the vessel great flexibility in working offshore. Additional work stations on the side ramp allowed for faster welding, giving the barge the capacity to lay up to two miles of pipe per day under good conditions. Two tire-type pipe tensioners on the ramp could be used to reduce sagbend stress. The conventional straight pontoon originally attached to the *Gordon* had to be replaced with an articulated pontoon several years later when the barge moved into deeper waters. All in all, the *Hugh W. Gordon* embodied the state-of-the-art in pipelaying equipment in the late 1960s, and the ship's construction was Brown & Root's proclamation that it meant business in the North Sea.

This was good news for companies in Europe, as well as for Brown & Root. Rotterdam yards built the *Atlas*, the *Hercules*, the *Hugh W. Gordon*, and later vessels required by Brown & Root and other offshore companies. The pipe for the West Sole line came from British Steel in Glasgow, and when it began to arrive in Yarmouth in February 1966, that city had already become a center of offshore-related activities. Shipyards throughout Europe had also begun to produce the mobile drilling rigs that were much in demand in the North Sea. As drilling and production expanded, a growing complex of associated industries spread along the shores of the North Sea, as had happened earlier in the Gulf of Mexico. The ability of the well-developed economies of the region to adapt to the demands of this new industry hastened the development of gas and oil in the North Sea.

Brown & Root geared up its operations for the completion of the West Sole line in the summer of 1966. The first job for the *Gordon* went smoothly. The pipe arrived on time and in good shape from Glasgow. The Great Yarmouth pipe-coating yard of Brown & Root (U.K.) wrapped the individual joints of pipe in fiber glass and bitumen and then covered them in reinforced concrete before transporting them out to the *Gordon*. The assembly line on the barge's deck then took over to lay the pipe from the shore at Easington out to the platforms in the West Sole field. With the completion of this 16-inch trunkline, Brown & Root laid a 24-inch pipeline for almost two miles under the Humber River. Then all that was left was to open the valves and celebrate the arrival of North Sea gas onshore.[25]

The entire system delivered gas from the West Sole Field into the British Gas Council's existing onshore pipeline grid in the fall of 1966, marking the first commercial development of North Sea oil or gas. Before the completion of the pipeline, the Gas Council had agreed to purchase from BP at least 50 MMcfd of natural gas for fifteen years, and the 200 MMcfd capacity of the new pipeline encouraged all involved to expect a steady increase in natural gas delivered to shore with the further development of offshore gas fields. Considerable excitement greated these developments, which symbolized a new era of energy independence for Great Britain. The practical result was evident in 1962 in Holland and in the late 1960s in Great Britain when those nations' domestic and industrial gas appliances were converted from coal gas to natural gas.

Meanwhile, Brown & Root carried out a critical search for a way to bury the West Sole line. As the *Gordon* laid the pipeline, the company towed in from the Gulf of Mexico the *M-228*, which one trade journal referred to as "a veteran bury barge." The *M-228* was too small for the North Sea. The hard clay bottom would not yield to the trenching equipment developed for the softer sandy bottom of the Gulf. George Brown acknowledged that burying was non-negotiable in the North Sea: "The pipelines must be buried to avoid ocean travel in that shallow water. The minute rough weather comes, boats drop anchor and the anchors drag right over the pipelines. So we have to dig a ditch on the floor of the sea and bury the lines." He also acknowledged the extraordinary difficulties of accomplishing this task with equipment not well suited for the harsh conditions. "When the winds blow for twelve hours running," he said, "it's hard to dig a ditch in a hundred feet of water and lay your pipeline and bury it, too. But that's what we have to do in the North Sea."[26]

Burying pipeline became a pressing problem for BP, which had contractual obligations to deliver gas in the winter of 1966. To meet them, BP decided to deliver gas through the unburied pipeline until it could be buried during the next construction season. Representatives of BP asked L.E. Minor directly, "What are you going to do about burying this pipe?" After explaining the problems with the *M-288*, Minor offered a solution: "We'll build a new barge and do it for you next year." When BP said they were not prepared to discuss this proposal unless they knew its cost, Minor and Wilson walked outside, discussed the situation, and decided on the probable cost of the new bury barge. They returned to the meeting, where Minor said simply, "We'll bury your pipeline for six million dollars." After excusing the Brown & Root representatives for closed door discussions, BP agreed.

When spring arrived in 1967, the *M-288* returned to work with upgraded, jet-suction equipment, which increased the pressure of the stream of water used to dig a trench for the West Sole pipeline. As work progressed, construction in Rotterdam moved ahead on Minor's new, improved bury barge, the *BAR-279*. In August the new barge joined the fray, combining with the *M-288* to complete the burial of the West Sole line in October.

Using German "Grassel" mining pumps and a Pratt & Whitney aircraft engine, the much larger new barge could generate two and one-half times the water pressure and work in deeper waters than its predecessor. Although the *BAR-279* represented a giant step forward in bury barge design, further advances would be needed. Indeed, even the combination of the old and the new bury barges could not conquer the conditions of the southern North Sea. In places, the West Sole pipeline could not be adequately buried, and barge crews had to bring in additional dirt to cover the pipe.[27] As with other aspects of the West Sole project, the pipeline was not covered with conventional equipment and procedures, but it was covered.

Brown & Root's contacts with oil and gas companies aided the company's expansion in the North Sea. One such "contact" particularly close to Brown & Root was George Brown, the chairman of the board of Texas Eastern Transmission Corporation. When the British Gas Council and Amoco jointly participated in the United Kingdom's first licensing round for exploration concessions in the North Sea, they included Texas Eastern, a diversified energy company with an extensive U.S. gas pipeline system, and Amerada Petroleum Corp. These concessions turned out to contain the massive Indefatigable and Leman Bank gas fields. The Indefatigable field—Block 49/18-1—was discovered in 1966 through a well drilled by *Mr. Louie*. This self-elevating drilling barge could drill in up to 150 feet of water and had drilled the first exploratory gas wells in West German waters. In the winter of 1966, *Mr. Louie* also found commercial quantities of natural gas on the Gas Council-Amoco-Texas Eastern site in the Leman Bank field, Block 49/27.

When the British government held a competitive bid for the contract to build a pipeline from these gas fields to the mainland, Texas Eastern's wholly owned subsidiary, Texas Eastern Engineering, Ltd., won the engineering and management contract and subcontracted the work to Brown & Root. In 1967, Brown & Root built for the Amoco/Gas Council group a 30-inch, 35-mile pipeline that connected the Leman field to Amoco's processing plant at Bacton, Norfolk, about one hundred miles northeast of London. Later, in 1971, it added 22 miles of 30-inch subsea line in 125 feet of water for Amoco to connect the Leman and Indefatigable Fields. During the same period, Brown & Root designed and constructed for the Shell/Esso Group a 30-mile, 30-inch pipeline from the Complex A in the Leman Bank field to Shell's gas processing plant at Bacton.[28] This project included a unique 4-inch, concrete-coated "piggyback" methanol line strapped to the larger natural gas line. Brown & Root also completed another 17-mile pipeline to Bacton from a platform in the nearby Hewitt field. This system of gas pipelines tied a growing complex of gas-producing platforms to processing facilities at Bacton, which became for a time the central collecting point for the offshore gas industry.

Construction of the pipeline system into Bacton called forth a new and improved approach to pipe-laying. The pipe had to be laid across ridges on the ocean floor parallel to the coast of England, which required that a dredging vessel "presweep" the path of the lay barge to eliminate sharp changes in elevation.[29] During the laying of the trunkline from Leman field to Bacton, gale winds twice interrupted operations, forcing the pipe and pontoon from the lay barge to be placed temporarily on the sea bottom while the barge itself took shelter in protected waters. Such conditions placed a high premium on precision in laying pipe along the route, and Brown & Root put to good use the microwave survey systems it had first used in the Middle East. With microwave positioning, operators could adjust the lay barge's position, allowing work to go forward despite fog. Its large size made the *Hugh W. Gordon* more stable in rough waters, further reducing time lost to bad weather.

In 1971 Brown & Root laid what was then the longest pipeline in the North Sea. The line connected the Viking Field to Conoco's onshore processing plant at Humberside. This 28-inch pipeline was 85 miles long, with a 3-1/2-inch diameter piggyback line strapped to the larger line. The line could transport an average of 550 MMcf/d with a maximum peak-load capacity of 920 MMcf/d.[30] Brown & Root completed this major project in the spring and summer of 1971 with careful planning, the use of its growing fleet of specialized equipment, and the application of all of the lessons it had learned in the North Sea. The company employed two lay barges, two dredge barges, and a presweeping barge to complete the pipeline within a single construction season.

One new departure—the use of the CRC-Crosse automatic welding system—helped speed the work on the Viking line. Brown & Root had acquired the exclusive rights to this process in 1970, and after a brief trial run in the Gulf of Mexico, it was put to use in the North Sea on the Viking line. The combination lay barge *L. B. Meaders* (400' × 100' × 30') was equipped with the CRC-Crosse system, which allowed for fewer welding passes and could thus weld faster than the traditional manual stick-rod approach. The overall process resulted in extremely clean and easy to inspect welds.[31] The *Meaders* also had the British Internal X-Ray (BIX) unit to X-ray the welding. The benefits from this new system could be seen by comparing the progress on the Viking line of the *Meaders* and the *Hugh W. Gordon*, which worked without the CRC-Crosse equipment. The *Meaders* welded almost twice as quickly as the *Gordon*. Subsequent analysis of conditions under which the two vessels worked concluded that automatic welding had improved productivity by at least 15 percent. Such gains meant greater speed of operations, a critical consideration in the race against the North Sea's weather.

Brown & Root's growing experience and new equipment helped work on the Viking line go smoothly. Two dredging passes preswept parts of the route, removing large boulders and knocking down underwater hills in the same fashion that a bulldozer would on land. The *Meaders* began laying the line from the shore in early May. Despite good weather by North Sea standards, shutdowns ranging from less than a day to six days slowed the work of laying the line. During these shutdowns, the repeated raising and lowering of the stinger caused substantial damage, forcing its replacement. In July and August, the *Gordon* laid pipe from the platform toward the shore. Then it used its 250-ton revolving crane to set the platform riser as the *Meaders* completed laying the pipe. Three slight bends in the pipe allowed it to skirt areas that could not be smoothed by dredging. After three days spent aligning the two sections of the pipeline, the system tie-in was completed on September 10, 1971. The *BAR-279* then buried the line under three to fourteen feet of cover.[32]

Despite the steady development of natural gas fields in the North Sea during the 1960s, the region entered the 1970s without producing the dramatic volumes of natural gas or oil predicted by oil companies and geologists. In May 1970, the *Oil & Gas Journal* sounded a note of pessimism: "Based on its experience of the past five years, the industry has come to look on the North Sea primarily as a gas target. Some operators, steeped in disappointment, have pulled in their horns, cut their budgets and staffs, and expressed little interest in further North Sea ventures."[33] Dry holes had far outnumbered producing wells in these early years of North Sea development,[34] and prospects for the discovery of commercial quantities of oil dimmed.[35] Years of futile searches for major oil fields introduced the phenomenon of "marginal fields" and left a growing suspicion that the much anticipated North Sea boom was not to be. Certainly the natural gas coming ashore in England from the fields in the southern North Sea was a welcome addition to that nation's energy mix, but the overall impact of the offshore oil and gas industry had fallen far short of the promise of the early 1960s.

This was not, however, the case within Brown & Root. The North Sea had been most kind to the company in these early years. Between 1964 and 1970, the firm had laid all five of the North Sea's major pipelines, totaling some 425 miles of line. It had constructed or engineered more than fifteen offshore platforms, and installed about 95 percent of the thirty-five offshore platforms erected.[36] By 1970, Brown & Root was the undisputed leader in construction and engineering in this new offshore province. It had earned this position by entering the region early, investing in equipment especially designed for conditions in the North Sea, and adapting its traditional procedures to this demanding environment. It had established and maintained close working relationships with the major companies active in the region, cementing its reputation as a company that could get things done by finding innovative ways to meet unique challenges. Whatever the future of the region, Brown & Root would play a leading role.

Notes

1. This chapter draws from interviews with A.B. Crossman, R.O. Wilson, Dirk Blanken, L.A. Starr, and Hugh Gordon.
2. R.O. Wilson and M.R. Martin, "Deepwater Pipelining for Central North Sea," OTC 1855, Offshore Technology Conference, Houston, TX, 1973.
3. "Supplying Rigs Proves Toughest Operating Problem," *Oil & Gas Journal* (May 10, 1965): 154.
4. "Supplying Rigs Proves Toughest Operating Problem," 156.
5. John Cranfield, "25 Years of Brown & Root (U.K.) Limited," *Petroleum Economist*, Advertising Supplement (November 1984): I. Also see Clive Callow, *Power from the Sea* (London: Victor Gollancz, Ltd., 1973), 44; B.A. Rahmer, "North Sea Oil—The Search Begins," *Statist* (May 29, 1964).
6. Callow, *Power from the Sea*, 45–6.
7. Callow, *Power from the Sea*, 39.
8. Chris Castaneda and Joseph Pratt, *From Texas to the East: A Strategic History of Texas Eastern Corporation* (College Station: Texas A & M University Press, 1993), 165–167.
9. *Thirty Years of Innovation*, (Den Haag, Holland: Moretus, n.d.).
10. See "News in Brief. . . ." *The Stock Exchange Gazette* (May 29, 1964), copy in R.O. Wilson clippings book, Brown & Root Archives.
11. "Western Europe's offshore play is on; production up," *World Oil* (August 15, 1965): 120.
12. J. M. Rainey, "Marine Builders are fast. . . .," *Offshore* (February, 1973), 91.
13. "Rig Construction Still Booming," *Offshore* (June 20, 1967): 166, 169.

14. "The North Sea . . . Oil's Biggest Gamble to Date," *Oil & Gas Journal* (May 10, 1965): 134. "Today - a trickle - tomorrow, a score of offshore rigs and a $100 million investment," *Oil & Gas Journal* (May 10, 1965): 134.
15. "Continental's fixed platform," *Petroleum Times - London* (November, 1965): 617.
16. See R. O. Wilson, "Movable Platform: 1964-1966," in Brown & Root History files. The story about moving the platform away from its originally intended position came from a Wilson memo but is not documented in any other source. Also, see "North Sea hopes spurred by two wildcat gas tests," *World Oil* (December, 1965): 119–20.
17. "The big North Sea gas rush," *The Sunday Times* (November 14, 1965).
18. R.O. Wilson binder, "West Sole Pit," Tab H, copy in Brown & Root Archives.
19. Robert A. Moorehead, unpublished manuscript on Brown & Root history dated April, 1968, 110.
20. Callow, *Power from the Sea,* 174.
21. "What British Petroleum Learned about North Sea Pipelining," *Oil & Gas Journal* (February 27, 1967): 135.
22. A.R. Desai and J.R. Shaw, "Marine Pipelaying Methods Defied Odds," *Offshore* (September 1979): 106.
23. *Ibid.*
24. "Underwater Pipeline," newspaper clipping in Blanken file, Brown & Root Archives.
25. "Two British Gas Strikes in Giant Class," *Oil & Gas Journal* (August 8, 1966): 46.
26. *Ibid.;* Moorehead manuscript, 110.
27. "Brown & Root's Bury Barge is Working in the North Sea," Brown & Root publication, n.d.; "North Sea," *Brownbilt* (Winter 1967–68): 2.
28. "To Shore at Bacton," *Brownbilt* (Fall 1970): 20.
29. "What's Ahead in the North Sea," *World Oil* (December 1967): 76.
30. "Conoco-NCB line is North Sea's longest," *Oil & Gas Journal* (January 10, 1972): 90–95.
31. "The North Sea: New Challenges and Accomplishments," *Brownbilt* (Winter 71–72): 22.
32. "The North Sea: New Challenges and Accomplishments," 22. "Conoco-NCB line is North Sea's longest," 90–5.
33. "Huge North Sea find has entire oil world vibrating," *Oil & Gas Journal* (May 25, 1970): 34.
34. ". . . No Bonanza Yet," *Oil & Gas Journal* (February 27, 1967): 5.
35. "Hopes Fading in Norwegian North Sea," *Oil & Gas Journal* (June 5, 1967): 82.
36. "To Shore at Bacton," 16.

CHAPTER 12

Ekofisk and Early North Sea Oil

The North Sea in 1969 and 1970 finally yielded two huge oil discoveries: Ekofisk in the Norwegian sector and the Forties in the U.K. sector. Both fields were in the central North Sea, farther north, in deeper water, and with harsher weather conditions than previous gas fields. To bring oil and gas from Ekofisk and the Forties to market, the offshore industry had to develop new techniques and tools for working in one of the most demanding environments yet encountered. But the size and location of these fields seemed to justify the effort—even in an era of $3 per barrel oil. The discoveries also posed an old question: How could these prospects be developed at the most reasonable cost and in the shortest time?[1]

Phillips Petroleum, the discoverer of Ekofisk, and British Petroleum, the discoverer of the Forties, both employed the most experienced offshore construction company in the North Sea, Brown & Root, to generate new answers. At Ekofisk, Brown & Root built a creative system for temporary production and then helped construct a sprawling permanent production complex. At the Forties, the company took a leading role in the design, building, project management, transportation, and installation of a complex of four fixed platforms that embodied stunning technological innovations. The success of both projects, after inevitable delays, affirmed both the viability of oil development in the North Sea and Brown & Root's leadership. Ekofisk and the Forties launched a new era in the history of the North Sea and the offshore industry as a whole.

After the 1973 oil embargo disrupted oil imports from the Middle East and almost quadrupled the prevailing world price of oil, North Sea development became frantic. Companies took advantage of the new economic and political realities to press the search for oil in this non-OPEC region with excellent access to large European markets. Said Brown & Root's Dick Wilson, "Once oil went to $12 a barrel, . . . the rush was on. And so, OPEC certainly gave a 'kick start' to the North Sea at a point when it might not have . . . if the price of oil had stayed where it was. . . . Deepwater North Sea oil from fields like the BP Forties was not commercial at $3 a barrel. And thanks to OPEC, it became more than that." Indeed, rising oil prices of the 1970s transformed the search for oil in the North Sea into one of the most frenzied booms in modern oil history.

Brown & Root stayed right in the middle of the boom. The company had become the leader in North Sea construction and engineering in the 1960s by extending and adapting Gulf of Mexico technology to conditions there. This process continued in the 1970s, but the oil boom required much more than new, improved versions of traditional technology. Departures in platform design called for new approaches to construction, transportation, and installation. The laying of larger pipe in deeper, rougher waters pushed the industry toward fundamental changes in the size and capacity of barges. The rising costs and growing complexity of offshore technology encouraged the creation of new approaches to managing the massive projects requisite to the North Sea. In responding to these demands, Brown & Root greatly expanded its presence, hiring engineers in London and other oil centers around the North Sea. They also built a massive fabrication yard in Scotland and acquired a fleet of heavy-duty vessels. This was an exhilarating time for Brown & Root and competitors attracted by the extraordinary opportunities.

Challenges of the North Sea

Ekofisk and the Forties presented technical, economic, and political challenges to the offshore industry. Jay Weidler, who joined Brown & Root as a senior marine engineer in 1969, the year of the Ekofisk discovery, later recalled the "vast technological gaps that yawned" in these early years of North Sea oil development. "All these problems were happening, and there was little prior work to find out what we were supposed to do. We had to resolve most of them ourselves in conjunction with the client and the authorities."[2] The high costs of North Sea development also heightened economic challenges, encouraging most companies in the region to speed up their work in an effort to produce oil revenues as quickly as possible. Brown & Root was no stranger to such pressures, but the vast sums of money involved in the North Sea gave them a special urgency.

The technical challenge began with the move from about 100 feet of water and wave heights of perhaps 45 feet in the southern North Sea, to depths of 200–500 feet and waves of 75–100 feet in the central North Sea. According to Weidler, "Not only did the wave heights increase signficantly, and with them the forces that were going to be applied laterally on structures, but also the deck weight increased by a factor of five." This yielded a five- to ten-fold increase in jacket size, creating the need to revamp almost every aspect of construction and installation. If previous offshore platforms had the look of tourist courts in the ocean, this new generation of North Sea structures would resemble luxury hotels. The scale was needed to ease supply problems in the harsh conditions of the region and to make use of every expensive square inch of each artificial island.

The North Sea presented another challenge: metal fatigue. Unlike mild environments, the North Sea was full of waves that could cumulatively wear down a structure. With no data from which to calculate the long-term impact of North Sea waves on the giant new structures planned for Ekofisk, Weidler worked with the Norwegian government to devise a practical formula for fatigue. "We really didn't have the tools in the early 1970s," he recalled, but the formula "seemed to have worked reasonably well at Ekofisk."[3] Designers had to rely on educated guesses, erring on the side of safety as they accumulated the data to build more sophisticated understanding.

One response to such considerations was the basic redesign of the traditional steel jacket. Two important new features emerged for structures in the northern North Sea: the "node" and the "cluster" pile. Both innovations were in response to the increased loads imposed by the deck and the wave. Engineers had long recognized that the most effective location for the pile was at the four corners of the jacket. With the exception of structures for Cook Inlet, prior designs had run the piles through each leg of the structure even when there were more legs than four. This inefficiency in pile use had been accepted as it eased both fabrication and installation of the structures. For structures in the northern North Sea it became clear that it was more efficient to move the deck and wave loads to the extremities of the jacket.

The four external legs then grew in size, such that the wall thickness of the joint can was in the five to six inch range. Recognizing that steels of this thickness were difficult and expensive to produce, especially with good notch toughness and through thickness properties, engineers fell back on internal ring stiffeners to reduce the required wall thickness. The difficulty in fabrication using previous procedures led to a construction method whereby the stiffened joint and portions of the incoming brace members were shop fabricated and then moved to the field. This assemblage was called the node.

As the force in the leg reached the bottom of the structure it was necessary to transfer it to multiple piles. Again the most efficient way to accomplish this was to cluster the pile sleeves around the leg in a circular pattern. The pile sleeves were connected to the leg using shear plates. The piles had to be a certain distance from the leg and a certain distance apart from one another so that the resistance offered by each pile was not diminished through too close a spacing. This assemblage, often weighing hundred of tons, was known as the cluster pile.

As fabricators faced transforming these designs into functioning platforms, construction facilities sprang up around the North Sea in the 1970s. In the early phases of the Forties project, Brown & Root entered the fray with the construction of the HiFab yard at Nigg Bay, Scotland. The scale of the new yard matched that of the new platforms, and the innovations applied to construction reflected their growing size and complexity.

After designing and building these giants, engineers and constructors also had to devise new ways of transporting them to their sites and installing them. The scale of these structures dictated modification or redesign of every piece of installation equipment, from derrick barges to steam-driven pile hammers. When enlarged equipment alone could not accommodate the demands posed by giant structures, more fundamental innovations had to be envisioned, as when Brown & Root launched the Forties platforms with reusable floatation tanks instead of building the giant barge that would have been required to launch the jackets more conventionally. At such moments every major construction company active in the North Sea faced vexing competitive questions: Did long-term prospects justify the incredible investments in equipment needed to keep pace with the needs of the oil companies and the competition? Which investments best fit the strategy and the profile of a specific company? How many major competitors could each sector of the market support?

Underlying such questions were tricky and difficult economic issues. Cost overruns plagued the best-laid plans, as conditions forced engineers repeatedly to reconsider designs. Even after debates over the magnitude of a "hundred-year storm" at Ekofisk led to the raising of the original height of the platforms and walkways after fabrication had begun, a photographer captured the image of the walkway awash in the spray from the high waves caused by a storm surge. "I think they had a hundred-year storm every year out there," said Larry Starr of Brown & Root in describing this dramatic photograph, which symbolized the need for humility—and re-engineering—in this environment. Brown & Root had thrived in such situations, when it could work out in practice what seemed probable in theory, but the high cost of experimentation in the North Sea often caused tension with clients, even after higher oil prices assured acceptable returns on investments.

Brown & Root understood profitability, but not politics in the North Sea nations. Sir Phillip Southwell, chairman of Brown & Root, U.K., Ltd., smoothed the way politically in England, and the company even learned to coexist with British labor unions. Norway proved less hospitable. The use of foreign workers, the lack of tax payments by employees working offshore, and the accidental death of a diver all plunged the company into public controversies. Brown & Root's image in Norway as an arrogant Yankee company made its work there more difficult, proving that technology alone cannot overcome all challenges.

In these exciting early years of North Sea oil development, Brown & Root's plate was filled to overflowing. Engineers grappled with cutting-edge design issues on many fronts at the same time. Giant platforms and ambitious pipeline projects absorbed the energies of the company's best employees. All that Brown & Root Marine had learned in its previous history was at once vital and obsolete. Beginning at Ekofisk and the Forties, the company had to build on the best of its long traditions of applied engineering and practical construction methods while also keeping pace with technical developments in a rapidly changing industry.

The Discovery of Ekofisk

The oil jackpot was long in coming, but proved to be worth the wait. In the 1960s, 200 exploratory wells in the region yielded no commercial quantities of petroleum. Norwegian waters alone witnessed thirty-three dry holes in these lean years. When the bit finally found oil at Ekofisk in late 1969 and early 1970, however, it found a giant field. Through 1980, the field produced more than 575 million barrels of oil and 1.4 trillion cubic feet (Tcf) of gas.[4] The cost to develop Ekofisk and the six nearby fields (West Ekofisk, Cod, Tor, Eldfisk, Albuskjell, and Edda) reached almost $6 billion, making this project second only to the Alaska pipeline as the largest private commercial engineering project completed up to that time.[5]

As in Alaska, the harsh environment contributed to this high cost. A Brown & Root engineering report noted that, "With the acceptance of the Ekofisk project, we encountered the most severe environmental conditions for design that our department had experienced up until that time. These conditions included water depths of 230 feet, a 78-foot design wave, and a 126-mile-per-hour wind. Additionally, it was estimated that an average of only 50 hours per month could be used for off loading of supplies during the winter months. . . ."[6] A Brown & Root engineer said it more simply: "One of the things we had to do early was to learn how to stay offshore and be able to handle the storms."

Ekofisk held out hope that Europe might some day achieve energy independence. The media described it as a "new, politically safe source of supply." It was also important for Phillips Petroleum Company, whose management had questioned the wisdom of spending more money on exploratory drilling in the North Sea after having nothing to show for $30 million spent there since 1966.[7] Seismological reports in Norwegian waters after the Groningen discovery had shown promise, but subsequent exploration had been fruitless. In 1968, the Phillips Norway Group (Phillips Petroleum Co, Norske Fina, Norske Agip, and Norske Petronord) thought it had found a significant field in the central North Sea, but hope failed when the Cod condensate field did not appear to be large enough for commercial production.[8] By the late 1960s, the once-heralded Norwegian sector was considered by many to be the least likely sector of the North Sea to contain oil.[9]

In late fall of 1969, Phillips Petroleum neared the end of its patience with the North Sea. As the weather worsened and the work season neared its end, Phillips decided to give it one last try before abandoning the site for the winter and perhaps for good. This last try proved historic. Drilling at Ekofisk 1-X during September 1969, Phillips's crew found traces of oil in the circulating drilling fluid at 5,500 feet. Technical problems forced the abandonment of the well. In December, a new well on Block 2-X about sixth-tenths of a mile from the original site found no oil at the 5,500 foot mark, but detected signs of oil at 10,000 feet. The volume of oil increased as the drill bit dug deeper.[10] These solid indications of oil came from a promising strata of chalky, highly fractured limestone with good porosity.[11] Phillips had tantalizing hints that it had found an epoch-defining discovery in the North Sea, but it would have to sit out the inclement winter weather before it could learn more about its find.

In the spring of 1970, Phillips returned to the site and confirmed that this new field contained a giant oil deposit, which measured eight miles by four miles and consisted of a 690-foot thick, porous, pay zone of carbonate rock. Initial estimates placed reserves in the seven billion barrel range, making this the largest oilfield ever discovered in Western Europe. The field, named Ekofisk, was a few miles east of the British-Norwegian offshore boundary and adjacent to a large block of uncommitted acreage. Its proximity to Stavanger made that Norwegian city the logical location for Phillips's Ekofisk headquarters. As oil and gas production in the Norwegian sector of the North Sea increased over the years, Stavanger became the center of the Norwegian oil and gas industry.

The Ekofisk discovery sent a lightning bolt through the oil fraternity. There was oil, and lots of it, in the North Sea. The pace of exploration quickened, and the focus shifted to the central North Sea. As development moved forward at Ekofisk, news came of additional oil strikes in U.K. and Dutch waters. What had been learned in the southern regions would have to be modified in response to harsher conditions in the central and northern North Sea.

It would be modified on the run, since Phillips Norway was in a hurry to secure production from the site. The company needed to recoup some of its investment in the region and generate funds for the extraordinary expenditures required to develop its massive field. Forsaking a traditional approach to development that seemed too slow to meet its timetable, the company chose instead a novel approach that promised to produce oil more quickly. Instead of relying on the European infrastructure, which had grown to support earlier projects in the North Sea, Phillips called on the resources of the worldwide offshore industry to push its project forward. Brown & Root was among those responding to the Phillips's call, and it played an important role in the initial phase of the field's development.

Phillips Norway followed a three-phase plan in its development of Ekofisk. The first phase included the construction of temporary production facilities capable of getting oil to market from the four wells previously drilled in the field. Funds generated in part by these temporary facilities helped finance the completion of Phase 2: the construction of a permanent central producing complex with three drilling platforms, a production platform, the first ever living-only quarters platform, a flare platform, a unique one-million-barrel concrete oil storage facility, and a long-distance subsea oil pipeline to shore. In Phase 3, pipelines connected seven outlying producing areas into an expanded central complex, complete with a central processing facility installed

Ekofisk concrete tank. Photo courtesy of Dirk Blanken

atop the concrete storage tank and known as Ekofisk center, a new pumping platform for an oil pipeline, a riser platform where flow lines could be interconnected, a second flare platform, and a natural gas pipeline. The complex resembled a small city, especially when viewed from the air at night, and it came to be known as "Ekofisk City."

The development was a marvel of modern offshore engineering. The *Oil & Gas Journal* noted that "the entire project has continually pushed against the frontiers of existing knowledge. Some of the innovative achievements have become commonplace in the offshore industry."[12] Brown & Root accounted for many of these innovations, particularly in its work on the temporary production facility.

Phase One: An Unusual Design for Temporary Facilities

Phillips had not yet proven the field at Ekofisk, but the company decided to "go ahead and produce the wells that they drilled during exploration, just to look at the viability of the field," said Bill Golson, Brown & Root's project engineer on Phase 1 of the Ekofisk project. Phillips hired Brown & Root to search for the most efficient way to get started at Ekofisk and suggested converting the 300-foot jack-up rig *Gulftide,* which they had under lease, into a temporary production facility. Golson and his colleagues at Brown & Root were pessimistic: "How in the hell are we going to make a production facility out of that thing?" Still, they forged ahead with feasibility studies, which indicated that a specially designed caisson might be used to convert the *Gulftide* into a temporary production platform. Golson said, "We were really out in the wild blue yonder with this one," but he also acknowledged that Phillips "was reaching on out there themselves." Time was money, and Phillips was willing to flaunt convention in search of a quicker route to oil production from Ekofisk. Brown & Root was more than willing to follow. As one Brown & Root engineer recalled, "We were given the go-ahead by Phillips to get after it."

Phillips wanted to "get after it" immediately, despite the obvious difficulties of working 200 miles from land in the central North Sea in the middle of the winter. When Brown & Root reminded its client that "this is the worst time of the year, particularly at this place. It's absolutely lousy weather and our downtime might be 60 percent," Phillips responded: "If your downtime is 60 percent and if we can produce for two months, we've paid for it all." For the first time, Brown & Root then ventured out into the North Sea winter and stayed out, far from shore, for months. On Ekofisk Phase 1, Phillips ordered Brown & Root to go full-speed ahead. Clyde Nolan remembered: "There were not enough hours in the day. You just worked your tail off and it didn't make any difference how many hours or whatever it cost, you just did it."

Work began on site in January 1971, when the *Gulftide* was towed to the site and jacked up on location at Ekofisk with the assistance of the *Hugh W. Gordon.* To prevent potentially disastrous scouring of the rig's legs, divers stacked thousands of sand bags against them on the ocean bottom. After helping set the mobile rig into position, the *Gordon* served as the base of operations for the entire conversion process.

Built for the North Sea, the *Gordon* had quarters for more than 250 workers and extra supplies for extended work periods. It had a reinforced hull and deckhouse, a 250-ton crane, upgraded lifesaving equipment, and a special radio for receiving short-range weather predictions. In addition, sections of the barge had been prepared for the use of divers. This was the first time that Brown & Root had used one of its barges out in the central North Sea for extended periods during the winter season, and some worried that the *Gordon* wouldn't survive. Model tests in a test basin suggested that the barge could survive a hundred-year storm in this section of the North Sea. After strengthening the vessel and modifying its deck layout and equipment to allow water to flow over the deck during storms, the *Gordon* moved out into the foreboding weather at Ekofisk. A hundred year-like storm occurred during the first winter, and the *Gordon* and its 200-man crew rode it out before being retrieved with an emergency line near the coast of Norway.[13]

Converting the *Gulftide* into a temporary production platform capable of supporting an array of production, processing, and transportation equipment was a daunting task. Golson recalled that "The challenge that we had using the jack-up rig, of course, was trying to get the oil from the seabed to the deck of the *Gulftide,* which was setting up at the maximum elevation of plus 89 feet above mean sea level. And it was set up that high because of the waves."[14] The removal of all drilling equipment from the *Gulftide*'s deck created space for custom-made separators, meters, pumps, and an elevated flare stack. Critical to the success of the conversion was the placement of wellhead equipment on four subsea wells, the connection of these seafloor wellheads to the production platform, and then the connection of production pipelines from the platform deck back down to the ocean floor and outward to two single buoy moorings for loading oil tankers. One Brown & Root engineer explained that the basic challenge of Ekofisk was to "perform constructive offshore pipelaying in fully exposed areas, where return to sheltered water is impractical in severe weather, even in the most hazardous season."[15] None of this had been done before in such deep water or under such harsh conditions.[16]

The oil flowlines, hydraulic lines, and control cables had to be connected at depths of more than 200 feet, which required precise positioning to assure secure connections. Brown & Root's engineers designed and fabricated a unique 44-ton subsea landing base which divers from the *Gordon* helped install

Gulftide during conversion

on the ocean floor between the legs of the *Gulftide* to form a base for flowlines and loading lines. A 42-inch pipeline riser caisson extended from the base up to the platform deck, providing a housing for all of the flow lines, hydraulic control lines, and oil loading lines.

Brown & Root engineers designed the long caisson to fit through the platform's drilling slot and extend down to the ocean floor. The caisson provided the conduit for the pipes and cables needed to take the oil on its journey from the subsea wellheads to the platform deck for processing and then back down and out to the single buoy moorings. By guiding, protecting, and supporting the subsea lines, the caisson rendered the conversion of the *Gulftide* a unique success. Two 4-1/2-inch coated flowlines carried the well streams from each wellhead to the platform. The inclusion of two flowlines, each with separate valving, allowed production to continue even if one flowline failed. This dual line design also allowed for efficient cleaning of the lines by closing the wellhead valves, opening a bypass valve, and injecting water into one line to displace the oil into the other one. The two lines also permitted pigging, if necessary in the future. In addition to the flowlines, a bundle of six hydraulic control lines connected the platform to each wellhead. All of the flowlines and their hydraulic bundles fit snugly into the space within the riser caisson, which one of the designers described, with an engineer's pride, as "rather full inside."[17] Hydraulic control lines controlled the wellhead valves, while chokes and manual valves located on the platform provided normal well control.[18] What Brown & Root designed "was not very pretty," admitted Rochelle, "but it worked."

Brown & Root employed fabricators from around the world to supply parts. Clyde Nolan states, "We spread this all over the place to get everything in time." Different suppliers were used "not because we wanted to, but because nobody could deliver everything in time." Rochelle recalled that Nolan was a "one-man band on getting the caisson and flowline into operation. He would be in the office one week, in Holland or Norway the next, on the barge the next, and then repeat the cycle." Nolan and Brown & Root "became the interface. . . . We took all of the equipment from people, matched it, put it together, and assembled it. And it worked out well."

Once the parts had been obtained, the fabrication and installation of the caisson went forward as smoothly as could be expected. The installation of this caisson—which was unique at this time, though later used more widely—required creative use of existing equipment. To pull a flowline from the barge to the base, workers installed a pull winch and cable specially outfitted for the *Gordon*. After extending the cable to the sea floor, divers threaded it through guide shoes on the concrete base. The winch then pulled it back to the stern of the lay barge, where the cable was attached to the pipeline pulling head. Operating at a tension of 70,000 pounds, the barge's winch then pulled the flowlines off the barge and down to the sea floor, where divers tied them in at the base. A short truss-type pontoon supported the flowlines as they were lowered off the barge to the sea floor for connection at the caisson's base.

The lines were weight coated and tied together with a one-inch wire rope. In addition to the flowlines, each well required a six-tube hydraulic control bundle. The lay barge reeled off the bundles in segments of 2,500 feet with steel weights to sink the lines.[19]

Connecting the wellheads to the *Gulftide* was only part of the job. The ten-inch flowlines had to be laid from the deck of the *Gulftide,* through the base, and to the single buoy moorings, which Brown & Root had never installed in water depths beyond one hundred feet or in waters as rough as those at Ekofisk. Six anchors ranging from 30,000 to 50,000 pounds held the single buoy moorings in position. The system could handle oil tankers in the 30,000- to 50,000-ton class and gave the Ekofisk field a substantial outlet for its oil during the years before completion of an oil pipeline.

The underwater connections at Ekofisk placed heavy demands on divers. Even before the *Gulftide* arrived on site, divers had prepared the way by making the first subsea wellhead completions in the North Sea. Connecting these wellheads to the pipes and cables in the caisson and tying in the single buoy moorings kept the divers busy throughout Phase 1. Winter weather made conditions extremely difficult, but Phillips had asked the divers "to stay out there during the winter and get every hour we could work." This made for a cold, lonely winter, with more waiting than working, "but we had to stay out there and get them hooked up, and it paid off in the end."

Taylor Diving Company completed this difficult and dangerous underwater work in almost unbelievably harsh conditions. The divers plunged to 230-foot depths to weld underwater pipeline connections, repair platform jackets, connect flowlines to platforms, and assist in pipeline trenching and burying. To make the underwater connections, divers used a combination of hydrocouples, hydraulic expansion sleeves, misalignment unions, and swivel joints. During later phases of Ekofisk's development, Taylor divers spent thousands of hours making saturation dives and hyberbaric welds to connect the system's pipelines.

Phase 1 at Ekofisk required much trial and error to create a temporary production system, but little time was lost to experimentation. Phase 1 began during the winter of 1970–71. By July 1971, the four original subsea completions at Ekofisk produced as much as 40,000 barrels of oil per day. These temporary facilities pumped oil into tankers for nearly three years. As these years passed, a massive permanent production and oil storage facility emerged at the Ekofisk field, paid for, in part, by the revenues produced by the *Gulftide.*

Some of the Texans working at Ekofisk felt that the Norwegians looked down at their "Jake Leg Cowboy Operation." When Brown & Root first moved its people and equipment out to the site in the winter, these skeptics said, "just watch those fools. We'll see what happens. They'll be home." Once the job had been completed, Brown & Root employees noted with pleasure the "amazement and chagrin of the Norwegians." Such almost good-natured joking thinly masked underlying tension that inevitably arose when an established American company aggressively "took over" major projects in the North Sea. In the years before Brown & Root-U.K. grew larger, the Norwegians and the

British viewed Brown & Root as a foreign company, and one with a certain swagger, at that. The company's work at Ekofisk gave it reason to swagger a bit. It could deliver creative solutions to difficult construction problems even in the winter months in the central North Sea. T. "Bo" Smith III summed up the company's achievements during Phase 1 of the Ekofisk project: "It was interesting, we did a hell of a job for them out there, both in the design and in the actual offshore construction, which I think set us up to be their contractor of choice from then on. And we kept equipment out there until that field was finally completed many years later."

Permanent Structures in Concrete and Steel

As oil from the *Gulftide* flowed to market, planning and construction of the permanent facilities of Phase II went forward. Brown & Root's work included designing, fabricating, and installing the U.S. portion of fabrication of the twelve-pile field terminal platform jacket, the "A" and "B" drilling and production platforms, the flare structure and deck, several bridges and connecting platforms, and the pipeline system.

To save time by making use of existing facilities, Brown & Root built the field terminal jacket at its Greens Bayou yard. The routine operation of loading the jacket on the barge for towing across the Atlantic Ocean encountered a slight hitch. Stan Hruska said, "As we were loading that jacket here in Greens Bayou, one of the verticals in the launch truss buckled. . . . You know, you just sat there and said, well, how could that happen?" The problem corrected, *BAR-267* towed the jacket the 5,000 miles from Houston to its installation site,[20] stopping at the Azores to inspect the jacket for fatigue. All the platforms were connected by bridges and walkways to the field terminal platform, a first-of-its-kind quarters platform, which could house 165 workers, and an emergency flare platform.

This episode illustrated at important point: supplying Phillips Norway and other companies active in the North Sea with the equipment to produce hydrocarbons would not wait for the maturation of local fabricating facilities. Until the establishment of such new yards and the expansion of existing ones, larger jackets would be built, if necessary, at yards thousands of miles removed from the North Sea. The infrastructure needed to support rapid development in the North Sea would emerge in Europe, but in those early years, the high costs of delays more than outweighed the high costs of long-distance shipments of much-needed products.

As such shipments arrived for installation, the severity of North Sea waves became ever more apparent. Without previous experience in this sector of the North Sea, oceanographers tended to underestimate the wave heights. Eight months after material lists had been prepared and four months after the bid drawings for four of the Ekofisk complex structures had been issued, the design wave was increased from 63 to 78 feet. This departure from design standards in the Gulf of Mexico and other offshore provinces compelled engineers to scramble to adjust designs.

Soil conditions also posed problems for platform installation. Larry Starr noted that, "In that boulder clay you couldn't drive the piles to grade. And so, we actually put in under-ream footings and it has worked rather well, although they have since had to raise those platforms because the withdrawal of the oil and gas caused a subsidence. . . . We actually evolved and developed the system for drilling those things out through the jacket legs and putting in concrete-filled under-ream footings on the platforms." Jay Weidler said that Ekofisk was a technical challenge because "the structures had to be underpinned by bell-bottomed foundations, which was a novel foundation ingredient for offshore platforms at the time. It was probably the first and only time such a foundation was ever used. This was because of the inability of the hammers of the time to drive the piles to what was considered an adequate depth."

Finesse, not strength, was required to install permanent flowlines without interrupting production from the four original wells. Phase 2 divers placed the permanent flowlines over the buried and still active temporary lines. After the activation of the new lines, divers cut and salvaged the original lines. Permanent 30-inch pipelines replaced the temporary 10-inch loadlines from the platform to the single buoy moorings, greatly expanding the capacity of the terminal. All other temporary facilities, including the *Gulftide* itself, were removed once the permanent facilities became operational.

As Brown & Root worked on a variety of steel structures at Ekofisk, a one-million-barrel capacity concrete storage tank was installed. This giant tank was the defining feature of the Ekofisk complex, and it symbolized a new design option for structures in the North Sea. Standing 270 feet tall and 302 feet wide and weighing 237,000 tons, it demanded attention. Its nine cylindrical storage chambers were surrounded by a perforated, prestressed concrete breakwater. Built in the Stavanger Fjord by as many as 1,000 workers, it was slowly and carefully towed in an upright position for 320 miles to Ekofisk, where flooding gradually pushed it to the ocean floor in June of 1973. During the early years at Ekofisk, the concrete tank provided ample storage for oil so that oil could be produced even when bad weather shut down the single buoy moorings. With the completion of an oil pipeline, the concrete tank was converted into the main base for Ekofisk's processing and production equipment. It became the center of the integrated oil and gas production facility at Ekofisk, where platforms connected by walkways (central complex) and pipelines (remote sites) produced one million barrels of crude oil and two billion cubic feet of natural gas per day.

Brown & Root had little to do with the design, construction, or installation of the first North Sea concrete structure. The company's experience and expertise was steel structures, and it had no interest in moving into concrete construction. The increasing use of concrete structures, especially in the northern North Sea, created a substantial market for this type of production platform, which could be installed without launching and held to the sea bottom without traditional piles. Brown & Root at times project managed the fabrication of such structures and their topside facilities, as well as for the towout, installation, and mating of decks to these structures. While doing so, Brown & Root retained its focus on what it had always done best, designing, building, and installing steel platforms and pipelines.

Laying the Pipelines from Ekofisk

Prospects for future work in the North Sea convinced Brown & Root to upgrade its fleet, and in 1972 Brown & Root announced plans to build two derrick/pipelaying barges on the Texas Gulf coast. *BAR-323* and *BAR-324* had the same dimensions as the company's most modern barges, 400 × 100 × 30 feet, but they incorporated improvements already tested in the North Sea on the *Hugh W. Gordon.* They contained more personnel quarters and special arrangements for the saturation diving crews. They had bows designed for heavier seas, fully enclosed pipe ramps, three high capacity pipe tensioners, space for dynamic positioning thrusters, and uprated anchor winches and cables. These barges expanded Brown & Root's pipelaying fleet to eighteen vessels, making it, according to the *Oil & Gas Journal,* "by far the largest in existence." Brown & Root also designed, constructed, and owned the first submarine for pipeline inspections in waters up to 1,300 feet deep in the North Sea.[21]

The surge of platform construction and pipelaying in the North Sea prompted Brown & Root in 1972 to enlarge the engineering staff in its London office, which enabled the firm to work more quickly and efficiently on its North Sea design and engineering work. As the staff grew from 40 to 400, Brown & Root U.K. took additional offices in London, Raynes Park, and, later, Wimbledon. A growing staff of experienced engineers in Southwell House and Olympic House could call on its own growing resources, as well as those of Houston, to interpret and meet the needs of an ever-changing North Sea marketplace.

When Phillips Petroleum sought contractors to build an oil and a gas pipeline from its Ekofisk field to shore, Brown & Root won portions of both contracts. Norpipe A/S, a 50-50 joint venture of the Phillips Group and Statoil, the Norwegian national oil company, owned the lines, but the Ekofisk pipelines would not go to Norway. The Norwegian oil market amounted only to 202,000 barrels per day of oil, and the combined capacity of Norway's three refineries was less than one-half the expected output of Ekofisk. Accordingly, a committee appointed by the Norwegian government decided that the United Kingdom was the most logical destination for Ekofisk oil. In addition to Norway's limited oil demand, the deep Skagerrak trench separating the Norwegian coast from much of the North Sea presented too daunting a technological challenge for a pipeline to Norway. Contractors feared that existing technology might not be able to lay a pipeline across this trench or repair a line once laid.[22]

In April 1973, the Norwegian Parliament approved the construction of a 220-mile, 34-inch line from Ekofisk field to Teeside, England. Brown & Root,

J. Ray McDermott, and Santa Fe International won the $200 million contract to build the line.[23] After dividing the work into three spreads, the contractors began work in late May 1973. Santa Fe's semisubmersible barge, the *Choctaw,* laid pipe from the Ekofisk field toward Teeside; J. Ray McDermott's lay barge *LB-24* worked from the mid-point; and Brown & Root's *BAR-324* laid pipe from Teeside seaward.

The Ekofisk oil line, which included two "booster" platforms en route, was the first major oil pipeline in the North Sea. The array of equipment used included five lay barges, three trenching barges, six dredges, thirty pipe-haul boats, one pipe-haul barge, eight tugs, four survey ships, eight supply boats, a submarine, a hydrostatic test crew with two work boats, six fabrication coating yards, a design contractor, and five inspection agencies. A diving team and equipment was included on each lay barge and trench barge. Also included were two tugs per barge, not to mention the helicopters for ferrying VIPs. A total of 2,300 persons worked on the line.[24]

The contractors hoped to overwhelm the North Sea through sheer force of numbers and complete the line in one working season, but severe weather thwarted these plans. Work finally was completed by May 1974, and in October, oil began flowing through the pipeline from Ekofisk into a tank farm at Teeside at the initial rate of 300,000 barrels per day.[25] At a ceremony marking the arrival of North Sea oil in England, the historic message was clear. With sufficient daring and financing, platforms could be installed in the central North Sea, large pipelines could be laid under its waters for hundreds of miles, and oil could be brought to shore at a market price that justified further development.

Gas production at Ekofisk began in the fall of 1974 at the initial rate of about 780 MMcf/d. Because natural gas could not be stored in large quantities in offshore storage facilities for pick-up by tanker, it was reinjected into the wells to spur oil production pending the completion of a gas pipeline. Ekofisk gas went to Emden, Germany. There, the gas would be processed and purchased by a four-company group of western European gas firms before entering the West German natural gas distribution network. The survey for the gas line was more involved than the one for the oil line, since the gas line crossed nine communication cables and traveled beneath one of the world's busiest shipping lanes. The communication cables had to be cut, folded back while the line was laid and buried, and then reconnected. Potentially more dangerous to the pipeliners were World War II mines along the path, which the German Navy swept clear of mines before laying the line.

The gas pipeline project consisted of six pipelay spreads, two on land and four offshore. Brown & Root again shared the project with several other companies. Brown & Root and Santa Fe each completed two of the marine spreads, while a joint venture of Preussag AG and Philipp Holtzmann AG laid the tidal-flats section, and IBU of Dusseldorf took charge of the land-line work. The 36-inch line had a capacity of up to 2.2 Bcf/d and at 275 miles long was the longest pipeline at the time in the North Sea. Work began in

April 1974. Plans to finish the line before the winter storms again succumbed to the realities of North Sea construction, and work had to be suspended in the fall with 85 percent of the trenching completed.[26]

Brown & Root encountered a new variant of an old problem in this case. The company's bury barges dug a suitable trench, but currents did not cooperate by pushing sand into the trench to cover the pipe. Divers had to guide thousands of sandbags into place, completing the burial of the pipeline and finishing the initial, three-phase development program at Ekofisk.

"Ekofisk City" continued to expand through the decade, but the ambitious program announced in 1970 had been accomplished. Despite delays, escalating costs, and engineering and construction problems, Phillips Norway won its gamble on two unproven technologies—the temporary production platform system and the concrete storage structure. Oil and gas had been produced more quickly than many had thought possible, and the field had more than lived up to its promise.

The arrival of North Sea gas onshore caused Europe to celebrate. A ceremony marking the opening of the gas line attracted royalty, politicians, and other dignitaries, who accurately proclaimed a new day in the energy history of Europe. The celebrants had reason for optimism. Five years of expensive, innovative development had brought oil and gas to market in impressive quantities, while serving notice that the North Sea could be tamed. During those five years, new discoveries had been made and companies had begun to develop other fields throughout the region. Even before the completion of Phase 1 at Ekofisk, Brown & Root had moved on to other high profile projects, where it could apply some of the lessons learned during its pioneering work there.

Notes

1. This chapter draws from interviews with R.O. Wilson, L.A. Starr, T. Smith, III, W.R. Golson, W.R. Rochelle, C.E. Nolan, A.V. Gaudiano, S.J. Hruska and J.B. Weidler.
2. "Improvision the Key to Early Structures Success," *Offshore Engineer,* U.S. Supplement (July 1985): 44.
3. "Improvision the Key to Early Structures Success," 44.
4. "Ekofisk innovations helped lead offshore construction development," *The Oil and Gas Journal* (February 11, 1980): 59.
5. *Ekofisk, Energy from the North Sea* (The Phillips Norway Group, n.d.), 4.
6. Larry Starr, "Phillips Ekofisk Project: Phase II."
7. Clive Callow, *Power from the Sea* (London: Victor Gollancz, Ltd., 1973).
8. "Norway claims Phillips has a big one," *Oil & Gas Journal* (November 10, 1969): 130.
9. "Norway claims Phillips has a big one," 131.
10. "Norway claims Phillips has a big one," 130–1. Also see "Phillips drills Ekofisk confirmation," *Oil & Gas Journal* (March 30, 1970): 70.

11. W.B. Bleakley, "Ekofisk goes on production," *Oil & Gas Journal* (May 31, 1971): 54.
12. "Ekofisk innovations helped lead offshore construction development," 62.
13. R.O. Wilson notebook, Tab K.
14. Bleakley, "Ekofisk goes on production," 54.
15. R.O. Wilson and M.R. Martin, "Deepwater Pipelining for Central North Sea," OTC 1855, Offshore Technology Conference, Houston, TX, April 29–May 2, 1973, 1.
16. Bleakley, "Ekofisk goes on production," 56.
17. Bleakley, "Ekofisk goes on production," 54.
18. Bleakley, "Ekofisk goes on production," 54.
19. Information for the technical aspects of the Ekofisk system was derived from Wilson and Martin, "Deepwater Pipelining for Central North Sea."
20. Brown & Root, Inc., *Marine* (n.d.), 27.
21. "Pipe Lay barges Beefed Up," *Oil & Gas Journal* (May 1, 1972): 38–9. Less than a year later B&R announced construction of a new bury barge, *BAR-331* with dimensions of 350' × 92' × 28', designed to bury pipe in 500 feet of water.
22. "Ekofisk-U.K. line gets preliminary nod," *Oil & Gas Journal* (February 21, 1972): 40.
23. "Lay Barges Standing by on Ekofisk Line," *Oil & Gas Journal* (May 7, 1973): 34.
24. "Pipe Laying: The Great Leap Forward," *Investors Chronicle* (September 5, 1975): 40.
25. "Work Begins on Ekofisk Crude-Oil Main Line," *Oil & Gas Journal* (May 28, 1973): 26. Also, see, "Phillips Completes Ekofisk Oil Line," *Oil & Gas Journal* (June 3, 1974): 87.
26. "Ekofisk-to-Emden gas line international affair, *Oil & Gas Journal* (June 3, 1974): 110, 113.

CHAPTER 13

Ring Master at the Forties Field

If Ekofisk tested Brown & Root's limits, British Petroleum's Forties field redefined them. The Forties was the highest of high-profile projects for Brown & Root. It was the first major oil field in U.K. waters and, at the time, involved the largest jacket in the deepest water. The British government, the press, and the populace watched intently in eager anticipation of their nation's arrival as a major producer of oil. As project manager, Brown & Root served as the ring master for a four-ring circus at the Forties. In the world of offshore construction, the company occupied the center ring from the time it began feasibility studies for this project in August 1971 until the arrival of Forties oil in England in November 1975.[1]

British Petroleum announced the discovery of the massive Forties field in October 1970, one year after the discovery of Ekofisk. Britain's first offshore oil field was approximately 110 miles northeast of Aberdeen on the Scottish coast. It covered 35 square miles and contained an estimated 1.8 billion barrels of recoverable oil. Bringing this oil and associated gas to shore presented unprecedented technical challenges to BP and the offshore industry as a whole. But the field's size and location assured that BP would seek to develop it, even at the prevailing price for oil in 1970. In retrospect, all involved greatly underestimated the costs of the technological system required to recover oil from more than 400 feet under the North Sea. Yet history proved kind to those developing the Forties. Skyrocketing oil prices in the 1970s gave them a windfall that paid escalating development costs, and the project succeeded despite the high costs to develop the field.

The discovery of the Forties marked the success of a sustained effort by BP to explore the British sector of the North Sea. After receiving good leases in the initial licensing of the U.K. sector, BP invested heavily in exploration. The centerpiece of its effort was the *Sea Quest,* one of the largest offshore drilling rigs in the world. The rig's tripod legs could be flooded to allow it to function as a submersible or semisubmersible platform. It could drill in 110 feet of water while resting on the ocean floor and in deeper waters while afloat. After helping to prove BP's gas fields in the West Sole area in the southern North Sea, the *Sea Quest* hit the bullseye in the central region with the discovery of the Forties.

From the earliest indications of oil, it was clear that this field was special. The discovery well flowed at 4,700 barrels per day. After a second well on the same block identified the northern flank of the Forties field, a third well four miles to the west of the discovery flowed at 3,260 barrels per day. A fourth well four miles west-southwest of the third well proved to be the largest of them all.[2] These wells in 350 to 420 feet of water tapped a substantial deposit of oil in Tertiary sands 7,000 and 8,000 feet below the ocean floor. They provided all the evidence BP needed to justify an ambitious development program. The company had labored long and hard to find oil in the North Sea, and it moved aggressively to claim the first fruits of its labor as quickly as possible.

British Petroleum chose Brown & Root as its project manager. The combination of the two companies created a talented team deep in North Sea experience. The extreme depth and demanding conditions at the Forties pushed the design of platforms and pipelines beyond conventional practices, forcing BP and Brown & Root to imagine new ways to build and launch the giant offshore structures needed for this field.[3] Dick Wilson, who exercised broad supervision from Brown & Root's London office, gave an understated summary of this adventure in applied engineering: "We were at the edge of technology and made mistakes, but eventually it all came together."[4]

Wilson concluded that "it was an exciting four years," and this statement, even more than the lengthy list of technological firsts at the Forties, captures the essence of the project within Brown & Root. There was a sense within the company that this was the project for which everyone had been preparing. Roy Jenkins, who oversaw the installation of the platforms, voiced a widely-held sentiment about the Forties project in a newspaper interview: "I was sitting out there (in Singapore) as Area Manager, reading the trade journals and afraid that I was going to miss it. I could see that this was going to be the big step forward in platform construction."[5] Those, like Jenkins, who got the call were not disappointed. Harris Smith's fond memory of the Forties captures a sense of the tone of the time: "We were full of the Brown & Root philosophy that we could do anything. These events occurred and we never looked back. We just changed our concepts. We did whatever it took to fix it. We never even considered failure. . . . We were going to do it." More than twenty years after the completion of the project, many of those who worked on the Forties for Brown & Root still consider it the highlight of their careers.

Departing from Traditional Designs

The most memorable aspects of the Forties project were the technological leaps in platform design, construction, and installation required to recover oil from under more than 400 feet of North Sea water. Matt Linning, who managed the project as head of BP Forties Development Group, acknowledged the great risks posed by the development of this field: ". . . It was an extremely courageous decision by the BP board to go ahead with the development of the Forties Field, bearing in mind the state of research and lack of precedent."[6] The phrase "on the cutting edge of new technology" is overused, but not in this case.

After deciding to go forward, BP aimed to minimize risks and speed the completion of the project by choosing Brown & Root as project manager. From the design phase through completion, Brown & Root, under the supervision of Linning, coordinated the work on the platforms while taking a significant role in building and installing them. Under a separate contract, the company also helped lay the Forties pipeline. By one estimate, the project management function alone absorbed the energies of 600–800 Brown & Root employees in the heyday of work on the Forties.

Project management was relatively new in the offshore industry. Brown & Root had taken such a role in projects in Africa and the Middle East in the 1960s, when it took full responsibility for the overall coordination of the work required to install loading terminals and pipelines. On most major projects in the Gulf of Mexico and the North Sea, however, the company developing the field directly managed the contractors and subcontractors responsible for the different aspects of development. Conditions on the Forties project convinced BP to employ as project manager an experienced construction company with well-developed contacts, which they hoped would hasten development on a project that would, of necessity, draw on the services of hundreds of companies based around the world.

The key to the success of this approach was the creation of a smooth working relationship between BP and Brown & Root and between Brown & Root and the other contractors. BP granted its project manager considerable autonomy, and Brown & Root executives later voiced appreciation for the leeway that they enjoyed. Bill Stallworth, Brown & Root's project manager at the Forties, said: "BP was extremely good to work for on the project. They turned the project management over to us. Didn't interfere. Generally acted on our recommendations." Heinz Rohde, a Brown & Root engineer with responsibilities for launching the platforms agreed: "There was absolutely no . . . interference from British Petroleum on this project." Such a relationship was possible only if BP had great confidence in Brown & Root.

This trust had been earned in the field on previous projects for BP—Iraq and Das Island in the early 1960s, the West Sole gas field in the late 1960s, and then on the North Slope of Alaska. Given this history and Brown & Root's extensive North Sea fleet, the company was a logical chice for British Petroleum. Dick Wilson recalls that BP "put out a fairly simple invitation letter to three contractors, Brown & Root, McDermott, and Earl & Wright, to present a proposal to manage the Forties project. . . . We were very familiar with the top management of BP plus their contracts group." After submitting the proposal, Wilson went on about his work on other projects until the summer, when he received a call from BP "saying that they were ready to proceed to award the project and could I come down to see them." When Wilson replied that he would come immediately, the BP representative added, "Oh, by the way, bring whoever you are proposing to have as project manager." This presented a momentary dilemma, since Brown & Root had not yet selected the person for this key job. The choice came quickly, since "at that time, Bill Stallworth walked into my office." As a project manager on one of the platforms at Ekofisk, Stallworth was a good choice for the new job, so he and Wilson took a taxi to BP's headquarters, where they soon found themselves discussing the organization of a project management team for the Forties.

If this selection process appeared somewhat casual, there was nothing casual about the job itself. BP borrowed $936 million from a consortium of twenty international banks to finance the project. The loan was one of the largest private loans negotiated to date. BP took this financial risk because it believed that the field could produce 400,000 barrels of oil per day for 20 years.[7] In turn, Brown & Root agreed to coordinate the work on the project, much of which was undertaken by Brown & Root itself. Stallworth later said that the company was well compensated for its jobs at the Forties. The profits reflected the technological challenges of the field development program.

The design of the first set of Forties jackets and flotation tanks was performed by a Houston team led by Dr. Max Koehler, a creative engineer. His leadership was essential in maintaining a schedule and devising a solution that worked. Innovative answers to a variety of environmental challenges had to be found.

The technological challenge at the Forties was demanding because of the convergence of three elements: the deepest water for an offshore project at 420 feet, the highest design wave at 94 feet, and the largest deckloads at 18,000 tons. By contrast, the water depth at Ekofisk was 230 feet. Even in the Gulf of Mexico, only a few structures had approached depths encountered at the Forties. From a design standpoint, deeper water depths increase the moment arm of the horizontal load produced by waves, wind, and currents on the platform. At the Forties, the moment arm would be nearly twice that at Ekofisk.

The design wave at the Forties was established at a height of 94 feet. This figure was only 20 percent greater than the design wave used at Ekofisk and about 50 percent greater than the normal hurricane wave factored into designing platforms in the Gulf of Mexico in the 1970s. But since maximum unit pressures increase with the square of the wave height, the maximum wave unit pressures at the Forties were 45 percent higher than at Ekofisk and 125 percent higher than in the Gulf. Thus, similar structures at Forties and Ekofisk would have different design criteria, since those at the former would experience horizontal forces about 45 percent greater and an increase in overturning moment of about 200 percent.

Finally, the deck weight at the Forties was more than four times greater than the largest deck ever placed on a single platform. This magnitude of load required a much heavier and larger supporting jacket than those used for previous platforms. Since the jacket would have much more area exposed to the unit pressures of the waves, the jacket would have to be almost five times heavier than what had been built previously.

Thus the Forties structures could not be designed simply as larger versions of existing Gulf of Mexico platforms. According to Bill Stallworth, "Prior to the Forties project, essentially everything in the North Sea had been designed as an extrapolation of the Gulf of Mexico criteria. The Forties was one of the first projects where a completely different set of criteria had to be developed for the environmental conditions of the North Sea."

These new design criteria began with the requirement to withstand a 94-foot storm surge and 130-mile-per-hour winds, but much more was involved than an extrapolation upward from traditional designs by factoring in *x* more feet of water and *y* more feet of wave height. As had been evident at Ekofisk, the high waves of short frequencies in the central North Sea forced fundamental design changes to take into account the much greater lateral forces at work on the platform.[8] Greater protection against metal fatique in the form of heavier jackets and stronger and thicker steel not previously used offshore also had to be factored into the new design equation.

Several factors were responsible for the extremely large deck sizes planned for platforms at the Forties Field. The difficulties of resupplying these platforms, especially during the fierce winter season, encouraged self-sufficiency and thus included a design with much greater storage space than a traditional platform. Hugh Gordon, who had worked offshore for Brown & Root in numerous areas, summarized the key difference between deck designs in other regions and in the North Sea: "[In the Gulf of Mexico], you have multiple jackets on the oil field . . . due to the hostile weather of the North Sea . . . you put everything, put the whole oil patch on one platform, so they were monstrous." This was particularly true at the Forties, where "platforms have more operational functions than any single platform ever installed offshore."[9]

A pressing concern for designers was the hitherto unprecedented deck loads. As an original estimate of the weight of the deck loads at the Forties, Brown & Root used "a nominal figure of 8,000 tons . . . which was . . . two times the biggest platform ever built. That proved to be grossly underestimated." As mentioned previously the actual weight turned out to be 18,000 tons. At the time, the largest deck going up in the North Sea was the 2,800-ton structure being installed by Brown & Root at Ekofisk. The heaviest operational deck load on Gulf of Mexico platforms in these years was around 4,000 tons.

The giant increase in deck weight for the Forties structures had ramifications for all phases of design, construction, and installation. Jay Weidler, who helped design the Forties jackets, pointed out the implications for jacket design of greater wave force and much higher deck weights: "Those two things happening simultaneously caused a five- to tenfold increase in jacket size, and in the problems attendent thereto."[10] Designers could not be very precise in analyzing the impact of deck load because "the fabrication of the upper 200 feet of the Phase I jackets was started before the deck load distribution was known."[11] When in doubt, the answer had always been to design conservatively. The first two platforms installed at the Forties were widely acknowledged to be heavier than necessary. Yet even in retrospect, it seemed clear to BP that the right choice was made, since structures designed on the run and conservatively built more than paid for themselves by going into production far sooner than the less expensive structures that might have been built after several more years of analysis. Such were the inevitable tradeoffs when design and construction went forward simultaneously.

Size was the most obvious determinant of the design of the Forties platforms, but speed of installation also influenced design decisions. The high costs and great risks of weather delays and storm conditions dictated a design "to enable setting the jacket on the seabed as rapidly and safely as possible."[12] Dangers abounded from the time the expensive jackets began their journeys in the open sea to the time they were secured to the ocean floor. So a "major concern in the case of the Forties field platforms was the offshore time required to install piling."[13] Designers thus had a strong motivation to think creatively about all aspects of installation.

Shaping all design considerations was the broad imperative to move fast in completing the job as a whole. The need for speed shaped every decision made about the Forties. In practice, this meant that only steel platforms in some variant of the tried-and-true design configuration were considered, because this was not the time to build a radically designed structure. It also meant that designers knew from the start that they would be expected to work under pressing deadlines and unusual uncertainties. Weidler acknowledged that "We had to design a structure from the top downwards, and any problems got worked out just shortly before the pieces were rolled."[14]

Brown & Root prepared a design feasibility study, which it presented to BP in April 1971. After meetings with BP to discuss the design criteria, Brown & Root did most of the engineering work in Houston; London personnel surveyed fabrication yards in Europe. A starting assumption was that "where a short engineering and construction schedule is necessary, the economics would . . . favor . . . a more conventional procedure of fabrication and installation." Brown & Root examined, but discarded, two relatively untested approaches: the Conitower concept that it had developed using conductors installed in jacket legs for piling, and the "Texas tower" concept that made use of three large diameter legs. The feasibility report studied in detail only two "more conventional concepts: an eight-leg tower, which is barge-transported, and a four-leg tower, which is self-transporting."[15] The eight-leg tower was a variant of the traditional design that "has always been the most economical type platform to fabricate and install." The four-leg tower was adapted from designs developed in the Cook Inlet in Alaska, where large ice forces dictated the use of structures with the fewest number of members or legs in the ice zone. Brown & Root acknowledged that this design adapted for a wave environment "generally requires more steel tonnages at a higher unit cost than a barge-transported, multileg jacket,"[16] but since a barge of this capacity was not available, the four-leg structure with its own floatation system appeared to be the best solution.

BP settled on the four-leg approach and Brown & Root began working on the design problems presented by this project. Four platforms would be used to develop the Forties, each of similar design but with variations in size.[17] Brown & Root had to adjust its traditional approach to platform construction to the new reality that on the Forties project, "practically everything we do is the largest, heaviest thing that has ever been done."[18] This was no idle boast. The Forties stretched existing offshore technology to a new scale and complexity. Brown & Root's design on the Highlands One, noted one company official, "was so far in advance of things that had been done here, it was unreal."

Several aspects of the platform design were noteworthy. To minimize danger from bad weather during installation and to secure the jacket against the worst assaults of the North Sea's waves and wind required innovations in piling. To set the jacket quickly, twelve piles located within the jacket framing were inserted in their guides in the fabrication yard, carried out in place, and released and allowed to run into the seabed under their own weight once the jacket rested on the seabed. Driving these twelve piles to design penetration and grouting them made the jacket secure against winter storms a short time after it entered the water. The remainder of the piles could then be driven from barges or from temporary decks constructed for this purpose. Between thirty-two and forty-four piles—each fifty-four inches in diameter by two inches thick—nailed each jacket to the ocean floor. Clustering the piling in concentrated groups around the four corners of the platform was one innovation. Clustering minimized the number of structural members in the jacket, "thus reducing the steel tonnage and the loads applied to the structure" while allowing for the development of a "streamlined diver-oriented system for installation," which "was found to be very successful."[19]

The most distinctive design choice at the Forties was the use of steel rafts instead of launch barges to carry the jackets from the fabrication yards out to the field. Such flotation tanks had never been used before to transport jackets, and they have never been used since. They were chosen out of necessity; the Forties platforms were too large for any existing launch barge. Indeed, the heaviest platform ever barge-launched weighed less than 4,000 tons—less than one-fourth the weight planned for the first two Forties platforms. The flotation tanks seemed less expensive, easier, and quicker to build than a super barge. They required innovations in telemetry and the development of a dependable clamp that could be released underwater once the tanks and jacket had been submerged, but both seemed doable in the short time available.

Forties platform at Nigg

These tanks were massive structures in their own right, stretching 450 feet long and 340 feet wide and weighing about 11,000 tons. The main support system consisted of cylindrical tanks 30 feet in diameter crossbraced by slightly smaller tanks. Plans called for the construction of the platforms in a graving dock atop these tanks, attached to them by huge hydraulic clamps. The flooding of the graving dock would set the tanks afloat, and they would then be towed out to the Forties field, an aquatic version of the Challenger spacecraft being transported piggy-back on top of an airplane. Once they arrived on site, the fun would begin.

Installation presented considerable uncertainty, although the plan was straightforward. Using sophisticated computers and radio equipment housed on a control vessel, valves would be opened within the tanks to flood predetermined chambers. Initial flooding would tilt the vessel slightly in preparation for the critical procedure, "crash flooding" of chambers in the jacket legs and flotation tanks to cause the structure to dive quickly to a 45 degree angle. After a pause to check out all systems, gradual flooding would lower the structure to the ocean floor. Once it had been pinned and pile driving had begun, the flotation tanks would be removed by opening the clamps via remote controls. Using liquid nitrogen—later replaced by compressed air—the flotation tanks would then be deballasted, whereupon they would be raised to the surface, brought back in the horizontal plane, returned to the graving dock, and reused to transport subsequent platforms. Because of its heavy reliance on telemetry, this system evoked many comparisons to the space program, which only three years earlier had succeeded in putting a man on the moon. The main justification for the use of this exotic, high-tech system was simple: the size of the jackets at the Forties field had outrun the existing equipment to launch them.

Deck design also required the rethinking of many traditional procedures, since the extremely heavy decks for the Forties platforms were a quantum leap beyond any previously installed. A three-level design with about 3.5 acres of surface emerged. Traditional deck construction utilized a module support frame on which different parts of the deck could be placed. The Forties design, by contrast, called for modules containing necessary equipment to be stacked directly onto two rows of six legs on the top of the jacket and then on top of each other. This modular design had its roots in the prefabricated structures used in Lake Maracaibo and on the North Slope of Alaska. The "deck" thus became the sum of these modules, with a savings of space and a decrease in the total weight lifted onto the platform.

Despite such creative efforts to save weight and make efficient use of space, the harsh reality remained that the 18,000-ton deck load of the first two platforms required numerous very heavy lifts. Five such "packages" weighed in at more than 1,000 tons each, with a maximum lift of more than 1,700 tons. Thirteen additional lifts ranged from thirteen to 555 tons. The largest heavy lift vessel in the world at the time was designed and built by Brown & Root's former partner, Heerema. The *Thor*, with a lifting capacity of 2,000 tons, proved its worth at the Forties field. Once lifts had been completed, each platform had space for all of the equipment and workers needed to produce as much as 150,000 barrels of oil per day.

The announcement of the plans for the design of the Forties platforms evoked inevitable comparisons between their size and that of familiar landmarks. The largest of the platforms "had more steel than the Eiffel Tower." It was "one and one half times taller than the Statute of Liberty." In the most unlikely contrast, it was "bigger than St. Paul's Cathedral and Big Ben on top of each other."[20] These eye-catching sizes reflected an ambitious design that combined the extraordinary strength of the platforms with innovations aimed at hastening their placement on site. Even before the completion of design work, fabricating had begun on many of the elements required in the design. If subsequent data altered assumptions that had affected planning, the design could be adjusted; likewise, lessons learned along the way could be incorporated in the later phases of the project.

Building the Jackets and the Nigg Bay Yard

The enormity of the structures designed for the Forties field meant that fabrication would require a giant construction yard. To save time, BP decided to use two fabrication sites. Laing Offshore in joint venture with ETPM of France would build two of the platforms (designated FA and FB) at a yard at Graythorp, England, which had to be remodeled and retooled to take on the job. A joint venture of Brown & Root and George Wimpey & Co. would build the other two (FC and FD). Early on, it was clear that Brown & Root's fabrication site should be in Europe, with good access to the Forties field and the many subcontractors on the continent.[21] As part of its initial feasibility study, a Brown & Root team searched for an existing yard capable of building jackets to be installed in 350 feet of water. Engineers visited shipyards in Belgium, Denmark, France, Germany, Great Britain, Greece, the Netherlands, Ireland, Italy, Norway, Portugal, Spain, Sweden, and the United States, but none of these yards met all of the necessary requirements. After concluding that "no existing facilities were located in northern Europe that possessed the capabilities for fabricating jackets for water depths ranging between 300 feet to 400 feet," the company began a second study "to examine and evaluate the 'grass roots' development of sites located in the European area for fabricating platforms for water depths ranging up to 700 feet."[22] In building the Forties jackets, Brown & Root would also build a fabrication yard for its present and future North Sea operations.

The search for possible sites began with surveys of locations surrounding the North Sea. Sir Phillip Southwell in the London office favored a site within the U.K. and George Brown agreed. Bill Golson and others responsible for finding a location for a pipe-coating facility needed for the Forties pipeline flew up the coast of Scotland to look for sites from the air. They saw excellent sites on Scotland's west coast, but the obvious problem was, "It's not the side of Scotland that the damned oil is on." Flying over the North Sea side as low as the pilot could safely go, "We flew over this area and . . . there it is. It opened right up into the North Sea. You could tell it was deep because of its color. And it was undeveloped. And, of course, that was Cromarty Firth." The combination of deep water relatively close to the Forties and other potential North Sea fields and large blocks of undeveloped land made the site attractive. After exploring the region from the ground, the team located a suitable 185-acre parcel of land at Cromarty Firth on Nigg Bay and received permission from the Scottish government to establish a construction yard there. In recognition of its location in the Scottish Highlands, the new site was named "Highlands Fabricators Ltd.," a name commonly shortened to "HiFab," and often referred to simply as "Nigg" or "Nigg Bay."

The work of building a construction facility began almost immediately for the joint venture of Brown & Root and George Wimpey & Co., an all-purpose British construction company.[23] This joint venture grew out of the two companies' complementary strengths and their past work together. In the mid-

1960s, the companies had teamed up to create onshore bases for offshore operations and work boats, including the facility at Great Yarmouth used heavily by Brown & Root during the North Sea natural gas boom of the late 1960s. Said Golson, "The deal that Brown & Root had with Wimpey . . . was that Brown & Root wouldn't do anything onshore in the U.K. without asking Wimpey to be their partner." As a partner in Highlands Fabricators, Wimpey built and maintained the fabrication yard, allowing Brown & Root to focus on the complicated process of constructing its first Forties jacket, Jacket FC or Highlands One.

As Brown & Root-Wimpey built a yard and recruited and trained a workforce, the Laing-ETPM joint venture began its work on Jacket FA, or Graythorp One. Laing-ETPM expanded an existing shipyard, equipping it with two of the largest cranes in the world and staffing it in part with experienced shipyard workers. Meanwhile, J. Ray McDermott formed a joint venture with the British company Redpath, Dorman, Long to develop a similar facility on a 100-acre site at Ardersier, Invernes-shire, on the Moray Firth.[24] A North Sea construction boom was under way, and its impact was increasing onshore with the expansion of existing fabrication yards and the creation of new and more specialized ones.

HiFab took shape quickly. Only twenty-six weeks after the creation of the Brown & Root-Wimpey joint venture, Wimpey had completed construction of Europe's largest graving dock at Nigg Bay; the site stood ready to fabricate jackets while work went on building the remainder of the yard. The 1,000 × 600 × 50-foot graving dock required the removal of giant mounds of sand and sandstone, which were deposited near the edge of the bay to create new acreage for the yard. In the two years it took to build the yard, Wimpey moved an estimated 1.5 million cubic yards of material. The completion of the graving dock involved driving pilings for new foundations, laying of crane tracks, and completing a sophisticated dewatering facility. The defining structure in the graving dock was a giant 16,000-ton concrete dock gate—426 × 50 × 45 feet in dimension—that held back the waters of the Cromarty Firth until the jackets were completed and ready for floatout.[25]

With the graving dock completed, work began on the jackets. The flotation raft, which had been fabricated in pieces in several other yards, was assembled on the floor of the graving dock so that workers could begin putting together the jacket. Meanwhile, work went on completing the remainder of the yard. During 1972 and 1973 a huge, modern yard sprang up at Nigg, complete with three spacious fabrication shops and a structural pipe mill capable of producing 200 tons of pipe per day—more than the pipe mill at Greens Bayou in Houston. With the huge graving dock and associated heavy lifting equipment, the pipe mill, and more than 200,000 square feet of buildings, the Nigg Yard rose from the sand dunes to become one of the leading fabrication yards in the world.[26]

Brown & Root imported many skilled workers from its operations around the world, but the core of the HiFab workforce came from the Highlands. The local government of Ross and Cromarty, Scotland, encouraged HiFab to hire as many Highlanders as possible, and hiring policy stipulated that 94 percent of employees recruited for training had to be local. Since most of these recruits had no experience in fabrication work, Brown & Root's Personnel Training & Development Department designed programs to train them. Twenty-four skilled supervisors began the training program in April 1972, instructing the new workers in flux core, stick, MIG and sub-arc welding; fabricating; rigging; layout; and pipefitting. The initial class of 580 trainees began work in field operations. The following year, phase two of the program began, with some of the original American instructors returning to work in the yard while local men trained in the first class took their place as instructors. In 1973, 632 more workers completed their training, and an additional 386 in 1974. A local newspaper headlined its article on the training program, "From Farm-Workers to Welders in Just a Few Weeks."[27]

This training program succeeded on several levels for Brown & Root. It delivered a much-needed new work force while also generating goodwill for the company in its new hometown. Between 1,500 to 2,000 workers completed the training program, in the process changing the mix of the workforce in a region where "crofters had been living on their farms for generations." Between HiFab and the pipe-coating yard built near Nigg Bay by another company, "We had sucked up all of the available labor. Anybody that wanted to work in that whole part of Scotland, we set them up." The rapid expansion of HiFab boosted the local economy, making Cromarty Firth the "emerging center of the oil-platform building industry."[28] While some lamented the death of the old way of life in this section of the Highlands, many others spoke with their feet in support of the growth of HiFab by joining the work force there.

One inevitable result of rapid industrial growth was labor unrest, which threatened the hasty completion of the jackets at HiFab. Joining the 2,000 or so local workers and welders at HiFab were about 1,000 more experienced union workers. This mass of new workers migrating to Nigg overwhelmed existing housing and social services. As much as half the work force moved to Nigg without their families, occupying any sort of temporary housing they could find. Brown & Root eased the housing shortage by acquiring two passenger ships, the *Highland Queen* and the *Hermes,* docking them in Nigg Bay and converting them into quarters. Several brief walk-outs by workers in February and May 1973 raised fears within the joint venture of longer delays due to strikes, but by the summer of 1973, Wolf Pabst, the general manager at HiFab, could report that "labor relations are not a major problem."[29] To boost morale during the long push to complete the first jacket, Brown & Root promised its workers 500,000 pounds (sterling) in bonuses if they met the deadline. Given the challenges of training, housing, and motivating a new work force, those in charge of the management of labor on the Forties project at HiFab did a remarkable job.

The fabrication of the giant platforms required innovations in manufacturing processes, most notably the use of node construction techniques. Bill Stallworth explained the fundamental change embodied in this approach to jacket construction: "In the past, each member had been coped and fitted into the main members in the field but on this project and other projects subsequently, the joints or the intersections of the various tubulars were actually done in a controlled environment in the shop, which was much more efficient." The use of nodes rolled and welded at other fabrication sites allowed the jacket to be pieced together in segments like a giant tinker toy. This was essential given the lack of lifting equipment capable of moving the entire jacket at once. The crucial welds on the major joints of the jacket could be made indoors, where they could be completed more efficiently and inspected more thoroughly than outside in the elements and high above the ground. This contributed to more uniform and durable welds, as did the practice of grounding the welds to smooth out the joints and discover possible weak spots. Even with nodal construction, however, the assembly of the jackets required that workers spend considerable time in the graving dock and up on the jacket several hundred feet in the air.

Stress relief equipment at HiFab unlike any previously used by the offshore industry reduced the susceptibility of the welds to brittle fracture, whereby a flawed metal subjected to less-than-design stress can fracture at low temperatures. Little was known about brittle fracture, but those building the Forties jackets reduced this uncertainty by using finer grain steel, improving welding processes, and subjecting welds to stress relief in specially built furnaces. Nodal construction and stress relief of welds were two manufacturing innovations that improved the strength of the jacket while also hastening its completion.

In the rush to complete the jackets, the Nigg yard whirled with movement. Giant cranes that had been removed from the derrick barges *L.B. Meaders* and *BAR-297* moved very large pieces of metal from site to site on the way to their ultimate destination, the jacket, which rose steadily in the graving dock. Stallworth said, "At one time on the Forties project, we had fabrication going on in, if I recall, about fourteen or sixteen different yards scattered throughout the U.K., Holland, Germany, and France." This made the project a "world-class-type project from the standpoint that you had people from all over the world . . . contributing to either equipment or labor . . . or brain power." As workers at HiFab assembled materials from many other yards, the deck modules were being built at places "throughout the U.K. and Europe." One of Brown & Root's most demanding tasks at HiFab was the coordination of all of the work being done at different sites so that the right parts of the project arrived at the appropriate time at HiFab and the second fabrication site at Graythorp.

A natural and friendly competition arose between the Graythorp and HiFab sites as to which jacket would be to first to float out to the Forties field. HiFab had the early lead, and its construction of deck modules for the Highlands One (FC), the larger of the two platforms, was given priority. But the more experienced workforce and larger cranes at Graythorp helped move Graythorp One (FA) ahead. Brown & Root had responsibilities as project manager for work at both sites and the installation of both jackets and the two to follow. After a series of troubling delays at HiFab, the Laing-ETMP jacket was the first to be installed in the summer of 1974.

More troubling throughout the fabrication of the first two jackets was a string of delays that kept pushing back the floatout date for both of them. BP's original plan for developing the Forties optimistically set the summer of 1973 as the target date for installation of the first two jackets, which had been ordered in January 1972. The summer of 1973 came and went, and all involved became increasingly concerned with the implications of further delays. Similar problems plagued most of the major projects in the North Sea in the early 1970s, so much so that the *Wall Street Journal* analyzed the reasons for these delays in a front-page story in August 1974. This article listed several problems, including "design changes, materials shortages, labor problems, bad weather, and the difficulty of the deep water work involved." Time was lost in 1973 because of a British coal miners strike. Most of those interviewed, nevertheless, agreed that early projections had been unrealistic, or as one unidentified source put it: "There's been a lot of wishful thinking and plain bull about how fast this work could be done." Dick Wilson stressed the difficulties of constructing the structures while still working on their basic design. "It was like trying to plan the moon shot while designing and building the spacecraft," he explained. With the experience of the first generation of projects, Wilson added, "future forecasts will be closer to the target."[30]

Delays spelled higher costs and postponed revenues. In general, the initial estimates of development costs in the North Sea proved far too low. The *Journal* article noted that Highlands One cost an estimated $200 million, whereas the initial estimate of the total cost of all four platforms at the Forties had been about $350 million. The escalation of costs greatly concerned BP. Wilson recalled one meeting where BP's representatives said that "this project can't continue on this basis because . . . the costs were just getting too high. It was a very disturbing meeting for us all." Events altered circumstances shortly after this meeting, since "the October War happened and the price of oil by the end of the year (1973) had moved up to where we did not discuss the overall project cost implications again with BP." The emphasis was then on speed of development, not cost, and expenditures in anticipation of bottlenecks were routinely approved. Rising costs still registered with BP, but soaring oil prices justified greater expenditures to bring the oil on line as soon as possible.

Finally the first two jackets left the fabrication yards and took their place in the Forties field. Graythorp One arrived first in June 1974, and as Highlands One prepared to join its twin on site, the celebration began. First, 12,000 people came to a ceremony at HiFab to christen the jacket. Then on August 14, 1974, immediately before the floatout of Highlands One, the Queen and other members of the royal family toured the facility. The day before, the royal yacht had cruised past Graythorp One on location in the Forties field. Government leaders—much less royalty—had never visited Brown & Root's platforms in the Gulf of Mexico. Something of national significance to the United Kingdom was taking place at the Forties.

"This Is It": The Moment of Truth at Installation

After eighteen months of hard work and delays, the Forties platforms were ready for float-out and installation. At dawn of the float-out day for Graythorp One, Heinz Rohde recalled "sitting there and contemplating the events that would be following, you know, now the real test came. . . . The sun came up and all of a sudden, the radios began to crackle, 'This is it.'" Roy Jenkins, the head of Brown & Root's installation team, said: "It was similar to a bull fight—when the matador kills the bull it's the moment of truth; well, when we floated out Graythrop One that was el momento de verdad for us."[31] Jenkins's supreme confidence that all was in order for successsful installations led him to distribute "Not To Worry" stickers to his crews. Beneath the confidence lurked a sobering truth. Jenkins and his coworkers would be forced to kill the bull four times over two summers, with jackets FA and FC installed in the summer of 1974 and FB and FD in the summer of 1975.

All four of the jackets followed similar installation procedures, with lessons learned at each launch incorporated into subsequent ones. Each journey began with the "memorable spectacle" of the float-out, when the gate opened to flood the graving dock and the raft and its valuable cargo slowly rose and was eased out of the dock. Spectators lined the path of the jacket to the sea, with views from some points inland giving the eerie impression of a giant structure gliding very slowly along the horizon. The float-out required careful monitoring of the weather. The jacket-carrying raft was as large as a battleship, but not as seaworthy, and the tow-out involved several days in the open sea. Each "package" of raft, jacket, piling, and ballast weighed more than 30,000 tons. A small armada of tugs and derrick barges assisted in the journey from the graving dock to the seabed at the Forties. Awaiting the raft at the field were the two largest derrick barges in the world, Heerema's *Thor,* which could lift 2,000 tons, and Brown & Root's *Hercules,* which could lift 1,600 tons.

The real moment of truth came when the "crash dive" sequence abruptly tilted the raft to a sharp angle in the water after preliminary flooding had eased the back of the raft down into the water. This breathtaking, ulcer-inducing procedure had never before been attempted except in test tanks with models. The earliest tests performed at Rice University in Houston in 1972 had produced a moment of horror, not of truth, when the model jacket had tipped over during the crash dive. To protect against this possibility, two large spheres, measuring fifty-one feet in diameter and weighing 540 tons, had been designed and fitted on the lower section of the jacket, where they could provide stability during the crash dive.[32] The memory of the wave tank test and the rising ten-foot swells of the North Sea added a sharp edge to the events which followed.

A floating command center on the supply vessel, *Oil Explorer,* carried the computer and radio equipment needed to control the precise sequence of flooding needed to usher the jacket to the bottom.[33] Operator Heinz Rohde described the various control systems:

We had four computing systems. We had manual operation where necessary. We had four radios to communicate between the jacket and the control boat. The control console, our position, was on the control boat. . . . A separate boat which had a little cabin on top contained all the computers, and the operating console. We had three people sitting at the operating console.

This was distinctly high-tech compared to anything previously used in launching platforms, but despite extensive shakedowns of the equipment, everyone knew that glitches could occur, demanding practical responses to correct any problems under intense pressure.

At the first launch, Graythorp One, the system worked well all the way through the dramatic crash dive. As crews prepared to continue flooding to ease the jacket down to the ocean floor, however, "that's when we had a little surprise. . . . All of a sudden, we had a failure. We could not communicate anymore." Later analysis of the failure showed that the telemetry equipment

Forties crash drive

housed inside the nosecone of the raft to receive messages from the control vessel had overheated during the tow-out after several unseasonably hot days in the North Sea. Whatever the cause of the problem, a practical way had to be found to finish the job.

Crews kept watch during the night over the platform, which remained stable at a 45-degree angle in the water. The next morning Heinz Rohde, Jack Morris, and Buddy Lehde were placed on the flotation tank using the basket on the *Hercules.* They shimmied up the main brace of the tank to the nosecone, which was high above the water, opened the small entrance to the cone, and crawled in to observe the equipment inside. It was obvious that the heat had shut down the computer system, and the men tried to bypass the computer with "a time-proven method, manual jumpers." This decidedly low-tech approach was successful. Using a make-shift jumper, they energized the solenoid valves that could not be energized with the computer. Thus the jacket was lowered "by hand" the rest of the way to the seabed. During prelaunch tests of the complex system of valves, Roy Jenkins had joked that he would bring along dynamite to the installation, so that if all else failed, he could blow holes in the rafts to flood them. Men with long experience in applied engineering understood that new technical systems often required more than a gentle nudge to get the job done out in the field.

Once the jacket had been pinned in place, the flotation tank would be removed and deballasted so that it could be returned to the fabrication yard for reuse on the second platform to be built at Graythorp. Plans called for using liquid nitrogen to force water from the tanks and refloat them. Problems arose, however, when the expensive arrangements for delivering the nitrogen into the tanks broke down when the propellor of the vessel that carried the tanks of liquid nitrogen cut the vital nitrogen hoses. Again, a practical solution was at hand in the form of a powerful air compressor on the *BAR-279.* Once compressed air deballasted the flotation tanks, the initial phases of installation of Graythorp One had been completed.

Two months later, the installation of Highlands One went more smoothly. The experience of the first launch had led to minor adjustments. Ventilation and a coat of white paint on the nosecones minimized the chance of overheating. Compressed air replaced liquid nitrogen for deballasting, and everyone felt more confident in the basic procedures used.[34]

The lessons learned in the construction and installation of the first pair of platforms were put to good use on the second pair. Once the flotation tanks had been returned to the graving docks, assembly could begin on Jacket FB (Graythorp Two) at Graythrop and FD (Highlands Two) at HiFab. Fabrication went much more quickly the second time around. The jackets were ready for float-out in about nine months, or roughly half the time required to build the first two. The HiFab yard was fully built before the start of construction on Highlands Two, which facilitated work there, as did the efficient fabrication procedures established during the first time through. In particular, workers made great strides in learning how to frame steel in the highly congested area around the bottom of the structures. Substantial savings in time and money came from platform design changes. By the time construction of the second set of structures began, the design had been completed. Knowing the final configuration and weight of the deck before the design of the jackets allowed the weight of the last two jackets to be reduced 20 percent.

When the second pair of platforms was launched in the summer of 1975, Brown & Root also had the installation process down to a comfortable routine. In launching Highlands One, however, as crews settled the last of the jackets down to the ocean floor, one of the spheres attached to the raft to stabilize the structure during its crash dive imploded under the water pressure exerted near the seabed. Later tests suggested weaknesses in the sphere due to imperfect sphericity—some of the interior beams were misaligned.[35] Such explanations were not of immediate concern to men in dinghies near the structure when the huge air bubble caused by the implosion burst to the surface. Benny Davis recalled that when the sphere imploded, "Air went everywhere up there but didn't bother the platform at all because it was

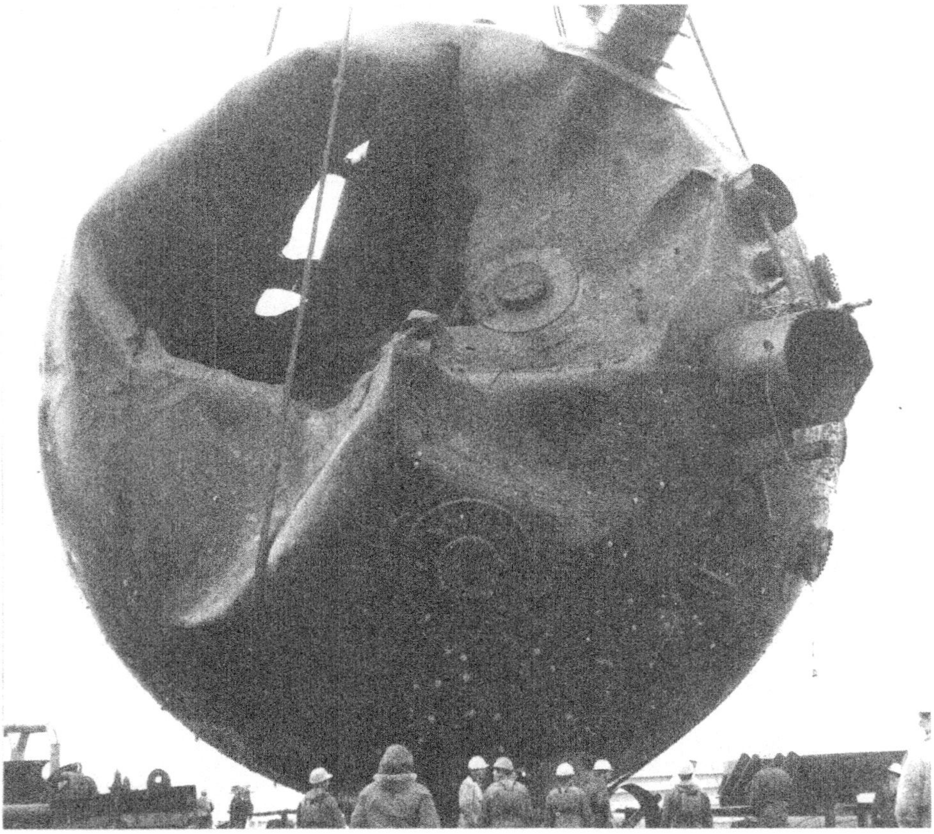

Imploded sphere. Photo courtesy of Dirk Blanken

already setting." Those at the Forties field could afford a nervous laugh, since they successfully had built the four platforms and set them in the field.

Now it was up to the heavy lifting cranes on the derrick barges to complete the platforms by installing the decks. Work began on the first two decks immediately after their installation. BP kept crews out during the winter of 1974–75 working on these structures. Original plans had called for the installation of temporary decks from which piles could be driven, but in practice crews found that they could more efficiently complete this vital work in the winter and spring using the "flying hammer" system, driving piles directly from the *Hercules*. The completed decks on the first two platforms weighed between 17,000 and 18,000 tons, and the *Thor* and the *Hercules* lifted them up onto the platform in a series of "packages," including several heavier than 1,000 tons. Because the topside designs of the four Forties platforms were nearly identical, modules could be swapped among the four platforms if called for by shifts in the work schedules on the four structures. Local newspapers followed the progress of deck installation in accounts describing "world record" lifts or presenting photographs of "the world's largest wire rope slings" used in these lifts. The three-tiered decks sat up on the platforms almost 100 feet above sea level, or 500 feet above the ocean floor, with the top of the platform's drilling derrick rising 690 feet above sea level.[36] Drilling began off Graythorp One in the summer of 1975 as work continued on the other three platforms.

Placing of Forties deck module. Photo courtesy of Dirk Blanken

The completion of the platforms for drilling involved more than lifting the deck modules into place. Establishing connections among the modules was a final essential task. Here Brown & Root called on its EUMECH (European Mechanical) division, which specialized in offshore mechanical, electrical, and instrumentation work. First created for Shell Oil's Leman project in the late 1960s. EUMECH's workers would swarm over the platform to prepare it for operations. EUMECH became known for two things, its profitability and its attachment to the color purple. Brown & Root veterans of the North Sea make similar points about how EUMECH worked: "[Hook-up] was done in a hurry at a very attractive rate." The company used purple equipment to keep track of its tools as it worked on platforms with other contractors. Before long, EUMECH's workers sported purple jumpsuits and its officials drove purple cars. After years of innovative nodular construction, exotic flotation rafts, and world-record deck lifts, the platforms could not be put into service until the men in the purple outfits had taken their purple wrenches and finished the work.

Struggles with The Forties Pipeline

The building of the Forties pipeline revealed the dramatic advances in pipelining since the first North Sea line a decade before. The West Sole line was 42-miles long and 16-inches in diameter in depths up to 120 feet. In contrast, the Forties line was a 32-inch, 115-mile long pipeline in water from 330 to 420 feet deep. One industry publication labeled this project "a pioneering effort in the installation of large-diameter pipelines in deep, open waters."[37] This was the deepest water in which a large-diameter line had been laid, and it was extremely rough going for the lay barges involved. One result was the escalating costs of pipelining. The daily rental of lay barges cost $50,000 per day in 1968 and $100,000 in 1975.[38] Moreover, the viability of traditional flat-bottom lay barges in the central and northern North Sea was uncertain. Even the heavy barges of the so-called "second generation" of lay barges were insufficient in the deeper waters of the North Sea.

In January 1972, Brown & Root and Saipem of Italy received the $39 million contract to lay the line connecting the Forties field to Cruden Bay, Aberdeen. From the line's terminus at Cruden Bay, CJB and Turriff Taylor laid a 130-mile pipeline costing more than $11 million to BP's Grangemouth refining complex in Scotland, which included a low-pressure oil and gas separator plant, gas-disposal facilities, and crude-oil storage tanks. In addition, a new crude-oil export terminal was built at the Firth of Forth to ship Forties crude to locations other than Grangemouth. On the offshore portion of the Forties pipeline, Saipem laid pipe from the shore toward the field. Brown & Root had the more difficult task of laying pipe from the field toward the shore.[39] The desire to bring the field on stream quickly led to the decision to use the summers of 1973 and 1974 to lay the pipelines before the platforms had been installed. The pipeline end at the Forties was capped, so that once the platforms were put in place, the line could be extended to them, the risers installed, and the entire system would be ready for production and transportation of oil.

The Forties pipeline proved to be Brown & Root's most difficult pipelaying job of this period. The route was farther to the north and in deeper water than any North Sea pipeline to date. In April 1973, Brown & Root's *Meaders* and *BAR-324* began trying to lay pipe from the Forties field in 430 feet of water. Bad weather forced numerous delays, but when the barges could work, ten 400–500 ton barges constantly supplied them with new pipe. The Morrison Knudsen Co.'s facilities near HiFab coated the offshore section of the line with a coal tar enamel overlaid with 2-1/2 inches of concrete.[40] When weather conditions improved, the *BAR-324* set performance records. Using CRC automatic welding on 32-inch diameter pipe on an around-the-clock basis, the *BAR-324* welded an average of 205 joints per day. During one 24-hour period, it laid 234 lengths of pipe, and 134 forty-foot lengths were laid in one 12-hour shift.[41]

The deep water of the central North Sea was a stern taskmaker, forcing the pipelay barges to cope with extreme broadside swells. The pontoon "Z-1," designed by Brown & Root for lay barges on the Forties line, broke on the first day of pipelaying and had to be replaced. Indeed, Brown & Root's primary lay barge at the Forties sat in the start-up position waiting out the weather for the first 25 to 30 days on site. According to *Offshore,* from May through October 1973, "One barge laying the Forties pipeline experienced a 57 percent downtime."[42] One July storm appeared too quickly for work crews to lower the pipeline, and the swells caused one of the barge's dragging anchors to rupture the pipe, causing extensive damage and delays. The crew had to lift the ruptured portion out of the water, expel the water by driving large rubber balls through the pipe with compressed air before it could be lifted, and replace the ruptured section.[43] Several buckles in the pipe further slowed work.

After Brown & Root and Saipem had finished laying their sections of line, the two ends had to be joined. Saipem was in charge of this operation, which took place in October 1974, 65 miles out in the North Sea in more than 300 feet of water. Saipem's lay/derrick barge, *Castoro II,* brought the ends of both lines to the surface by attaching four buoyancy tanks to each line. The two ends were clamped to the side of the barge and welded together in a delicate operation that lasted more than 14 hours.[44] During the following summer, the pipeline was connected to the first two structures in the Forties field. After "pigging" and testing, the line was put into operation. In September 1975, production began and oil finally flowed onshore from the Forties.

To mark the opening of its field, BP held an elaborate ceremony in Aberdeen in November 1975 at which Queen Elizabeth officially inaugurated the flow of oil through the Forties pipeline. The companies involved received the Queen's Award for their technological innovations in developing the field. Forty Brown & Root employees had the honor of being presented to the Queen, Princes Phillip, Charles, and Andrew, and representatives of the British government. The memory of this meeting remained long after the end of the Forties project, as did the memory of the sense of teamwork and achievement. Heinz Rohde spoke for many on the Forties project: "I remember very well . . . the close relationship we had with our team. . . . We were sort of by accident thrown together . . . and we turned out to be very compatible with each other and supporting each other. So, that was a good team effort."

Brown & Root emerged from the Forties field and Ekofisk with a dramatically expanded presence in the North Sea. By the mid-1970s, the company owned an 18-vessel North Sea fleet, including seven lay barges, six bury barges, and two combination pipeline/derrick barges, along with three spud barges and 40 enhanced supply boats. Brown & Root's fleet was engaged in all aspects of platform and pipelaying in the summers of 1974 and 1975. Yet even as these vessels completed their work at the Forties and Ekofisk, their performance raised questions about their capacity to operate effectively farther north in the North Sea. No such questions arose concerning the capacities of the thousands of engineers and skilled workers who greatly expanded the ranks of Brown & Root, U.K. The London office put the company in a much stronger position for engineering and construction work on future North Sea projects.

BP invested over one and one half billion dollars to construct the Forties; a lot more than initially thought necessary. Fortunately, with oil at fifteen dollars a barrel in 1976 and beyond, this investment was rapidly repaid.

A fraction of those profits accrued to Brown & Root for its path-breaking work on the project, but in addition to the excellent financial returns for its work, the company came away from the Forties well-prepared to assume new challenges in the fields developed in the North Sea in the remainder of the 1970s and the early 1980s. Frank Frankhouser, Brown & Root's project manager for phase two at the Forties, summed up the project simply and well: "Yes, the heaviest loads, the deepest water, the severest weather. At the end of the day, it was a good job, and it went quite well."

Notes

1. This chapter draws from interviews with H.P. Smith, B.E. Stallworth, H.K. Rohde, Hugh Gordon, W.R. Golson, R.O. Wilson, Dirk Blanken, B.L. Davis, Bob Weatherly, H.S. Frankhouser, and J.B. Weidler.
2. "Forties field development planning involves new wrinkles," *Oil & Gas Journal* (January 10, 1972): 78. P.J. Walmsley, "The Forties Field."
3. See Albert M. Koehler, Stanley J. Hruska, and Robert C. Walker, "Development of the North Sea Forties Field Platform Concept," OTC 2245, Offshore Technology Conference, Houston, TX, May 5–8, 1975, 9–20.
4. R.O. Wilson binder, "BP Forties," Tab L, copy in Brown & Root Archives.
5. "Roy Jenkins, the Offshore Man," *BP Forties News* (November 1975).
6. David Telfer, "Well Up to Schedule for Initial Oil Flow," newspaper clipping in B&R Archives.

7. "BP Sets Sights on 400,000 b/d from Forties," *Oil & Gas Journal* (December 27, 1971): 50. "BP borrows $936 million to Develop Forties," *Oil & Gas Journal* (July 3, 1972): 25.
8. "Improvision the Key to Early Structures Success," *Offshore Engineer,* U.S. Supplement (July 1985): 44.
9. Koehler et al, "Development of the North Sea Forties Field Platform Concept."
10. "Improvision the Key to Early Structures Success," 44.
11. Koehler, et al, "Development of the North Sea Forties Field Platform Concept," 13.
12. Koehler, et al, "Development of the North Sea Forties Field Platform Concept," 9.
13. Koehler, et al, "Development of the North Sea Forties Field Platform Concept," 13.
14. "Improvision the Key to Early Structures Success," 44.
15. "Fixed Base Platform Feasibility Study," (April 1, 1971): 3–1; copy in B&R Archives.
16. "Fixed Base Platform Feasibility Study," 3–2.
17. "Forties Field Development Planning Involves New Wrinkles," 78. "First Two Platforms set for Forties," *Oil & Gas Journal* (January 24, 1972): 38.
18. "North Sea oil and gas extraction brings a construction boom," *Engineering News-Record* (November 21, 1974): 21.
19. Koehler, et al, "Development of the North Sea Forties Field Platform Concept," 12.
20. "Platform 1 to North Sea's Oil Treasure," *Evening Standard* (November 27, 1973).
21. "The Gift Horse Gallops By," *The Economist* (July 26, 1975), 21–4.
22. "Presentation to BP Petroleum Development LTD. on Plans for Fabrication of Deepwater Structures at Cromarty Firth, Scotland," November 1971, copy in Brown & Root Archives.
23. "Nigg Yard Meets North Sea Challenges," *Brownbilt* (Fall, 1972): 25. "Europe's largest contractor in venture with U.S.' largest," *Engineering News-Record* (April 11, 1974): 63–4.
24. "BP Sets Sights on 400,000 b/d from Forties," 50.
25. "Tall Platforms, deep pipelines harvest North Sea oil reserves," *Engineering News-Record* (June 7, 1973): 23–4. *Ocean Engineering,* 54.
26. Highland Fabricators Information Brochure, copy in Brown & Root Archives.
27. Peter McCallum, "From Farm-Workers to Welders in Just a Few Weeks," *The Press and Journal* (August 2, 1974): 18.
28. "Nigg Yard Meets North Sea Challenges," *Brownbilt* (Fall, 1972): 26.
29. David Teller, "How Early Problems were Ironed Out," news clipping in Brown & Root Archives.
30. Neil Ulman, "Equipment Problems Dim Hopes of Getting North Sea Oil Soon," *Wall Street Journal* (August 15, 1974): 1.

31. "Roy Jenkins, The Offshore Man."
32. "Delayed Drilling," *Wall Street Journal* (August 15, 1974).
33. Vincent T. Coffman, "Instrumentation System for Installation of the North Sea Forties Field Platforms," OTC 2249, Offshore Technology Conference, Houston, TX, 1975.
34. "Big Production Platform is Installed in Forties Field," *Oil & Gas Journal* (September 9, 1974): 54–5.
35. "Billion-dollar-plus Forties project in full swing," *Oil & Gas Journal* (June 3, 1974): 106–9. "Imploding floats plague oil production platforms," *Engineering News-Record* (February 6, 1975): 10. "Ocean floor pockmarks may limit oil platform design," *Engineering News-Record* (June 19, 1975): 41.
36. "North Sea," Brown & Root publication, no. date, 29. "Largest offshore platforms are set for U.K. North Sea," *World Oil* (May, 1972): 84–85. "Success! Monster North Sea production platform positioned in the Forties Field," *The Press and Journal* (July 4, 1974): 5. "North Sea Plans turned into Tangibles," *Oil & Gas Journal* (January 8, 1973): 66–7. Also see "Forties Field Faces Severe Weather," *Oil & Gas Journal* (December 10, 1973): 87.
37. J.R. Calvert, "North Sea Progresses Despite Hazards," *Offshore* (September 1979): 178.
38. "Forties Field Pipeline on the North Sea's Toughest, says Hugh Gordon," *Oil & Gas Journal* (September 9, 1973): 47–50. "Pipe-lay barges beefed up," *Oil & Gas Journal* (May 1, 1972): 139.
39. Hugh W. Gordon, Jr., "Concepts refined for deepwater lines," *Oil & Gas Journal* (September 1, 1975): 125–6.
40. "BP Chooses Contractors for Forties Line," *Oil & Gas Journal* (May 15, 1972): 54.
41. "Highlands Fling!," *Brownbilt* (Fall 1974): 16.
42. "Highlands Fling!," 16; Calvert, "North Sea Progresses Despite Hazards," 180.
43. "Forties Field Pipeline is the North Sea's Toughest," 47–50.
44. "Above-water tie-in techniques completes BP pipeline," *Oil & Gas Journal* (May 5, 1975): 184–6.

CHAPTER 14

Project Management in a Boom Era

Brown & Root's success as project manager for the development of the Forties complex heralded an era of involvement in a series of major North Sea projects. The discovery in this region of numerous large oil and gas fields after the early 1970s focused international attention on this emerging center of offshore production. The dramatic rise in oil prices brought about by the energy crises of 1973 and 1979 made these fields economically attractive despite their high development costs. Because of its established presence in the region, Brown & Root was well-positioned to take part in a North Sea boom that began in the early 1970s and continued until the precipitous drop in oil prices in the mid-1980s.[1]

Brown & Root's clients, the major operators in the North Sea, needed large, flexible management organizations to control and supervise the fabrication and installation of the concrete and steel mega-structures built in this era. The size and complexity of North Sea projects encouraged the use of project managers to foster efficient organizational relationships among the field operator, the project contractor, and the numerous subcontractors. H.S. "Frank" Frankhouser, vice president of Brown & Root (U.K.), described the project manager's goal as the efficient coordination of engineering and procurement, supervision of fabrication and installation, project control, and administration. "The project management organization," Frankhouser noted, "must be designed to provide the performance, coordination and control of these elements." The term "project management" was used to include "people that plan, schedule, cost estimate, process vendor data, procure, and collate documentation."

At the height of its involvement at the Forties, Brown & Root employed some 750 people on its project management and design staff. Subsequent

North Sea projects required even larger contingents of project managers, who were recruited from the United States and Europe. Organizing such large groups into coherent and efficient management teams required considerable planning, but the resulting rewards were great. During the boom years in the North Sea, a series of some half dozen large projects, each lasting from three to four years, enabled Brown & Root to establish firm support bases around the region. New offices in Bergen and Stavanger, which already supported the extensive pipe-lay and lift installation activities, joined established bases at Rotterdam, Great Yarmouth, Aberdeen, and Peterhead. Numerous field offices also sprang up to support smaller projects around the North Sea. Shortly before the sharp fall in oil prices in 1983, the Europe and Africa division of Brown & Root employed thousands of designers, engineers, and project managers.

Project management services offered the most efficient administration system for full-fledged contractors like Brown & Root. Field operators, of course, retained ultimate responsibility for field development and production, but their involvement in day-to-day project activities varied greatly, ranging from minimal to high. Project managers preferred near autonomy, as Brown & Root had enjoyed during its work for BP on the Forties platforms. But during the heyday of the late 1970s and 1980s, operators often assigned hundreds of staff members to oversee and participate directly in all phases of the project.

Project management held within it a potential for duplication of functions, which could work against efficiency. "I guess the best way to understand it," recalled Jay Weidler, "is that in the worse case you would have more people watching the work being done than those doing the work—more people keeping score than people actually producing things." Brown & Root was not immune to problems inherent in what became known in the industry as the "glorification of project management." Huge North Sea projects during the boom years often required up to five years for completion and included 2,000 to 3,000 personnel for project management functions. With high oil prices cushioning the impact of the cost and consequences of duplication of management functions, contractors, operators, and regulators all contributed to the growth of top-heavy layers of managers.

The key technical consideration during the North Sea boom remained the design of the platform best suited for production at a particular site. The decision to use a particular type of platform came after careful analysis of field size, water depth, oil storage, and transportation requirements, and the availability of the equipment and expertise to build and install the needed platform quickly and economically. In this era of rising oil prices, operators explored numerous new platform designs in their quest for the rapid development of oil and gas deposits. Brown & Root focused on variations of traditional steel jackets with innovative designs of decks, jackets, and flotation systems, but the company also did substantial work on deck designs for concrete-gravity or Condeep platforms. Its most innovative design work was on the development of the first Tension Leg Platform (TLP) for use in Conoco's Hutton field.

As a major offshore construction company, Brown & Root was directly affected by one of the key considerations in jacket design, the equipment needed to build and install platforms. Along with its major competitors, the company had to make hard choices concerning how the future direction of platform design might affect the market for its construction equipment. Derrick and lay barges had to be capable of handling the jackets, decks, and pipelines required by offshore oil companies. Brown & Root thus faced a difficult question in planning its investments in equipment: What was the most efficient size and the appropriate technology for its new barges? At a critical juncture in its North Sea work, the company invested in larger and larger versions of its traditional flat-bottomed barges as several of its competitors chose ship-shaped and semisubmersible designs that proved better suited to conditions in the North Sea.

Such strategic missteps did not, however, prevent Brown & Root from participating in the design, engineering, and project management of many of the major platforms and pipelines built during the North Sea boom. In addition to its work before and during the Ekofisk and Forties projects, the company worked on aspects of the development of the Frigg, Brent, Beatrice, Brae B, Tern, Hutton, Maureen, Heimdal, Valhall, Morecambe, Odin, Magnus, and Statfjord fields, including the installation of more than 1,700 miles of North Sea pipeline. The firm's success reflected its ability to adapt to the harsh conditions of the northern reaches of the North Sea.

Deep-Water Steel Structures

Brown & Root Marine had specialized in the design and construction of steel structures throughout its history, and it remained at the forefront of this technology in the North Sea. In stretching the company's capacity to build and install large steel platforms in the rugged North Sea, the Forties project had emphasized the need for creative approaches in design and for more powerful equipment for installation. In the decade after the completion of the Forties platforms, Brown & Root worked on a variety of North Sea projects that embodied further advances in steel jacket technology, most notably the fabrication of self-floating jackets for Chevron's Ninian field and BP's Magnus field and the HIDECK design for the Maureen field developed by Phillips Petroleum. Although the company designed, constructed, and installed numerous other steel platforms in the North Sea in this era, these three projects illustrate how Brown & Root continued to adapt steel jacket technology in response to the new challenges posed by discoveries deeper in North Sea waters.

The decision to use steel jackets rested ultimately upon financial concerns as well as the requirement to get the fields in production as soon as possible. When Chevron Petroleum discovered the Ninian field in the far northern waters of the U.K. sector during January 1973, the company and its eight partners decided to try one concrete platform and two traditional steel structures in a $2.55 billion development plan. The field contained estimated recoverable crude reserves of 1.1 billion barrels, making it the third largest offshore oil field in the U.K. sector. Chevron's desire to have the jackets in place by the summer of 1977 dictated its choice of jacket design. When planning began in 1974, Chevron considered both a barge-launched jacket and a self-floating steel substructure. With no assurance that a barge capable of launching a 22,000 ton jacket would be available, Chevron chose a self-floating design for the Ninian-Southern field. As had been the case with most of the jackets used earlier in Alaska's Cook Inlet, two of the four main legs of the jacket were oversized, allowing it to float on its own legs instead of on dedicated flotation tanks similar to those used for the Forties jackets. Chevron contracted with Highland Fabricators to build this structure, which stood more than 550 feet tall.

The arrival of the largest launch barge in the world, Brown & Root's *BAR-376,* a year later changed the calculus of decision making for Chevron. Completed in April 1978 at Hyundai in Korea, the barge was towed to Hifab in time for use on Chevron's Ninian-Northern development. With the promise that this giant barge would be available, Chevron opted to use a barge-launched jacket at Ninian-Northern. The reason was simple. A barge-launched jacket capable of supporting roughly the same topside weight as the 22,000-ton self-floater used in the southern sector of the field weighed only 14,300 tons. Cutting more than one-third of the weight produced dramatic cost savings. Brown & Root designed the new platform, which was fabricated at HiFab and installed in 470 feet of water in June 1978.[2]

The installation of this large jacket from the giant new barge had its moment of high drama. *BAR-376* was equipped with hydranautics hydraulic launching jacks, which momentarily failed during the launch of the jacket, stranding it halfway down the launchways. Frantic jacking, a bit of luck, and the ice-cool direction of Roy "Not to Worry" Jenkins, moved the jacket farther down toward the end of the barge, where the tilting beams did their work and tipped the jacket off and into the water. Heinz Rohde and many others of the successful Forties team used sophisticated telemetry and a simplified ballast valve upending system that evolved from those used on the Forties project to place the structure safely on the seabed. As the launch took place in June, pile driving and module lifting to complete the platform could be carried out in the summer weather window.

With the timely arrival of a new generation of giant launch barges, the barge-launched jacket had been more economical than the self-floater at the Ninian field. But self-flotation had a place in the design and fabrication of much larger versions of the Ninian-Northern jacket that could not be launched from existing barges. The most significant of these for Brown & Root was the huge jacket built for BP's Magnus field, discovered in 1973 in more than 600 feet of water in the northernmost U.K. sector about 125 miles northeast of the Shetlands. Design criteria reflected BP's requirement to begin oil production quickly. Also, the structure did not require offshore oil storage facilities because oil would be pumped to a nearby underwater trunkline. The designer,

CJB-Earl and Wright, developed a self-floating, traditional-style steel platform. But because it was designed for 614 feet of water, the design was far from traditional in size. The massive steel substructure weighed nearly 38,000 tons, making it the heaviest steel jacket built to date. The entire platform stood 695 feet high, the equivalent of an 80-story building. Its $2 billion cost made it the most expensive single platform yet installed in the North Sea.[3]

In late 1979, HiFab won the $193.5 million contract to build the structure, which weighed 44,000 tons including its main piles which were in place during the float-out. This size far outstripped the capacity of available launch barges, and the Magnus was designed to float on two main legs 35 feet in diameter with a flotation chamber between them. Six temporary buoyancy tubes in the pile guides provided additional buoyancy. In April 1982, five tugs pulled the Magnus jacket on a smooth journey from the HiFab dock. But a serious mishap occurred during installation, underlining one of the problems of steel jacket installation. During the ten-step lowering procedure, several of the 84-inch diameter, 330-ton piles broke from their retaining clamps and slid into the water until their lower ends rested on the seabed. This threw the jacket out of balance, forcing a halt to the installation process with the jacket floating precariously at a 20 degree angle. Installation engineers pumped water into the platform's legs and towed the jacket away from the angled piles, causing them to fall out completely. Using the piles that remained, crews secured the structure to the seabed. Replacements for the lost piles were quickly fabricated and installed. After all piles were in place, Heerema's *SSCCV Balder* lifted 19 topside modules weighing a total of 34,000 tons into position atop the jacket.[4]

After installation, the large diameter legs served as storage for potable water, drilling water, and diesel fuel. The platform itself contained 20 standard well conductors. The jacket also included 11 flow-line risers. Seven of these connected the satellite wells to the platform and four of the risers were available for additional wells. Because the Magnus field was long and narrow, and a single platform could not therefore fully exploit the reserves, BP developed the field's outlying areas with subsea satellite wells tied back to the central platform. To reduce the demands on divers, an elevator positioned down the center of the jacket allowed a remote controlled inspection vehicle to work exclusively within the jacket members.

With a production capacity of 125,000 barrels per day of oil and 9,000 barrels per day of natural gas liquids, the Magnus field was large enough to justify its own pipeline connection to shore. Fortunately, this field could be tied into the nearby Ninian field trunkline to Sullom Voe terminal in the Shetland Islands. Contractors installed a 24-inch, 56 1/2-mile connecting line for pumping crude and dissolved gas liquids from Magnus. A trencher remotely controlled from the diving support vessel *Seaway Sandpiper* assisted in laying these flow lines. Brown & Root's *Semac 1* laid 42 miles of 20-inch gas line in the summer of 1982.

Counting mishaps and delays, the completion of the hookup and commissioning of the Magnus required several million offshore man-hours. Oil did not begin to flow until August 1983, more than a year after installation. The huge expense of offshore hookup work on the giant Magnus platform demonstrated the need for innovative thinking about deck design and installation. Most North Sea construction had featured a limited number of platforms with relatively large decks, and the traditional approach of lifting modules onto jackets at location and then making all necessary hookups on site was costly in time and money. In some sense, this was the equivalent of the "stick-building" process long abandoned in jacket construction in favor of onshore fabrication. Brown & Root made an important contribution to the evolution of new deck-building technology with its work on a path-breaking deck design that came to be known as HIDECK.

The design was originally developed as an alternative to offshore hookup in the later 1970s by Larry Farmer and Phil Abbott. The first application of HIDECK was in Phillips' Maureen field. HIDECK was an integrated deck fabricated fully equipped, outfitted and tested in a graving dock on shore, then barged as one unit to a specially designed, pre-installed steel substructure or jacket for mating and final start-up. The HIDECK system sharply reduced the time needed to activate a deck on a large platform, cutting a typical job from 10 months to less than two months. In addition, this new approach did not require the hiring of expensive crane barges for installation. Dramatically lower overall installation costs resulted, and drilling and production could begin as many as six months sooner.[5]

Design engineering for HIDECK began on July 3, 1978. Brown & Root (U.K.) received the contract for structural and installation design of HIDECK, while Technomare of Italy received the award to design the 40,000 ton (later increased to 93,000 tons) Technomare Steel Gravity base (TSF) substructure, which was built by Ayrshire Marine Constructors at an old concrete site at Hunterston on the Scottish west coast. This was the first and the last steel gravity substructure in the North Sea, and it was the world's first unpiled steel platform that combined flotation, ballasting, and oil storage. Howard Doris-NAPM won the contract to fabricate the HIDECK at its Loch Kishorn yard in Scotland.

Unlike the traditional jacket, which is usually narrow at the top and wide at the bottom, HIDECK required the jacket to be wide at both the top and the bottom. Its innovative design included an inverted arch so that a large cargo barge could position itself underneath the HIDECK structure within its supports, allowing it to be towed out to sea. The HIDECK design eliminated the costly and difficult process of hooking up the separate modules on site, thus cutting costs, speeding installation, and enhancing safety. The HIDECK's deep open truss permitted a more orderly layout of facilities and straighter piping and cabling runs, which reduced onshore construction costs when compared with the construction costs of modules.[6]

The completed Maureen HIDECK weighed 18,600 tons and measured 225 × 246 feet. It included a flarestack towering 389 feet above the bottom of the lower deck. In addition, an intermediate mezzanine covered about half of the deck. The upper deck contained accommodations, power generation, workshop, control room, compression and drilling modules, the flare tower helideck, and three deck cranes. The process and utility equipment was on the lower deck, and the majority of this equipment attached to skid packages. The primary steel frame of HIDECK consisted of five main trusses supported on ten legs.

Maureen Hideck

The February 1983 tow-out of HIDECK from the Howard Doris yard to the inshore mating site was a delicate operation. To accommodate its transportation by barge, HiDECK was designed and built in the form of a bridge, with one side resting on the quay and the other supported by a line of dolphins placed 170 feet into Loch Kishorn. To tow the structure out to the mating location, workers positioned the large launch barge *H-114* underneath HIDECK and then raised it by deballasting. This was the largest operation of its type yet attempted. Three 3,300 HP Voight Schneider tugs positioned the barge within the HIDECK arch. To assure perfect alignment, four vertically mounted lasers were focused on selected targets, ensuring that the mating points on the HIDECK matched those on the barge. During this operation, a computer monitored and automatically adjusted the flow of 10,000 tons per hour of water ballast.[7]

The next phase of the operation called for the mating at Loch Kishorn of the HIDECK with the substructure, which had been floated out earlier from Hunterston. The mating procedure, described by the engineers as both an art and a science, occurred in deep water. The deck weighed approximately 19,000 tons and the TSG weighed some 93,000 tons, including its solid ballast of concrete and iron ore. Only about the top 26 feet of the ballasted TSG was visible above water. Tugs maneuvered the cargo carrying the HIDECK into the substructure slot, where winches and wires made more precise adjustments to its position. After aligning the ten mating cones in two rows of five, the TSG was deballasted and the structure rose into the HIDECK. The withdrawal of the barge then transferred all weight to TSG. Further deballasting gave the structure the 177-foot draft required for tow-out.[8]

Since the economics of the Maureen field did not justify laying a pipeline to shore, tankers collected the oil production from the platform. The TSG included three large storage tanks with a total capacity of 650,000 barrels, so that oil production could be stored during periods when rough weather prevented tankers from reaching the platform. Tankers collected oil production at a concrete single-point mooring station located near the platform. The HIDECK design allowed the field to come on stream two weeks ahead of schedule after only 80,000 man-hours of offshore hookup and commission work. This design moved the industry a giant step forward in understanding how to limit the need for extremely costly mating operatings at sea. Although gravity-based structures made of steel never caught on in the North Sea, the gravity-based structures made of concrete were quite common. In both cases, as well as with traditional steel jackets, the reduction of offshore hookup and commissioning work had become one of the new frontiers of platform design.

Concrete and Politics in the Norwegian Sector: Statfjord

Concrete gravity structures offered a promising alternative to steel structures in the North Sea. They absorbed tremendous topside weight even though, in the opinion of some, their construction costs were often less than those steel platforms. Because these structures remain upright during transit tow to their offshore sites, deck mating and hookup could take place in sheltered water near the fabrication site. Concrete structures had certain technological advantages over steel in water as deep as 1,000 feet, and in particular where the field properties required that extensive equipment be located in the topside design or offshore storage was desired.[9]

Most of the first generation of concrete gravity structures used the Condeep design, with the deck modules transported from Europe and mated atop the concrete substructures. Slipforming of the bottom part of the concrete structures, which contained the ballast and oil storage cells, took place mainly in dry dock in fjords near land. Once the bottom had been completed and floated out, the remaining two-, three-, or four-legged columns were erected and then towed to an inshore, deep-water mating site. The Norwegian coast, with its many protected inlets and fjords, lent itself very well to this purpose. There, the concrete columns were crowned with a large module support frame, or later, with an integrated deck.

Norway took the lion's share of the fabrication of the concrete gravity-base structures (GBS) mainly as a result of the deepwater, yet sheltered, building sites along the coast. Attempts were also made to fabricate GBSs elsewhere, including the site at Hunterston on the Scottish west coast, the cradle of the Chevron Ninian Central platform. Another site was on the shallow west coast of Holland, where ANDOC put together the Dunlin A platform substructure, although the GBS was again mated with its topsides off the island of Stord in Norway. Still another successful attempt to build concrete structures was made by Howard Doris, who blasted out a small graving dock from the surrounding rock face at Loch Kishorn on the east coast of Scotland and fabricated Total's MCP1 and two manifold platforms.

The Statfjord field sported the largest concrete gravity platforms of the period in the North Sea. This huge field was located in the far northern section of the North Sea on the dividing line between the U.K. and Norwegian sectors, with the bulk of the field on the Norwegian side. Brown & Root participated extensively in the development of this field, providing project management, installation, and hook-up services for the Statfjord A platform, which became the tallest platform of any type then to be installed in the North Sea. Brown & Root also provided project management services, and designed the topsides in joint venture with Norwegian Petroleum Consultants (NPC) for Statfjord B, which at tow-out in the early 1980s was the world's largest offshore structure. In 1982, *Scientific American* called this platform the "most advanced concrete platform yet constructed." Platform C, which was modeled after B, followed quickly and was installed in June 1984.[10]

The Statfjord projects were especially complex and problematic. One problem was geopolitical in nature. Because Statfjord B rested on the median line separating the Norwegian and U.K. sectors, it had to meet the regulations of both the Norwegian Petroleum Directorate (NPD) and the U.K. Department of Energy (DoE). Joint ownership of the field also posed problems. Mobil, which had only a 13 percent interest, was the project manager and designated operator. Norway's Statoil, however, had a 50 percent interest and also played a major role in project management. Brown & Root's project management for this platform included conceptual and detailed engineering of the topside as well as site management and supervision of the fabrication of modules and the concrete gravity base. Brown & Root's Frank Frankhouser, project director on Statfjord B, recalled that "We had made a schedule and estimate of the job and four years later, it ended up almost exactly on both the schedule and the cost estimate that we made. So, in that sense, it was a successful job." The responsibilities of project management involved employing a large workforce and that meant providing for the families of this workforce as well. During its work on Statfjord A, Brown & Root, which had set up an office in Bergen, Norway, bought a local building and established an American high school for the many foreign children of the project management staff.

Brown & Root's joint venture partner, the Norwegian Petroleum Contractors (NPC), consisted of ten of the largest engineering and construction companies in Norway. "They were all kind of bitter rivals commercially. And we, Brown & Root, were the sponsors of the joint venture. So, it was quite a management challenge," recalled Frankhouser. "On the Forties field Brown & Root was alone at the engineering and project management. Statfjord B was a joint venture, so everything we did, we had to check with our venture partners and so forth on major decisions. The Norwegians initially were quite suspicious of Brown & Root . . . because we were a foreign multinational company. It took a long while to overcome that and get to working together as a team but, ultimately, we did."

Thad "Bo" Smith III, of Brown & Root had a similar view about work on the earlier Statfjord A: "All these minority partners giving them [Mobil] hell! And we were the high-profile, high-exposure project management and hookup team that had to go out there and put all that stuff together. But we were very fortunate. We had a lot of good guys, like Roy Jenkins and Randy Davis and his team, and we found good Norwegians that worked with us. And finally we molded a pretty good crew out there." The successful completion of the project required 5.2 million man-hours. At Statfjord B, Brown & Root had as many as 2,200 people on the job spread out to London, Oslo, Stavanger, Bergen, and other fabrication sites throughout Western Europe.

The Statfjord B platform consisted of 335,000 tons of concrete plus a massive two-level steel module support frame. The base, built in dry dock in Stavanger, consisted of 24 cells in a honeycomb configuration. Each cell was 66 feet in diameter with three-foot thick walls and 20 of the cells at the base of the structure were domed over and used to store crude oil and diesel fuel.

The politics of labor threatened Brown & Root's timely completion of its work at Statfjord A and B. The Norwegian government required its citizens to be given preferential treatment in staffing the projects. Brownaker, Brown & Root's Norwegian joint venture with Aker, was responsible for hiring more Norwegians and letting go of other workers. "So we started changing the crews out," recalled

Bo Smith. An offshore strike followed at both the Statfjord and Ekofisk fields. Brown & Root's Spanish workers decided to strike in sympathy. "We had never had a bit of labor problems from those kids from northern Spain in our life," said Smith. "They were great, great people. They knew their jobs. They were being replaced by Norwegians. They went on strike." Smith contacted Paco Montero, Brown & Root's Spanish personnel manager stationed in Rotterdam, who took a helicopter out to the platform and managed to settle the labor dispute.

Work continued and the 41,000-ton topside unit was assembled at a yard near the construction site for the Statfjord B structure. Workers at the Moss Rosenberg yard at Stavanger built a 7,000-ton support frame for the deck modules. After installation of the modules on this frame near the shipyard, four piggy-backed and ballasted barges moved into position underneath it. After deballasting, the barges lifted the entire topside unit over the waiting concrete substructure and crews connected the two massive structures with post-tension rods and grout.

In August 1981 five of the world's largest tugs slowly towed the 900,000-ton structure 230 miles to a pre-leveled offshore site in 472 feet of water. The weight and shape of the massive concrete structure meant that piling was not necessary. Upon installation, the structure measured 910 feet from the seabed to the top of its derrick, with 450 feet rising above the sea. The Statfjord structures became models for numerous major concrete gravity-based structures subsequently fabricated and installed in the Norwegian sector.

Brown & Root's image in Norway caused problems for the company during its work there. The death of a diver during tests of deep-water hyperbaric welding in a Norwegian fjord in preparation for possible welding of a pipeline across the Norwegian trench brought a strong public outcry. An Oslo newspaper reported the incident under the headline, "American Vessels Using Human Guinea Pigs for Diving Tests." Later, when the Norwegian government embarked on a highly publicized investigation of Brown & Root's method of arranging to pay its workers in ways that minimized their Norwegian taxes, newspaper accounts proclaimed "Offshore Oil Firm 'Dodging Taxes'." Norwegian socialists kept a close eye on Brown & Root, the embodiment of American capitalism, making the political waters about as rough as the North Sea in the Norwegian sector.

Tension Leg Platform: Conoco's Hutton TLP

The Tension Leg Platform (TLP) represented the next important innovation in platform design originating from the North Sea. This was a steel structure of a different sort, with a design that promised great benefits in waters up to 2,000 feet deep. Like the HIDECK, the TLP had a fully integrated topside with all deck components installed and hooked up prior to the final tow-out. Conoco chose to pioneer this new technology in its Hutton field in the North Sea, and the company recognized that, "it is significant for the support industries involved in design, supply, and construction. They also will get the chance to be pioneers."[11]

Conoco discovered the Hutton field in 1973 approximately 90 miles northeast of the Shetland Islands in 485 feet of water. Because Hutton had reserves of only 175–250 million barrels, reservoir engineers considered it marginal in comparison to existing North Sea fields. Working with an eight-company consortium, Conoco looked for an efficient way to develop this relatively small field which had several distinctive characteristics. Because the field held little natural gas, the deck would not need space for gas reinjection equipment and could thus be smaller than a normal North Sea deck. The field had a relatively short projected lifespan, making a TLP particularly attractive, since—at least in theory—much of the structure could be moved from loca-

Hutton TLP

tion to location. This promising new technology seemed particularly suited to the Hutton field, and Conoco decided to build the first full-scale TLP for drilling and production there. The estimated cost of the platform was approximately the same as a steel jacket design for the site, and the TLP design was worth testing.

Conoco initially estimated that the project would cost $1 billion and require four years to complete. The original design plan announced by Conoco called for twelve tubular steel lines or cable tethers to anchor the TLP to piled foundation templates on the sea floor. Designers expected some horizontal movement to occur, but heaving could be regulated by tightening the tethers and discharging ballast, which could create a tension of about 1,000 tons on each of the twelve lines. "The main problem to be solved," wrote a team of Brown & Root engineers, "is how to obtain sufficient compliance to enable the structure to oscillate with the waves without overstressing the foundations."[12]

In the summer of 1981, Conoco awarded the design of the TLP to Brown & Root (U.K.) in association with Vickers Offshore Projects and Development. Brown & Root assembled a team of 800 design and engineering personnel in the Colliers Wood office in London for this high profile project. During the four-year design phase, the team produced approximately 10,000 fabrication drawings and 675 procurement packages. During fabrication, the team also monitored the weight of over 40,000 individual components to meet strict weight requirements. Altogether, Brown & Root put more than 2.5 million man-hours into the project.

McDermott Scotland received the award to build the huge 243 × 260 feet integrated deck at its Ardersier yard. HiFab at Nigg Bay landed the contract to build the semisubmersible hull for the TLP. At the time of letting the contracts, a Conoco official admitted that the total cost of the Hutton TLP venture would far exceed the original estimate of one billion dollars, although the project still remained financially attractive.

HiFab began preparations to build the six-column hull structure similar to a semisubmersible barge, but with two to three times the displacement. Conoco had changed its original design from an eight-column structure to a six-column one to facilitate the mating of the hull and the deck. With eight columns in place, a cargo barge could not slide under the deck during mating. Ten different fabrication yards, including HiFab's own Nigg Bay facility, separately constructed 32 segments of the hull and delivered them to Nigg by spring of 1982. The hull was made with relatively thin steel plate strengthened with internal stiffeners. The corner legs measured 59 feet in diameter and were 187 feet long, while the two center legs of the same length measured 46 feet in diameter.

Problems occurred during fabrication in the summer of 1982. At the Nigg yard, a routine magnetic particle inspection (MPI) of a weld joining a ring stiffener to the inside of a column block revealed a defect, even though the weld had passed an earlier NDT examination. Further investigation revealed that almost every weld on the ring stiffeners fabricated at HiFab suffered from hydrogen embrittlement, which results when a weld receives moisture from the atmosphere or from welding rods that have not been fully baked. The release of hydrogen from the water forms high-pressure pockets that cause cracks. Because these defects did not usually appear on the surface of the weld, they were very difficult to detect. Welding teams searched for the defective welds, ground them out, and replaced them. The embrittlement problem added thirteen months to the project.

Once these problems were corrected, mating took place. In May 1984 the cargo barge *Oceanic 93* transported the deck to the inshore mating location at Moray Firth, where it was placed over the ballasted hull and anchored securely just above it. Deballasting then raised the hull to meet the mating points and Aker Offshore Company (AOC) completed the hookup work. On July 8, five tugs towed the structure to the Hutton field. Heerema's first generation semisubmersible crane barges (SSCV) *Balder* and *Hermod* helped secure the TLP to the seabed on July 15.[13]

This required a critical new operation, attaching the tubular steel tethers to the pre-installed template foundations. First, Brown & Root engineers directed the lowering of four of the sixteen tethers from the platform and secured them to the foundations with an anchor connector. During the next several days, divers connected the remaining twelve tethers to the foundations. The next step included tying the ten pre-drilled wells to the platform. After installation, the platform crew drilled more wells as required. The TLP required only 22 days to be brought into full readiness after installation, compared to the 9- to 14-month period required for the hookup of a traditional steel jacket with modular topsides. Conoco also laid a four-mile underwater pipeline to Amoco's Northwest Hutton platform for oil flow through an existing line to Shell's Cormorant platform and then to the oil terminal at Sullom Voe in the Shetlands via the Brent oil trunk line. Conoco's confidence in the new TLP technology was evident three months after initial production at Hutton, when the company announced plans to build a second TLP at Green Canyon field in the Gulf of Mexico.

Platforms and Pipelines: Frigg Field

One of the largest North Sea projects for which Brown & Root served as project manager was the Frigg field. Discovered in 1971, this was one of the largest gas fields in the North Sea, with a projected capacity of one billion cubic feet per year. Elf Aquitaine of Paris was responsible for the development of the field and Total was responsible for the pipeline transportation system. Brown & Root served as project manager for the pipeline and various other structures at the production complex. Rick Rochelle commented on the challenges of this project: "Frigg was much more complicated than Forties ever thought about being . . . Setting risers around that thing and putting all of those Z-bends on the bottom in what at that time was extremely deep water."

To facilitate its work on the project, Brown & Root moved its design team to an office in Paris at the La Defense Complex. Jamie Dunlap, project manager for the Frigg work, recalled that, "The Paris office was a real showplace. We had originally brought in a computer line, direct access to Brown & Root's mainframe in Houston. We had earlier set that up and had it operating in London. But there had never actually been one from the USA to an engineering office on the continent. And so, we were able to bring one across to the Paris office and we set up a Data 100 system." The project management staff of approximately 1,300 people included both designers and engineers. The project management team worked on fabrication of modules, deck pallets, and various other components." Dunlap noted that, "We had fabrication work going on eleven different fabrication facilities in seven different countries around the North Sea area."

Like the Statfjord field, the main Frigg field crossed the boundary line between the UK and the Norwegian sectors of the North Sea. "And as such," recalled Dunlap, "part of the design had to be done to Lloyds standards and part had to be done to DnV Standards (Det norske Veritas Standards). At that time, the Frigg project was probably one of the most complicated projects, politically and technically, that we had ever encountered or worked on."

Another unique aspect of the project was the module support frame that sat on top of the concrete gravity base. The steel frame was too large for an existing crane barge to lift in one piece, so Brown & Root designed a tandem lift using its *L. B. Meaders* and the French *ETPM 1601*. Dunlap remembered that after completing the two-legged concrete structure at Andalsness, tugs towed it out into a Norwegian fjord. "We deballasted it down to a level where we could take this tandem barge lift and set the structure up on top of the two concrete columns." Despite much "plotting and planning," the company encountered "a real communications problem when we started up, which was something that we hadn't anticipated." On the *Meaders* barge, "we had a Spanish-speaking operator running the crane and we had a Scotsman . . . giving instructions. And the ETPM barge . . . was operated by a Frenchman." After delays put workers behind schedule, they deballasted the structure at ebb tide and had to wait to start the lift until the tide started changing. The tidal currents then caused the concrete columns to rotate until they were about six or eight inches outside the diameter of the steel frame that was going to be set down on top of them. After several tries, the cranes successfully "stabbed" the legs.

The Frigg field pipeline project presented even greater challenges. Total Oil Marine managed the construction of dual 32-inch diameter pipelines about 200 feet apart, but both used a terminal on Scotland's east coast. The Frigg lines were more complex than either Ekofisk or the Forties field lines. According to Dunlap, "We were responsible for everything that happened on the concrete structure. Plus we were responsible for designing into the structure the scheme for installing, pulling in the pipelines into the base of the structures through tunnels and then pumping the tunnels that were connected by caisson to the top of the structure." Installation required innovation: "the pipeline itself was pulled in through a winch system operating from the top of the platform. We pulled the pipe in from the lay barge and they then laid

Frigg field complex. Photo courtesy of Dirk Blanken

away from the structure. The seals were activated remotely and the water was pumped out and the riser welds were made down on the sea bed. This was the first time that this approach had ever been taken."

Brown & Root joined Oceanic Contractors, a subsidiary of J. Ray McDermott, and the French firm ETPM to lay the Frigg field pipeline. Brown & Root began laying pipe from the Frigg platform toward the intermediate point in the North Sea; Oceanic, from the intermediate point toward shore; and ETPM, from the shore toward the Frigg field. Construction began in 1974 and continued through 1976. The 230-mile, 32-inch gas pipeline from the Frigg field to St. Fergus in Scotland was the longest pipeline yet laid in the North Sea. Previously, the deepest North Sea pipeline was 420 feet deep, but this one would be 500 feet deep with one fifteen-mile segment as deep as 550 feet. It was also located the farthest north in the North Sea.

Brown & Root laid the deepest section of the pipe using conventional flat-bottom barges beefed up for North Sea service. It laid the portion of the line in 550 feet of water using its lay barge *BAR 324,* which measured 400 × 100 × 30 feet and could handle pipe up to 60 inches in diameter using an in-house designed semi-articulated stinger. Due to the size of the pipe and the deep-water conditions, Brown & Root called on Taylor Diving for hyperbaric welding over a twelve day period in July 1975. Taylor used its patented Submersible Pipe Alignment Rig (SPAR) and the Brown & Root pipelay-derrick barge *BAR-297* to connect two previously completed sections of 32-inch pipe at a depth of 382 feet.

The Frigg project moved Brown & Root into several new directions. The firm opened a large high-tech office in Paris and participated in the project management of a large gas field crossing national boundaries. The hyperbaric welding technique proved its commercial usefulness in this project and for the first time in the North Sea, and a tandem, two-barge lift was carried out. But Brown & Root had to continue to upgrade its barge fleet. On the Frigg pipeline project, a French barge did what Brown & Root needed to be able to do, lay double-jointed pipe in deep water.

Equipment for the Future: The Semisubmersible

The Frigg project demonstrated the importance of maintaining a large and efficient fleet of derrick barges, lay barges, and trenching equipment. Higher oil prices in the 1970s had propelled an offshore boom around the world, but the focus of the action in this era was in the North Sea. As companies scurried to develop one giant field after another, their designs often were shaped by the size and efficiency of available construction equipment. North Sea jackets and decks grew much larger very rapidly in the boom decade, and contractors hustled to keep pace with the offshore industry's needs by building newly designed barges. Brown & Root had enjoyed a lofty competitive position in the early years of North Sea development, but the boom years witnessed the emergence of numerous aggressive competitors—led by the company's one-time partner Pieter Heerema. The extraordinary opportunities evident in these years in the North Sea convinced Heerema and others to invest heavily in new and larger vessels with innovative designs that allowed them to outperform more traditional work vessels such as the flat-bottom barges favored by Brown & Root. These new vessels posed serious competitive questions for Brown & Root, and by the late 1970s, the company faced difficult strategic choices concerning the design and expansion of its fleet.

In its early years in the North Sea, Brown & Root had made good use of the *Global Adventurer,* the crane barge *Atlas,* the lay barge *Hugh W. Gordon,* and numerous other vessels constructed in Europe or borrowed from the Gulf of Mexico to establish its preeminence in North Sea construction. But as larger projects in the harsher areas of the region moved forward in the 1970s, the handwriting appeared on the wall. Larger and more sophisticated equipment would be needed to work efficiently and competitively in the North Sea. In 1972 and 1973 during the fabrication of the Forties complex, cargo barges used in the North Sea were of 250 × 75 × 16 feet, Gulf of Mexico dimensions. Originally designed in the early 1960s, they proved too small for the violent North Sea. Brown & Root operated six of these workhorses in the North Sea. To satisfy the requirements for heavier marine equipment, the Forties team persuaded subcontractor marine equipment operators to build bigger cargo barges measuring 300 × 90 × 20 feet. Initially, twelve of these were built and used at Forties. At the end of the 1970s more than fifty such "Standard North Sea Barges" were available in the North Sea, creating an enormous surplus and lowering rental costs.

As all major offshore construction companies in the region explored their options for the design of the next generation of larger vessels that would be needed to meet the demands of the North Sea boom, the semisubmersible design quickly became the focus of the debate. When buoyant members of a semisubmersible vessel are ballasted below the area of wave action, only the widely spaced columns receive wave-induced forces. Work vessels of this new design proved capable of remaining relatively stable in the rough waters of the North Sea and could work in weather conditions that would shut down traditional barges. With sufficient size, semisubmersibles also promised dramatic increases in the operational capability of onboard cranes and equally impressive increases in the speed and efficiency of pipelaying operations. Although the technology underlying the semisubmersible design had been applied to mobile drilling rigs since at least Project Mohole, the companies that led the way in attempting to build effective semisubmersible derrick and lay barges faced significant risk—and extraordinary opportunities.

After considering in the mid-1970s the option of investing in this new and as yet unproven technology, senior staff at Brown & Root decided to stick with traditional designs for the company's new construction vessels. Thus as Santa Fe and others introduced semisubmersibles into the North Sea, Brown & Root announced the construction of a new "superbarge," the *BAR-347.* This mammoth vessel dwarfed everything in Brown & Root's fleet. "The world's largest pipelaying barge" was launched amid much fanfare at Rotterdam in early 1976. To work in the deeper water conditions of the North Sea and elsewhere, *BAR-347* was 650 feet long by 140 feet wide with a depth of 50 feet. Making use of a centerline, elevated pipe ramp with three pipe tensioners, *BAR-347* was designed with the capacity to lay double-jointed 80-feet lengths of 36-inch diameter pipe in water depths up to 1,100 feet without the use of a pontoon. In theory, this barge could lay pipe across the Norwegian Trench, a job which seemed in the offing in the late 1970s. With accommodations for a crew of 350 and storage for up to 20,000 tons of pipe, the superbarge was proclaimed by Brown & Root to be the prototype of the coming third generation of lay barges.[14]

Despite the impressive size and quality of equipment on the *BAR-347,* Brown & Root pipeline superintendent "Buddy" Hoke recalled that "Nobody wanted a *347.* . . . And I know it seemed like those old barges had made Brown & Root fortunes and we had taken them all over the world and they had done real good, and if a little bit was good, a whole lot would be real good. . . . When you ran that *347,* you never got a sixth sense of where you were at. It was kind of like an arm that didn't have any feeling in it. The communication was horrible on it. It had another bad flaw that nobody ever wanted to talk about and that was that touchdown point down there, that would buckle the line. At the touchdown, where the pipe touches the bottom."

Unfortunately, by May 1976, only months after its much heralded launch, the superbarge had already departed the North Sea to work in the tamer waters of the Gulf of Mexico. By the end of 1980 most of Brown & Root's flat-

Superbarge (*BAR-347*)

bottom derrick and lay barges joined *BAR-347* in leaving the North Sea. The *Atlas* left in summer 1977, and *Hercules* followed in December 1977. The older *H.W. Gordon* had already left in December 1975. The *Global Adventurer* left for Angola in May 1973. *BAR-228* went back to the United States in October 1975. The *BAR-323/324* (lay) and *BAR-316/331* (trench) remained throughout 1979. The era of the traditionally designed barge had come to a sudden end, and Brown & Root suddenly found itself scrambling for competitive position in a new era of semisubmersibles.

The limitation of even giant flat bottom barges in the North Sea were exposed by a cluster of semisubmersibles owned by Brown & Root's competitors. In response, the company reevaluated its previous decisions and purchased the shipshape monohull *Ocean Builder* in 1978 and the semisubmersible lay barges *Choctaw 2* in 1979 and *Semac 1* in 1982. The former was purchased from Santa Fe for $17.2 million, and extensively modified for another $6 million in Rotterdam. However, the vessel was never very successful. Brown & Root found little market for its services amid the general decline of business in the North Sea, and it moved the vessel to the Gulf of Mexico in 1983.

In 1976, Brown & Root had the opportunity to participate in the construction of the *Semac 1,* but the firm declined for several reasons including cost. "The primary thing," said Joe Lochridge later, "was not only to get something that had better motion characteristics than our smaller lay barges, but to try to lay double joints. And so, I kept telling Hugh [Gordon], we're never going to be able to lay double joints until we are ready to build a bigger barge . . . or come up with a revised welding process, a single or double station welding process, which we have not been able to do successfully." Discussion continued, but Brown & Root did not want to spend the money necessary to build a semisubmersible. Eventually, it became clear that not having a semisubmersible meant not being competitive, and the cost of building or buying one became a secondary consideration.

In 1982 Brown & Root purchased the semisubmersible *Semac 1* in order to maintain its position as the dominant pipe layer in the North Sea. Brown & Root's decision to purchase the *Semac 1* proved its worth many times over, and the vessel remains in the 1990s a centerpiece of EMC, Brown & Root's joint venture with Saipem.

Following this, Brown & Root briefly entered into a courtship with the monohull shipshape crane vessel concept and bought *Ocean Builder 1* from Amoco and *Sarita* from Ugland in, respectively, 1983 and 1984. Neither worked in the North Sea; but they did find uses in other environments.

The debate on the merits of the semisubmersible flared up again, for the last time, in May 1983. In order to regain the lost ground on marine construction, a team of specialists was set up in the London office to investigate the technical and commercial merits of investing in a second generation semisubmersible crane vessel (SSCV). The vessel was to be bigger and more advanced than Heerema's *Balder* and *Hermod,* with a projected capacity to lift integrated decks weighing in excess of 11,000 tons using dual cranes.

The building specifications were developed and Götaverken (GVA) in Gotenburg, Sweden, was contracted to prepare a conceptual design. Building yards worldwide were invited to bid on the $600 million Super-SSCV and more than 70 presentations were given to oil companies on how such a vessel would reduce offshore construction and hookup schedules. Also, extensive hydrodynamic model tests were held and subcontractors, including a crane manufacturer, were signed up. Nearly $12 million was spent on this conceptual phase of the work. While Brown & Root conceptualized, however, several other companies acted, and the announcement of the construction of several additional semisubmersibles altered Brown & Root's economic calculations, convincing the company that the existing market for the use of advanced semisubmersibles was becoming glutted.

This was an unsettling hint of things to come. The sharp rises in oil prices in the 1970s had unleashed a decade of extraordinary expansion in the North Sea and around the world, presenting the offshore industry with unprecedented opportunities to apply new technology. But at least in part, the industry became a victim of its own success, since the large offshore reserves developed in the 1970s undermined the ability of OPEC to sustain high prices. Sharp price declines of the mid-1980s marked the end of an era of almost frenzied expansion marked by far-reaching engineering advances. Brown & Root had been a pioneer in the North Sea boom, and as the boom subsided, it would be forced to become a pioneer in charting a sustainable path in an era of lower prices.

Notes

1. This chapter draws from interviews with Dirk Blanken, H.S. Frankhouser, T. Smith, III, W.R. Rochelle, Jamie Dunlap, Bob Weatherly, Buddy Hoke, J.C. Lochridge, and J.B. Weidler.
2. "Ninian Reserves Hiked, Third Platform Planned," *Oil & Gas Journal* (February 14, 1977): 63; "More Ninian Platform Contracts Let," *The Oil and Gas Journal* (January 5, 1976): 43.
3. Fred S. Ellers, "Advanced Offshore Oil Platforms," *Scientific American* (April 1982): 39–50.
4. "First Oil Flows from Magnus," *The Oilman* (September 1983): 24.
5. "Hideck Comes Out into the Open," *Offshore Engineering* (April 1978); "$2 Billion Oil Platform Uses Innovative Concepts," *Engineering News-Record* (January 11, 1979): 18–19.
6. "HIDECK Saves Time and Millions," *Brownbilt* (Summer 1979): 8–11.
7. Phillip A. Abbot, Larry Farmer, Graham J. Blight, and David Osborne-Moss, "A New Integrated Deck Concept," OTC Paper 3879, May 5–8, 1980.
8. "Maureen Tow-Out," *The Oilman* (March 1983): 28; "Maureen Platform's Deck, Base Mated," *Oil & Gas Journal* (April 18, 1983): 46.
9. Finn Rosendahl, "Concrete Structures Offer New Options," *Offshore* (September 1979): 63; "Two Views of the Future," *The Oilman* (June 11, 1973): 20–21.
10. Ellers, "Advanced Offshore Oil Platforms."
11. Robert R. Steven and Diane Crawford, "Plans Set for TLP Debut," *Offshore* (March 1980): 55.
12. Andrea Mangiavacchi, Phillip A. Abbot, Shady Y. Hanna, and Rudolf Suhendra, "Design Criteria of a Pile Founded Guyed Tower," OTC Paper 3883, May 5–8, 1980.
13. Robert Steven, "Conoco Details Hutton TLP Delay," *Offshore* (September 1983): 43.
14. "'Superbarge' to be Ready in '76," *Brownbilt* (Summer, 1975): 28–29; "*BAR-347:* The Quiet Giant," *Brownbilt* (Winter 1977); G.H.G. Lagers and C.R. Bell, "The Third Generation Lay Barge," OTC Paper 1935, Houston, May 6–8, 1974.

EPILOGUE

A New Dawn After the Storm

Oil prices did not continue their upward trend after 1981. Prices for crude leveled off in 1982 and then began to decline slightly. Most people in the industry thought that this change was an aberration and that oil prices would soon begin to rise again. In retrospect, the stabilization and subsequent decline of the value of oil resulted from the success of two ongoing initiatives: the avalanche of new oil from offshore and the success of energy conservation measures in the developed nations.

Brown & Root Marine initially was not affected by this change. Work continued in Mexico, while the North Sea appeared oblivious to the price of crude and the U.K. and Norway continued their quests for energy independence and full employment. The company even embarked on a plan to build a heavy lift, semisubmersible derrick vessel to take on Heerema in what was perceived to be a lucrative North Sea installation market. In the United States, company engineers researched deepwater platform design for the anticipated market beyond 1,000 feet and in formulating platform solutions for the Beaufort Sea. A mechanical underwater trencher was designed, built, and tested by the company.

An important barometer of the health of the offshore industry, attendance at the annual Offshore Technology Conference, peaked at 120,000 in May of 1981, but no one knew it at the time. By late 1982, the count of active rigs began to decline as well as the demand for offshore vessels. Those companies that had invested heavily in assets to meet the expected increased demand became concerned for their balance sheets. Oil companies became hesitant to continue exploration and development in the face of declining oil prices.

Although engineering work remained strong in the North Sea, opportunities declined in Houston. The population of the company's marine engineers shrunk by two-thirds in the 1983-1987 period. When the price of oil bottomed in 1987, even North Sea companies retrenched, while the once active Gulf coast fabrication yards had weeds growing in empty parking lots. An internal study in 1987 showed that there were ten construction barge days available for each day of anticipated work over the next three years. Profits in the oil service industry became a contradiction in terms.

Things were not all dark for Brown & Root Marine in the mid-1980s. The company won the Queen's Award for its efforts in the design of the world's first tension leg platform for Conoco's Hutton field in the North Sea. Three-dimensional computer-aided design and drafting (3D-CADD) was first used in the offshore industry for Hamilton's Esmond platform in the North Sea. Numerous important engineering and construction projects were completed in all parts of the world in this time period.

Internally, the marine group worked to rationalize the situation. It wrote down the bulk of the marine fleet (some thirty plus major vessels) as a non-performing asset and eventually sold it, primarily to Offshore Pipelines, Inc. The company also sold the fabrication yard in Bahrain, closed the yard at Labuan, and mothballed fabrication facilities at Sunda Straits and Harbor Island.

Brown & Root and Saipem formed a joint venture in 1988. Each company contributed a semisubmersible lay barge to the venture, which was called European Marine Contractors (EMC). Brown & Root also formed an underwater construction venture with Smit International, combining the diving assets of Wharton Williams Taylor with those of Smit to create Rockwater. Fabrication facilities at Nigg in Scotland and Greens Bayou in Houston remained open, while engineering kept locations in Stavanger, Aberdeen, London, Houston, Singapore, and Kuala Lumpur.

Oil prices stabilized at sixteen to twenty dollars per barrel in the early 1990s, stimulating oil companies to plan exploration projects around fifteen dollar oil—or about half of what the price had been only a decade earlier. While projects in the 1970s and early 1980s had been schedule-driven, those of the 1990s would be driven by cost.

The marine group's strategy as the 1990s began was to build on its strengths: engineering, project management, pipelaying, fabrication, and underwater construction. The company chose to exit most of the commodity end of the construction business and enter the offshore maintenance and logistics support business. Fortunately, this strategy coincided with a relatively stable oil price, a modified tax policy in the U.K. that encouraged development, a decision to bring more gas into Belgium, and the downsizing of oil companies.

This strategy greatly accelerated a trend that had been obvious within Brown & Root Marine for several decades, the growing importance of design over construction. What had begun as an offshore construction company with a talent for engineering evolved into an engineering company with a proven capacity in construction. The difficulties of the late 1980s forced the company to chart a future as an engineering and design specialist with construction capabilities primarily in pipelines and onshore fabrication.

The decision to keep a stake in fabrication reflected the need to maintain the all-important "feedback" loop between those who designed offshore structures and those who built them. This loop kept designers in touch with the practical implications of their designs. The Greens Bayou yard on the Houston Ship Channel, the historical center of Brown & Root's presence on the Gulf Coast, remained, though reduced in size. The HiFab yard in Scotland was well positioned to benefit from any rebound in the North Sea. It continued to attract enough work to stay busy even at the depths of downturn. Brown & Root obtained a 100 percent interest in this yard in the early 1990. Then in 1996, J. Ray McDermott, a long-time rival in offshore construction, combined its yard at Ardeshier with Brown & Root's Nigg yard into a new fabrication venture known as BARMAC. Rationalization of yards in the U.K. had been overdue, and BARMAC played a vital role in the resurgence of North Sea construction in the late 1980s and 1990s.

Pipelaying in the North Sea continued for Brown & Root even during 1985, when the company laid over 250 miles of North Sea pipeline in the midst of

Semac I

the offshore depression. The creation of EMC in 1988 greatly strengthened the company's position in pipelaying by bringing together two strong competitors with two of the most advanced pipelay vessels in the world, Brown & Root's *Semac I* and Saipem's *Castoro Sei.* Both vessels were giant semisubmersibles with state-of-the-art equipment to lay large diameter pipe in very deep water or to complete infield projects and tie-ins. Several of the traditional flat-bottom barges that had served Brown & Root in offshore markets throughout the world continued to lay pipe in the 1990s, but for other companies that had purchased them from Brown & Root.

EMC went forward with its two mammoth semisubmersibles to complete ambitious projects around the world. Two high profile jobs spotlighted the venture's prominent place in the pipelay business. Zeepipe, the longest marine pipeline yet laid, was completed by EMC during the 1991 and 1992 work seasons for Norway's Statoil. In addition to the *Semac I,* the *Castoro Sei,* with more than 1,000 people and some 60 support vessels, worked on this 670-mile line from the Sleipner field in the Norwegian sector of the North Sea to Zeebrugge, Belgium. Built at a cost of more than $350 million, the line carried 1.1 billion cubic feet per day of natural gas to a growing European market. In 1994, EMC completed a 440-mile line for ARCO from the Yacheng 13-1 field in the South China Sea to Hong Kong. This pipeline was second only to the Zeepipe line in length, and it was the longest line ever laid in a single work season by a single vessel.

Closely connected to Brown & Root's strength in pipelines was its growing strength in diving and subsea work. After its creation in 1990, Rockwater quickly emerged as a dominant force in the rapidly expanding underwater construction industry. Headquartered in Aberdeen, Scotland, Rockwater was well-positioned to take advantage of the boom in subsea work in the 1990s. As oil companies moved into deeper waters in the North Sea and elsewhere, subsea completions gained favor as an economically effective alternative to larger platforms. Rockwater quickly became an industry leader in all phases of subsea work, including pipeline and riser construction. In 1994, Brown & Root bought out Smit's interest in Rockwater, and this subsidiary played a critical part in the company's growth.

The expansion of subsea work presented Rockwater with excellent opportunities for innovation. For example, in response to the growing number of interconnections between subsea templates and collecting platforms, Rockwater developed new ways of prefabricating bundles of pipe containing all the equipment needed for the operations of a subsea manifold. By towing bundles of prefabricated pipes several miles long to the installation sites, the company dramatically cut the time and expense of hooking up the subsea connections. Rockwater also became expert at such tasks as installing the tethers on tension leg platforms and the numerous other underwater jobs required by the offshore industry. As the company grew, it acquired a small fleet of work vessels, including the semisubmersible *Regalia,* which it used as an offshore platform from which to complete its varied subsea work.

Subsea template at BARMAC

Brown & Root also responded to the growth of subsea completions in its work as a designer and fabricator. The design and construction of subsea templates and floating production systems became increasingly significant to the company. The use of systems other than traditional fixed platforms exploded in the 1990s, as companies sought cost-efficient ways to produce oil and gas from smaller fields in an era of lower prices. Various floating systems had been heralded since the 1960s as the wave of the future offshore, and in the mid-1990s, the future finally began to arrive.

This did not mean, however, that the demand for platforms disappeared. Instead, companies began to survey more carefully the range of options available for maximizing profits from a given field. In some cases, innovative designs for permanent platforms were used. Such was the case with the Santa Ynez unit expansion project undertaken by Exxon in the Santa Barbara Channel in the mid-1980s through the early 1990s. This ambitious project included the construction of two new platforms, onshore processing facilities, and offshore pipelines. Brown & Root worked on the design of the pipelines, the process facilities onshore and offshore, the onshore infrastructure, and the underwater power cables for the Santa Ynez unit expansion, which resulted in an additional 150 million cubic feet of natural gas and 150,000 barrels of emulsion per day. Brown & Root helped Exxon cope with the more stringent environmental requirements in California. Three million engineering hours were spent on the design of one of the last platforms to be installed on the West Coast. Throughout the 1990s, the company worked on similar projects throughout the world, as the offshore industry expanded the production of existing structures and found and developed new fields in areas ranging from offshore China to West Africa to Trinidad.

In the North Sea, a new market opened after a disastrous fire at the Piper Alpha platform killed 167 offshore workers in August 1988. In the wake of this tragedy, North Sea companies moved to make their giant platforms safer. Brown & Root showed the way in this area with its work on Piper Bravo for Elf. This redesign of the complex that had been destroyed in 1988 employed more than 1,100 people and required over four million man-hours of work by Brown & Root, making it the largest engineering project in the company's history. As all companies active in the offshore industry saw the need to rethink the design of platforms to assure that workers were as safe as possible, Brown & Root's work at Piper Bravo set a high industry standard in designing safer platforms.

In all of its design work, Brown & Root embraced the computer revolution that swept through the world economy in the 1990s. The company had been a leader in adapting computer technology to offshore engineering dating back to the 1960s, when it helped develop computer programs to analyze the stress placed on structures by various environmental forces. It continued to upgrade and expand such programs in the 1970s and 1980s, by which time computer-aided design had revolutionized the engineering of offshore structures. But the 1990s brought a second, more visible, revolution, in the form of 3D-CADD.

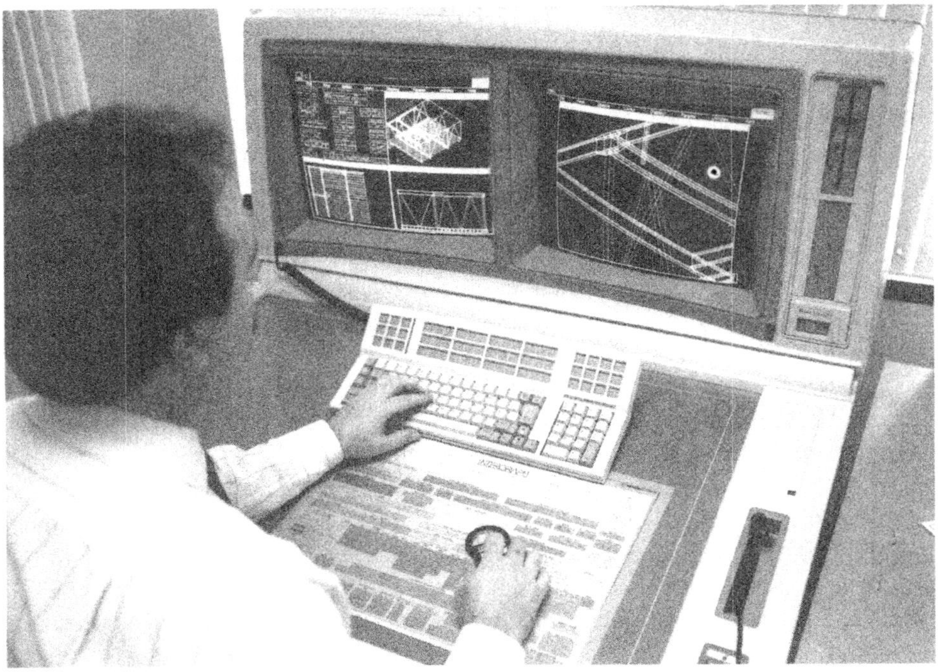

3-D CADD designer

Again, Brown & Root took the lead in applying this astonishing new technology to the practical problems of the offshore industry. Three-dimensional computer programs allowed the user to visualize both the choices being made and their consequences. Using this technology, designers could construct a virtual structure and subject it to virtual interactions on the computer screen, making design choices more concrete and understandable. These sophisticated programs could be used to interrelate all aspects of a structure, as well as to isolate any aspect of the structure and correlate information about it.

Computers also hastened advances in other aspects of the offshore industry. They allowed companies such as Brown & Root to manage information more efficiently, providing real-time feedback to clients about the status of specific projects. The creative application of computer technology also allowed companies to streamline the design and production process. One particularly important example of this came in fabrication, when computer models developed by designers could be taken into the fab yard and used to program machines to cut steel, thereby sharply reducing the cost and time of projects while improving their quality. All such applications of computer technology required innovative thinking and practical work by highly trained specialists. In all phases of offshore work, the advent of more powerful and subtle computers in the 1990s brought opportunities for stunning advances in the management and analysis of information.

Equally striking advances in management came from creative changes in the organization of projects. Brown & Root had pioneered the use of project management offshore, in which it worked with the operator to determine the projected time and cost of a project and then helped coordinate the work of the numerous specialized firms needed to complete the project. In the 1990s, the imperative to cut costs encouraged the creation of alliances, a fundamentally different way to organize offshore work. Again, Brown & Root pioneered this new approach in which clients put together a team of companies to complete a project and to work together to reduce the total cost and time. Once the operator had developed a target cost and schedule for a project, the members of the alliance met to pool their ideas and experiences in search for lower cost alternatives. Each alliance member assumed some of the risk for making these ideas work; each stood to reap some of the rewards if costs were indeed lowered. This cooperative approach required each company to view its long-term self-interest as identical to that of the alliance as a whole. This represented a stark change from traditional practice. The underlying assumption was simple and persuasive: with stable, historically low prices, the key to the successful development of new offshore fields was the systematic use of new technology and creative management to reduce costs. The future of each specific company ultimately was tied to the growth of offshore development as a whole, and the industry could no longer assume that higher oil and gas prices would open new opportunities.

The benefits of alliances and partnering were quite evident in Brown & Root's work for BP in the Andrew field in the North Sea. The Andrew Alliance, which included several prominent engineering and construction companies involved in platform development, joined two other alliances organized for this project: the Cyrus Alliance for subsea completions and BP's subsurface alliance. The results were striking. The original target cost of about $600 million was reduced by more then twenty percent, and oil came onstream in June 1996, more than six months earlier than initially projected. Given such results, it was hardly surprising that the alliance approach spread rapidly throughout the industry in the mid-1990s, producing a closer working relationship among offshore specialists and pointing the way toward the creative interchange of ideas and technology.

Implicit in the alliance approach was the view that the individual companies had strong incentives to work collectively toward the maximization of the revenue stream from offshore reservoirs. The added efficiency from this approach was good for the bottom line, since it increased the value of each reservoir while also increasing the number and size of offshore reservoirs that could be profitably developed. Not coincidentally, it was also good for society as a whole, which benefitted from an expanded supply of energy at the lowest possible price.

One additional step in the same direction was the life-cycle approach to planning reservoir development. Even the best alliance focused on reducing construction and installation costs might make choices that ultimately increased operating costs. By encouraging careful analysis of trends over the life of a field—from initial development to final production—life-cycle planning sought to assure the most productive and cost-effective use of each reservoir. An early step in this direction came in Brown & Root's work for BP in redeveloping the Forties field in the early 1990s. But even greater benefits would have been possible had planners in the 1970s taken a longer term view of the profitability of the Forties field. Of course, the climate of the 1970s emphasized speed of production over costs; twenty years later, much lower petroleum prices and improved technology dictate the new emphasis on costs during the life of a field.

As Brown & Root and its parent Halliburton help create a new calculus for maximizing long-term profitability in the development of reservoirs in such proven offshore provinces as the Gulf of Mexico and the North Sea, they also continue to expand around the world in emerging "frontier" areas. In the late 1990s, this includes new initiatives in offshore Australia and China, as well as new ventures onshore in areas as diverse as Colombia and Algeria. In such contemporary ventures, as in its past, the company relies on the creative adaptation of techniques previously proven in other regions.

The future of offshore development holds extraordinary promise. Already in 1997 oil has been produced at depths exceeding 5,000 feet in the Gulf of Mexico. Innovations in technology and organization have opened opportunities that seemed impossible in the gloom of the depression in the oil industry only a short decade ago. Numerous potential paths are evident. The expansion of subsea completions seems certain, as does the growth of related facilities such as floating production systems. Traditional fixed platforms and a variety of floating platforms have an enduring claim to efficiency within certain water depths, and their future also seems assured. Continued advances in related technologies, notably 3-D seismic for finding oil and directional drilling for producing it, promise great benefits to offshore companies. Yet the past also teaches that new technologies that are now taking shape in the minds of engineers/entrepreneurs will continue to burst unexpectedly on the scene, redefining the range of what is possible and forcing everyone in the industry to reevaluate their traditional assumptions.

Long experience in the industry gives traditional leaders such as Brown & Root the confidence spawned by past achievements, while also reminding them of the imperative to adapt to new challenges. On the fiftieth anniversary of offshore development, it is appropriate to congratulate the industry on its historical accomplishments. Comparison of the Kermac 16 and such contemporary projects as the massive Hibernia platform off Newfoundland yields amazement at the sustained pace of technological change in the industry. But the celebration of the past must be tempered by the certain knowledge that those who look back on the occasion of the centennial of the offshore industry in 2047 will no doubt shake their heads in amazement at the fundamental advances made since the birth of the new era in the offshore industry in the 1990s.

Index

Printed in Dunstable, United Kingdom

91380918R00181